Research Reports ESPRIT

Project 26 · SIP · Vol. 1

Edited in cooperation with
the Commission of the European Communities

G. Pirani (Ed.)

Advanced Algorithms and Architectures for Speech Understanding

Springer-Verlag
Berlin Heidelberg New York London
Paris Tokyo Hong Kong Barcelona

Volume Editor

Giancarlo Pirani
CSELT
Via Reiss Romoli, 274
I-10148 Torino, Italy

ESPRIT Project 26 "Advanced Algorithms and Architectures for Speech and Image Processing (SIP)" has the objective to develop the algorithmic and architectural techniques required for recognizing and understanding spoken or visual signals and to demonstrate these techniques in suitable applications.

The work was planned in three parallel areas: speech analysis, image analysis, and pattern recognition and understanding. Work on speech processing took two approaches, one statistical and one knowledge based. In image processing, various algorithms were analyzed, compared and implemented, and different layer approaches to the architecture for image feature extraction were considered. Medical and industrial applications were used to test the tools developed and to study the issues involved.

ISBN-13: 978-3-540-53402-0 e-ISBN-13: 978-3-642-84341-9
DOI: 10.1007/978-3-642-84341-9

Publication No. EUR 12821 of the
Commission of the European Communities,
Scientific and Technical Communication Unit,
Directorate-General Telecommunications, Information Industries and Innovation,
Luxembourg

Softcover reprint of the hardcover 1st edition 1990

2145/3140 – 543210 – Printed on acid-free paper

Preface

This book is intended to give an overview of the major results achieved in the field of natural speech understanding inside ESPRIT Project P. 26, "Advanced Algorithms and Architectures for Speech and Image Processing".

The project began as a Pilot Project in the early stage of Phase 1 of the ESPRIT Program launched by the Commission of the European Communities. After one year, in the light of the preliminary results that were obtained, it was confirmed for its 5-year duration.

Even though the activities were carried out for both speech and image understanding we preferred to focus the treatment of the book on the first area which crystallized mainly around the CSELT team, with the valuable cooperation of AEG, Thomson-CSF, and Politecnico di Torino.

Due to the work of the five years of the project, the Consortium was able to develop an actual and complete understanding system that goes from a continuously spoken natural language sentence to its meaning and the consequent access to a database.

When we started in 1983 we had some expertise in small-vocabulary syntax-driven connected-word speech recognition using Hidden Markov Models, in written natural language understanding, and in hardware design mainly based upon bit-slice microprocessors.

At that time the USA and Japan were starting big projects with very ambitious objectives in speech recognition and understanding: the former was supported by the Advance Research Project Agency of the US Department of Defense (DARPA), while the latter was related to one of the fundamental issues of the Fifth Generation Computer Project.

Given this scenario, the task we were undertaking was very challenging, and some fear stemming from the comparison with these giants was unavoidable. However, at the end of the project, the results obtained were quite satisfactory and allowed us to look at new developments and applications in the second phase of ESPRIT with a significant degree of confidence. We felt that we had given a small contribution to the fundamental goal of the ESPRIT Program of reducing the technological gap that divided Europe from USA and Japan.

Three partners guaranteed the stability and continuity of the project for its entire duration: AEG (now Daimler Benz), CSELT, and Thomson-CSF. They formed the core of the Consortium which allowed it to cope successfully with the departure of some partners and to find some new ones with a remarkable technical skill and speculative attitude.
The success of the project was not only determined by the skill of the researchers who took part in it but also by the sound management of the committee that steered it: this committee was chaired with invaluable efficiency by G. Perucca (Project Manager) who represented the Prime Contractor (CSELT) and was composed of the Technical Management Committee Coordinator, S. Giorcelli (CSELT), and of the Technical Management

Committee Members representing the other partners: H. Mangold (AEG), T. de Couasnon (Thomson-CSF), G.E. Hirsch (École Nationale Supérieure de Physique de Strasbourg), and, for the first half of the project period , P.V. Collins (GEC). Finally the valuable supervision effort of the CEC Project Officer, T. Van der Pyl, is worthy of particular mention.

List of Contributors

Gian Paolo Balboni, CSELT, Italy
Piergiorgio Bosco, CSELT, Italy
Robert Breitschadel, Daimler Benz, West Germany
Carlo Cecchi [1], CSELT, Italy
Alberto Ciaramella, CSELT, Italy
Davide Clementino, CSELT, Italy
Luciano Fissore, CSELT, Italy
Roberto Gemello, CSELT, Italy
Egidio Giachin, CSELT, Italy
Alfred Kaltenmeier, Daimler Benz, West Germany
Pietro Laface, Politecnico di Torino, Italy
Riccardo Melen, CSELT, Italy
Giorgio Micca, CSELT, Italy
Corrado Moiso, CSELT, Italy
Roberto Pacifici, CSELT, Italy
Roberto Pieraccini [2], CSELT, Italy
Giancarlo Pirani, CSELT, Italy
Jean Pierre Riviere, Thomson-CSF, France
Claudio Rullent, CSELT, Italy
Giorgio Sofi, CSELT, Italy
Giovanni Venuti, CSELT, Italy

[1] died in 1990
[2] now with AT&T Bell Labs

Table of Contents

Chapter 1

Introduction to the Book

Giancarlo Pirani (CSELT)

1.1 Historical Notes

ESPRIT Project P26, "Advanced Algorithms and Architectures for Speech and Image Processing", started in October 1984, after a one-year feasibility study (since October 1983 to September 1984) in the ESPRIT Pilot Phase, and ended in September 1988 with an overall effort of about 130 man-years.

Contractors involved were:

CSELT (Prime Contractor), Italy

AEG (now Daimler Benz), West Germany

Thomson–CSF, France

ENSPS (Ecole National Superiéure de Physique de Strasbourg), France, from October 1986

Subcontractors involved were: Polytechnic of Torino, University of Torino, Italy, and HITECH, Greece.

In addition, Plessey, United Kingdom, participated in the pilot phase, and GEC, United Kingdom, participated until September 1986.

The main objective of the project was to develop a coherent set of techniques, both algorithmic and architectural, for speech and image recognition and understanding, and to validate the performance of these techniques by means of suitable demonstrators, in order to assess their usability for industrial and commercial applications.

Two basic technical approaches were used: the former consisted in the integration of low-level, statistics-oriented recognition algorithms with higher-level, knowledge-based understanding techniques; the latter relied upon an extensive use of parallel processing architectures.

Although one of the goals of the project was to investigate the feasibility of a commonality (at least conceptual) between the speech and image understanding architecture, there was a distinct activity relevant to the development of the speech processing "branch" of the system. In particular, it is possible to identify a self-contained "sub-project" which was devoted to the implementation of a continuous speech understanding system.

The bases of the speech understanding project were laid down in the early 1980s; these were the years in which the new DARPA project was being launched, after a five-year hiatus, to re-think objectives and techniques after the results of the first program ended in

1977 [1]. In that sense we could exploit the lesson drawn by that experience together with the background in speech recognition, natural language understanding, DSP hardware, and parallel architecture available in our ESPRIT Consortium.

This expertise, together with knowledge of the state of the art in the world, has steered us to choose the large-scale objectives of our project in order not to be too ambitious, to be realistic enough but to have at the same time sufficiently advanced contents in algorithms and technology to justify a research effort of five years.

The system we decided to develop had to show the following characteristics:

- accept continuous speech from co-operative speakers
- accept natural language with limited syntactic coverage
- be trainable by the user with a vocabulary independent of the application
- use a close-talking microphone in a computer terminal room
- exploit a vocabulary of 1,000 words
- work in a constrained semantic domain, relevant to the inquiry of a geographical database
- achieve nearly real-time performance

One of the basic decisions that was taken at the very beginning of the project was that the logical architecture of the overall speech understanding system should rely upon a clear separation between low level (recognition stage) and high level (understanding stage). This is shown in Fig. 1.1, which gives also some anticipation of the general structure of the system.

This type of architecture was decided also in the light of the difficulties experienced during the first ARPA project, mainly stemming from its integrated approach to the problem of speech understanding. In fact, one of the major advantages that we expected in decoupling the problems was to emphasize the activities for improving the performance of the two blocks separately.

The aim of the low-level stage is to find out a lattice of word hypotheses; each item of this structure consists of an element of the lexicon (word), its acoustic likelihood, and its estimated temporal boundaries. The lattice is the interface between the low-level speech processing and the high-level linguistic analyzer. The main goal of the understanding level is to process the lattice of word hypotheses and to produce a representation of the utterance meaning. The meaning representation is used to access the database and to generate the answer to the inquiry in natural language.

According to this decoupling philosophy a big effort was made to obtain a lattice as accurate as possible in order to improve the overall performance of the system; this involved a deep study of the acoustic-phonetic decoding techniques to obtain highly reliable scores for the word hypotheses, with the principle that what is lost at the lower level can hardly be recovered by the higher one.

The separation between the two levels of processing has also implied a corresponding architecture from the hardware viewpoint, as Fig. 1.1 shows. This allowed us to define the global hardware structure of the recognition stage, which was based upon two different

[1] D.Klatt, "Review of the ARPA Speech Understanding Project", *J. Acoust. Soc. Am.*, vol.62, no. 6, pp.1345-1366, Dec.1977.

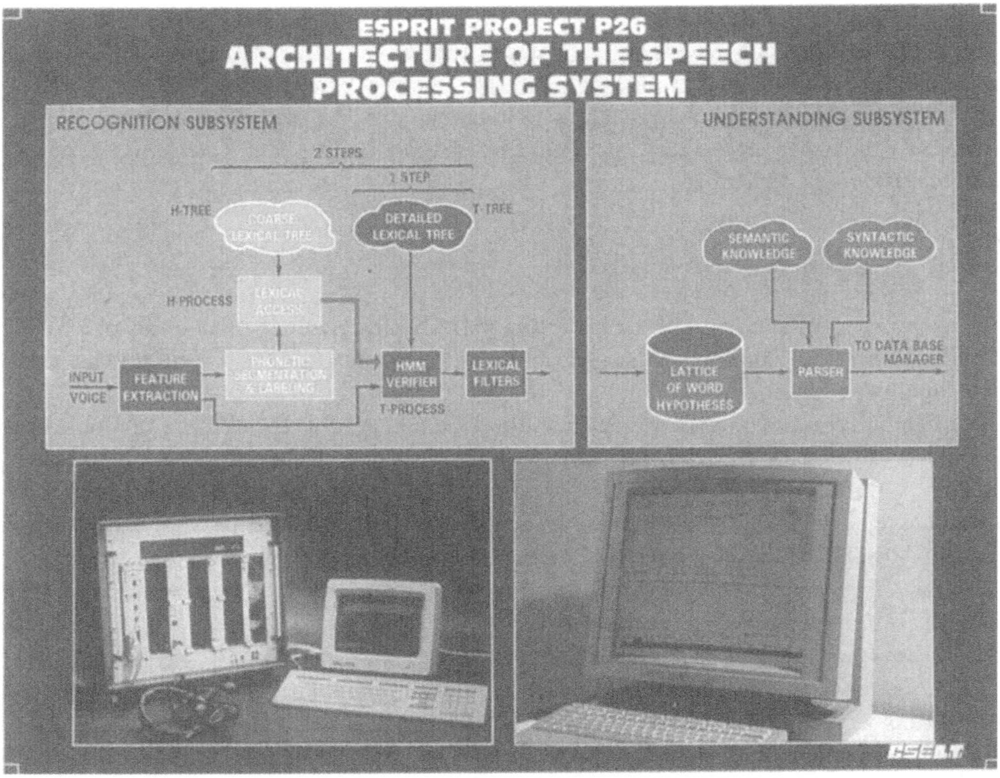

Figure 1.1: Global architecture of the speech understanding system.

types of boards, expressly developed for the project: the former performs the A/D conversion of speech together with analog amplification and filtering; the latter is equipped with a DSP (TI TMS32020) and can perform both digital signal processing and pattern matching operations. In this way we could concentrate from the beginning of the project on the most appropriate structure of the front-end processor to produce the lattice in real time.

On the other hand, the high-level algorithms of the understanding stage could be implemented on a LISP Symbolics workstation, with a view to improving its efficiency through a re-design and a re-writing in C language, and to studying the feasibility of a parallelization of the algorithms that would be suitable for implementation on the parallel architecture being developed.

1.2 Overview of the Book

The objective of this book is to describe the activities carried out in the part of ESPRIT P26 devoted to developping the speech understanding system, and to present the results that have been obtained.

The main body of the book is divided into four main chapters that address the four different tasks (or work packages) of the project workplan that had to be carried out to develop the final system: recognition algorithms, real-time hardware implementation of the recognition stage, understanding algorithms, and parallel languages for parallel architectures. For this reason the chapters are to some extent self-contained and autonomous.

Chap. 2 gives a detailed description of the recognition algorithms developed with the main objective of designing an efficient technique to extract an accurate lattice of word hypotheses from a continuously spoken utterance.

As the size of the vocabulary was medium-large, the first approach tried was to exploit a two-step strategy: the first step consisted in reducing the whole lexicon to a subset through a segment classification into gross phonetic classes and a subsequent lexical access; the second step was to give an acoustic score to the surviving words belonging to the subset. Although a partly knowledge-based approach was attempted, relying upon some skill at the University of Torino, the approach finally chosen was completely statistical. In this way the Consortium could benefit from the expertise of AEG in statistical pattern recognition for gross phonetic classification, of the Politechnic of Torino for lexical access, and of CSELT for automated training of sub-word units HMMs (Hidden Markov Models) and their use in the recognition process, through a suitable version of Viterbi decoding.

In the first phase of the project, this approach proved itself quite effective for a large-vocabulary isolated recognition task; therefore the first outcome of this research was an isolated word recognition system with vocabulary sizes in the range of 1000 – 20000 words. This was a quite interesting by-product of the project, but when we switched to the real goal of the project (1000 words and continuous speech) an extensive experimentation showed an important result: even if very efficient control strategies were invented to perform the hypothesization and the HMM scoring processes in an integrated and interactive way, the one-step approach was still slightly more appropriate in terms of accuracy and computation time.

In fact, the size of about 2000 words seemed to be the threshold that has to be exceeded to make the two-step approach become more convenient.

All the choices relevant to the hardware architecture of the recognition stage are described in detail in Chap. 3.

The general philosophy was to have an architecture based upon specifically developed boards. These boards had to exhibit enough computational power together with the possibility of being connected as intelligent peripherals to a commercially available family of microprocessor boards (Motorola 68020) through standardized VME and VMX buses. This type of architecture showed a noteworthy modularity that made it suitable for tasks of different complexity and computational requirements.

The speech acquisition board was designed by AEG, while the DSP-based ones were developed by Thomson-CSF. The choice of the TMS32020 processor was determined by its larger addressable area with reference to both internal and external memory, at the time when the selection was made (1985).

With the exception of the classifier module, all the firmware implementing the recognition stage was written by CSELT, which also took on the task of integrating all the hardware blocks into a unique system.

Whereas the recognition stage was the result of a joint effort of AEG, CSELT, and Thomson-CSF, the understanding-stage algorithms and architectures were developed almost entirely by CSELT.

As Chap. 4 describes, the philosophy that led to the developed understanding algorithms was to follow a knowledge-based approach. In fact, this approach was felt more suitable for the integration of syntax and semantics as well as for the achievement of a certain commonality between speech and image architectures.

When the project started we believed that we could take advantage of the new representational tools developed after the end of the ARPA project, that were more powerful and expressive than context-free grammars, even if more complex. In this way it was possible to exploit the language constraints more efficiently for a system with its purpose not limited to recognition, but including also meaning comprehension.

Conceptually, the basic philosophy was to start from a state-of-the-art system for written natural language understanding and to extend its capabilities to deal with word lattices instead of definite word sequences.

In fact, the rôle of the understanding stage is to parse the word lattice to find the most likely word sequence and to provide a formal representation of its meaning. We chose to have the parsing strategy based upon a score-driven search that is capable of avoiding the bottlenecks that are likely to arise when a lattice with some low-quality hypotheses is to be processed. In this way the system can evolve better towards applications where the quality of speech is impaired or speaker independence is required.

The developed approach started from the decision to use two separate representations, for semantic and syntactic knowledge respectively. These representations are then combined into common parsing rules through an automatic compilation which exploits also the property that the "caseframe" formalism, chosen for semantics, is characterized by structures which are similar to those of the "dependency grammar" formalism chosen for syntax.

An additional characteristic of the developed parser is its ability to perform correctly, when necessary, without having to rely upon hypotheses regarding words that are both short and semantically irrelevant, like determiners and some prepositions.

Finally it was decided to derive a formal and non-ambiguous representation of the meaning of the spoken sentence in a connected caseframe format at the end of the parser process, which is used both for completing word recognition and understanding.

One of the most challenging objectives of the project was also the design of a parallel architecture that would allow the understanding task to achieve real-time performance, both for speech and image. Actually, at the end of the project the parallel architecture was available, but there was not enough time to insert the speech understanding algorithms into such an architecture.

Although this book is devoted essentially to the description of the speech understanding system, we decided to make it more complete, including also a chapter which describes the basic principles of the design of the parallel architecture, focusing on the specification of a parallel symbolic programming language that permits its efficient exploitation.

Therefore, Chap. 5 points out how the specific objective of the activity on the architecture was to show the feasibility of a parallel machine with about 1 GIPS, medium cost, small volume, standard paackaging, and no communication bottlenecks; from the software viewpoint the goal was to show the feasibility of a problem-oriented parallel PROLOG on a distributed-memory machine, parallelizing significant understanding tasks.

From the technological point of view, the choice of INMOS transputers was made in order to have a very modular, powerful structure based upon a network of these PEs (Processing Elements) with an outstanding ease of communication.

Chapter 2

The Recognition Algorithms

Luciano Fissore (CSELT), Alfred Kaltenmeier (Daimler Benz),
Pietro Laface (Politecnico di Torino), Giorgio Micca (CSELT),
Roberto Pieraccini (CSELT)

2.1 Introduction

Subtask 2.1 of the P26 project was devoted to the study of the problems related to the development of the front-end of a speech understanding system. In the early stages of the project it was decided to separate the front-end, referred to in the following as the *recognition* module, from the *understanding* module, that deals with *syntax* and *semantics*. This decision was drawn taking into account several considerations mainly based on a practical point of view: the research groups working on Subtask 2.1 were at their first experience with speech understanding systems and their background was mainly in developing systems for small-vocabulary isolated and connected word recognition. Approaching the speech understanding problem required a strong effort both in knowledge acquisition and software development. For instance, methodologies for dealing with phonetic transcriptions of lexical items had to be developed from the beginning. More important was the lack of any practical feeling about the problem. Nobody knew (and very few in the world did at that time) what performance could be realistically achieved using a 1000-word vocabulary with a system based on sub-word unit modeling, hence which integrated strategy should be planned to attain a reasonably good understanding of the spoken sentences. The choice of a two-module system with a one-way interaction seemed the most appropriate for starting to acquire the proper knowledge on the problem. Besides, as people working on the two modules belonged to different groups and used different techniques as well as different programming languages (stochastic modeling and FORTRAN for the *recognition* group, knowledge-based parsing and LISP for the *understanding* group), the best solution looked like the one by which the development of the two modules did not have to suffer from unavoidable mutual time dependencies. The decision to consider the *recognition* and the *understanding* as two clearly separated modules with a bottom-up interaction was followed by the decision to make a time-wise separation among the two sub-systems. The understanding process should start only when the sentence has been completely analyzed by the recognition module. Again, it is advisable to stress the fact that such a strategy could not be the most suitable for a speech understanding system. It is well known that perceptual experiments, like those reported in [38], demonstrate, for example, that the linguistic analysis of an unknown utterance by human beings begins as soon as the first words have been perceived but, again, the practical choice of designing, testing and improving each module separately was considered more realistic. Furthermore we had to avoid the risk of replicating some errors made in the late ARPA SUS project [27] where most of the systems never met their project goals and where most of the care was taken in the design of the interaction between the modules rather than in obtaining the best performance by every single subsystem.

Once the kind of interaction between recognition and understanding was decided, it had to be taken into consideration which form such an interaction had to have. Given the above preliminary remarks we considered only two possible reasonable forms: a string or a lattice of word hypotheses. The first solution was attainable by implementing a standard decoder [23] that, given an acoustic/phonetic model of the words, yields the best sequence of lexical items according to a defined optimality criterium (like for instance maximum likelihood). This solution presents some disadvantages when used within a speech understanding module. First it must be noticed that every error generated by the recognition module (substitution, deletion or insertion of a word) is propagated to the understanding module without any possibility of being recovered. Hence the accuracy of the recognition part must be high enough to obtain a good understanding rate; this can be achieved by introducing some kinds of linguistic constraint at the acoustic/phonetic decoding level. Linguistic constraints can be used either under the form of a regular grammar or of a n-gram stochastic model, like for instance word trigrams [24]. The regular grammar has the disadvantage of being very rigid about the allowed sentences, while the trigram model, to be effectively usable, needs an enormous database of text for the trigram probabilities to be estimated. Moreover a linguistic constraint at the recognition level represents a replication of a knowledge source already present in the understanding module. The second solution, that interfaces the recognition and the understanding subsystems in terms of a lattice of words, seemed to be a more general approach that includes, as a byproduct, also the best-string-of-words solution. A lattice of word hypotheses is a database whose items consist of 4 pieces of information: vocabulary word identifier, time location of the hypothesis, explicated into beginning time and ending time, and its likelihood score. A sample word hypotheses lattice is shown in Figure 2.1. This strategy is in compliance with the generally adopted criterion in speech recognition that attempts to delay every decision to the moment when enough knowledge is available. Furthermore, if the lattice of words is properly computed and relevant hypotheses are not purged before they are given to the understanding module, it does not need to be built using linguistic constraints. The use of linguistic constraints during the generation of the lattice simply changes the scores of the word hypothesis (again if no purging is done). That operation can be done at the understanding level, giving the very same results but avoiding an unnecessary duplication of the knowledge sources.

An additional issue of the overall project was system effectiveness. Good performance can be obtained only as a result of intensive experimentation. When the response time of an experiment exceeds reasonable values, tuning a module or comparing different techniques becomes impossible. Thus, most of the choices, like discrete density HMMs versus continuous density ones or lexical preselection based on coarse phonetic classes, reflect the need to trade higher accuracy for reasonable computational load. Furthermore, as a real-time system for continuous speech was the target of the project, several design choices were made on the basis of the technology at hand.

Although the goal of this Sub-task was continuous speech recognition on a 1000-word

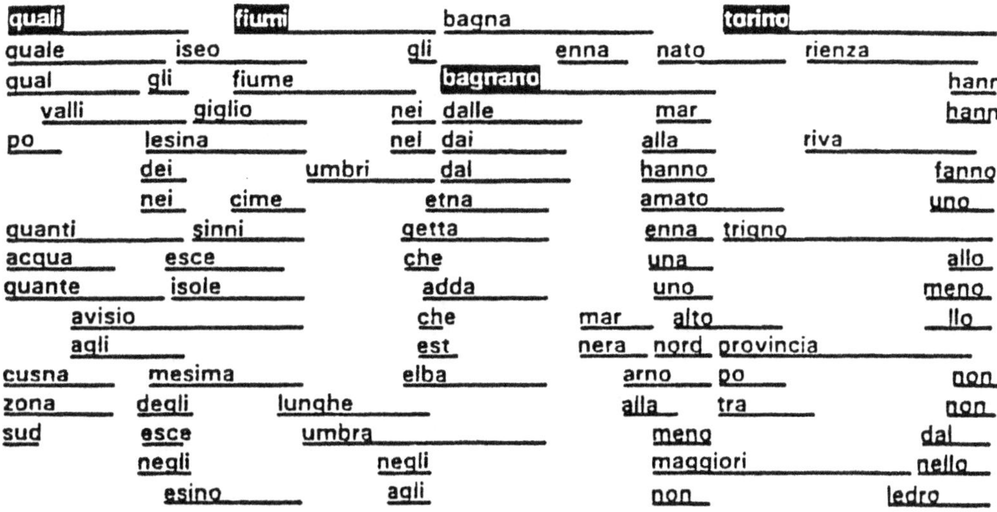

Figure 2.1: A sample word hypotheses lattice

vocabulary, nevertheless many resources were devoted to the task of isolated word recognition because most of the problems are common. It has been demonstrated that the techniques developed for the isolated word task can be extended to continuous speech. Hence particular attention has been devoted to the use of algorithms that do not take advantage on the constraints given by the simpler isolated words recognition task. After the above discussion we must conclude that the architecture of the speech understanding system developed in P26 is not to be considered as "the solution" to the speech understanding problem, but rather as a test-bed of different modules and algorithms for the investigation on two major problems of speech understanding: how to generate word hypotheses in continuous speech and how to parse a lattice of scored word hypotheses. These modules can be considered as a baseline for building more sophisticated architectures and control strategies.

2.2 System Description

In the automatic speech recognition area significant results have been achieved in research projects by using pattern recognition and stochastic modeling methods [33]. Following these paradigms, several commercial products have been developed and marketed that perform very well for simple tasks and in constrained conditions (single speaker, limited vocabulary, isolated words) [6, 37]. Nevertheless, several difficult tasks and applications still exist that need further research and engineering efforts to achieve systems that are really useful and widely acceptable by the end users.

Office dictation systems, voice-activated telephone dialing and information access with large vocabulary are emerging as realistic and useful applications.

These applications share the need of fast accessing large vocabularies of several thousand words, a difficult task even for speaker dependent systems. As the number of words to be discriminated is large, it is not practically feasible to collect thousands of templates, thus it is mandatory that lexical knowledge is built from a phonetic transcription of the orthographic form of the words. To this aim, sub-word recognition units must be defined, that can be trained from a reasonably small size learning vocabulary and used as building blocks for the words of any lexicon. Furthermore, in order to reduce the computational complexity of the pattern matching process, the search for the best matching words must be as far as possible focused. The reduction of the searching space can be obtained by carefully exploiting the structural constraints that a lexicon imposes at the phonologic level [49, 43] by using the hypothesize and test paradigm. According to this strategy, elsewhere called the *two-step approach*, a vocabulary subset to which the utterance is estimated to belong is hypothesized on the basis of a description that allows a fast search to be performed. Second, a more detailed and time-consuming verification process is activated only for words belonging to that subset [24, 26, 29, 21]. Different approaches can be used in the preselection step. The search can be carried out for all words in the vocabulary through a very simple and approximate description designed on the basis of heuristic and pragmatic knowledge [26]. As this kind of approach relies on the detection of word boundaries, it cannot be directly applied to continuous speech.

A less heuristic method is reminiscent of perceptual models of word recognition such as those introduced in the Cohort Theory and in the Phonetic Refinement Theory [43]. It avoids matching all words by characterizing each lexical entry by means of a partial phonetic description, so that acoustically similar words are clustered together [22]. From the automatic recognition point of view this is important because broad phonetic classes can be hypothesized more reliably than detailed phonetic segments. The effectiveness of the latter approach, in terms of preselection capability, has been evaluated by examining the statistical properties of large vocabularies under the assumption of a correct partial description of the words [54, 47, 16, 8]. For instance, as far as Italian language is concerned, describing a 13747 word vocabulary by using only six broad phonetic classes, 7225 words can be uniquely identified, while the maximum and average size of the subset of words bearing the same description is 34 and 1.5 respectively [20]. The results of these statistical analyses, however, do not take into account segmentation and classification errors. These errors depend on the inherent variability in speech and occur even if the acoustic-phonetic module must discriminate among a limited number of gross phonetic categories. Moreover, lexicon specifications made on the basis of a reduced set of symbols can lead to small redundancy, that is, a small distortion occurring on a string of symbols corresponding to a set of words is likely to perfectly fit the representation of a different set of words. Lexical access must be performed, therefore, through error correcting procedures that face the problem of high confusability of partial descriptions of words by generating a suitable set of likely candidates. Although word subsets larger than those predicted by an error free analysis are hypothesized, the results of several experiments, referring to different languages [31, 52, 47, 22, 7] show the substantial preselection capability of the method even in the presence of classification errors.

This hypothesize and test paradigm has been chosen first for developing a speaker-dependent isolated word recognition system with a vocabulary ranging from 1000 to 20000 words. Then the possibility has been analyzed of using such a preselection stage in a continuous speech recognition system. An original strategy has been devised for using the *two-step* approach in the word lattice generation, even if, in the final demonstrator, the *single-step* approach proved to be more effective due to the low size of the application vocabulary (1000 words). Nonetheless we will report in detail the work on the the stage approach because we think it is of practical interest when the size of the vocabulary increases.

In the *two-step* approach a first words preselection is carried out by segmenting and classifying the input signal in terms of broad phonetic classes (plosives, fricatives, vowels, etc.) To achieve high performance, a lattice of phonetic segments is generated, rather than a single sequence of hypotheses. The lattice can be organized as a graph in a structure referred to as "micro-segmentation". Words are hypothesized by matching the micro-segmentation graph against the models of all vocabulary words. A model is a phonetic representation of a word in terms of a graph accounting for deletion, substitution, and insertion errors. A modified Dynamic Programming (DP) matching procedure (three-dimensional DP or 3DP) gives an efficient solution to this graph-to-graph matching problem.

Hidden Markov Models (HMMs) of sub-word units are the basis of a more detailed knowledge in the verification step. The word candidates generated by the previous step are represented as sequences of diphone-like sub-word units, and the Viterbi algorithm evaluates their likelihood by observing sequences of labels, associated with each centisecond of the input signal, obtained by vector quantization of 18 cepstral parameters.

To reduce storage and computational costs, lexical knowledge is organized in a tree structure where the initial common subsequences of word descriptions are shared, and a beam-search strategy carries on the most promising paths only.

This strategy of lexical access has been applied to vocabularies of different size and complexity. Large-scale experimentation has been possible because all models can be trained without hand labeling or segmentation, allowing a ready adaptation to new vocabularies and to new speakers.

2.2.1 System Overview

The modules developed at the recognition level can be divided into two groups: those involved in the development and training phase and those involved in the recognition phase. Some modules, like for instance feature extraction, are used in both phases. Figures 2.2 and 2.3 give sketches of the two phases. The **FEATURE EXTRACTION** module is in charge of computing a parametric representation of speech at each 10 msec frame.
The **VECTOR QUANTIZER** additionally reduces the redundancy of the patterns by associating a symbol with each speech frame through a **codebook** of spectral vectors.
The **PHONETIC CLASSIFIER** associates with each frame one or two coarse phonetic labels, while the **PHONETIC SEGMENTER** detects segments belonging to a given phonetic class and represents them through a lattice of phonetic hypotheses.
The **LEXICAL ACCESS** matches the segment lattice with the coarse phonetic representation of the words that are arranged into a **PCL tree** (Phonetic Class Lexicon) and gives a list of candidate words.

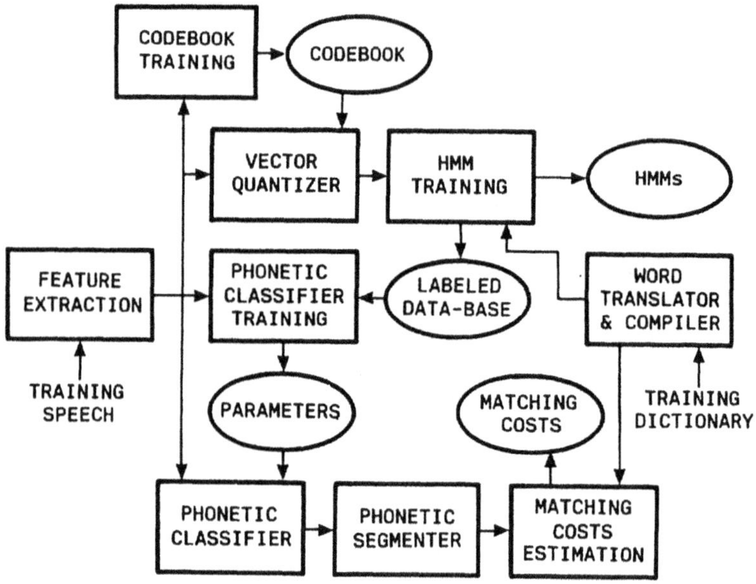

Figure 2.2: Modules active in the training phase

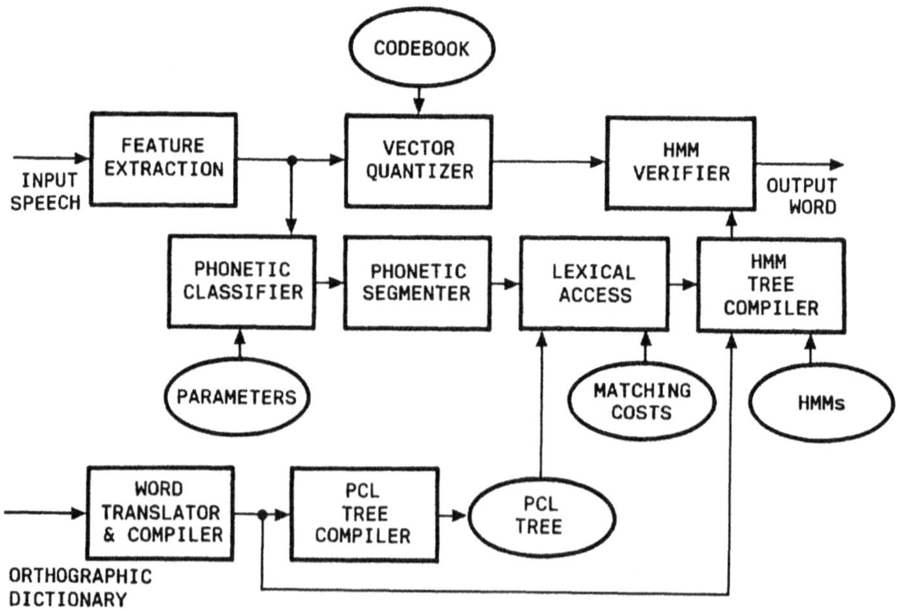

Figure 2.3: Modules active in the recognition phase

The **WORD TRANSLATOR** gives a phonetic representation of words starting from the orthographic description, while the **WORD COMPILER** gives a word representation in terms of sub-word units.
This representation is used to build the PCL tree, by the **PCL TREE COMPILER** or a HMM (Hidden Markov Model) state tree through the **HMM TREE COMPILER** since each sub-word unit is acoustically represented by a HMM.
The **HMM VERIFIER** performs a matching between the symbols given by the vector quantizer and the HMM state-tree representation of the words. Finally the **CONTROLLER** controls the information exchange between the lexical access and the HMM verifier. The information flow is bidirectional between the modules in the sense that, depending on the application (isolated words or continuous speech), constraints can be propagated from the lexical access to the verifier and vice versa.

Training the system from scratch requires four steps:

- **Codebook generation**: the Feature Extraction module performs a Mel-based cepstral analysis of the signal.The signal is collected through a head-mounted microphone, low-pass filtered at 6 KHz, and sampled at a 12 KHz rate. An FFT analysis is performed on each 10 ms frame, over 20 ms overlapping Hamming windows. At each frame, a cosine transform is applied that produces a vector of 18 cepstral coefficients. A simple endpoint detector extracts the portion of the signal corresponding to the uttered words on the basis of the energy of the frames. A fixed amount of the initial and trailing silence is kept to prevent occasional deletion of initial and final weak consonants. The Vector Quantization (VQ) module associates with every speech frame a label belonging to a finite alphabet of acoustic symbols (codebook); these symbols are used as an observation sequence by the HMMs training and verification modules. The VQ codebook is generated using the LBG clustering algorithm [35, 4, 3]. All experiments were performed using 7-bit speaker-dependent codebooks (128 codewords).
- **Sub-word units training**: the Word Translator rewrites, according to a set of phonologic rules, the orthographic description of a each word into a sequence of sub-word recognition units. This sequence is then compiled, by the Word Compiler module, into its corresponding HMM chain that is trained through the Forward-Backward algorithm [23]. The transition and emission probabilities of each sub-word unit model are obtained by processing all words of a properly designed training vocabulary. Trained units can be used as building blocks of the words of any vocabulary, and the Viterbi algorithm can estimate the likelihood that a given utterance corresponds to a word in the vocabulary. It is worth noting that none of these procedures need labeled speech, or human interaction. On the contrary, an important byproduct of stochastic modeling of sub-word units is that a speech database can be automatically segmented and labeled. In fact, once the models are trained, the Viterbi algorithm can estimate the best path through the states of the HMM chain corresponding to a known utterance, and the boundaries of the units composing the word or the sentence can be detected by a traceback procedure.
- **Phonetic Classifier training**: this module computes, from a previously labeled speech data base, the parameters of the frame by frame Phonetic Classifier. The

Phonetic Classifier estimates the likelihood that a cepstral vector belongs to a set of broad phonetic classes.

- **Estimation of Phonetic Segments matching costs**: adjacent frames with the same phonetic label are collapsed into segments by the Phonetic Segmentation module. A statistical estimation procedure generates the costs for the substitution, insertion and deletion of segments (matching costs). As will be detailed in Sect. 2.3.1, this module describes an utterance in terms of a lattice of phonetic hypotheses rather than by a single sequence of segments.

2.2.2 Feature Extraction

The selection of a good parametric representation of acoustic data is a crucial task in the design of any speech recognition system. Most parametric representations described in the literature may be divided into two groups: those based on the Fourier spectrum and those based on the linear prediction spectrum. The first group comprises filter bank energies and cepstral coefficients derived from those energies. The second group includes linear prediction coefficients (LPC). In [13] a number of parametric representations have been compared, and a clear performance advantage of the Mel-based cepstrum over all other parameter sets has been demonstrated.

2.2.3 Mel-based Spectral Analysis

A Mel-based spectral analysis is performed using a filter bank centered on the critical bandwidths of the human auditory system. Acoustic information, at the primary perceptual level, is analyzed into so called frequency groups. These groups are nearly logarithmically spaced, starting with small bandwidth at low frequencies. The spacing of these frequency groups defines a scale for the frequency selectivity of the human ear and is called the Mel-scale.
The frequencies and bandwidths of the 18-channel filter bank used in this project ([4]) are illustrated in Figure 2.4 and Table 2.1 respectively.

Speech signal is collected through a close-talk microphone and linearly digitized with a 12-bit accuracy at a 12 kHz sampling rate. Spectral analysis is performed every 10 msec using Fast Fourier Transform over 128 samples with an overlapping Hamming window of 256 samples. Finally, cepstral coefficients C_i are computed according to the following formula:

$$C_i = \sum_{j=1}^{N_F} Log(E_j) \cdot \cos i(j - \frac{1}{2})\frac{\pi}{N_F} \ , \ i = 1, \ldots, N_F - 1. \tag{2.1}$$

where N_F is the number of filters and E_i are the log-energies of each band.

Figure 2.5 further illustrates the Mel-based analysis. The first plot is related to the power density spectrum of a vowel segment computed using FFT, and the second one shows the corresponding Mel-based approximation. Channel vectors are not optimal parametric representations of speech signals, because their components are highly correlated. The first cepstral coefficients have a straightforward physical meaning: C_0 is proportional to the sum of the logarithms of the energies of each band, while C_1 represents the ratio

Band No.	Freq. Cutoff (Hz)		Central Freq. (Hz)
1	187	280	229
2	280	374	324
3	374	476	422
4	476	588	529
5	588	710	646
6	710	850	777
7	850	1009	926
8	1009	1186	1094
9	1186	1382	1281
10	1382	1606	1490
11	1606	1868	1732
12	1868	2167	2012
13	2167	2522	2338
14	2522	2942	2724
15	2942	3456	3189
16	3456	4110	3769
17	4110	4950	4510
18	4950	6071	5482

Table 2.1: Sub-bands of the filter bank according to Mel scale

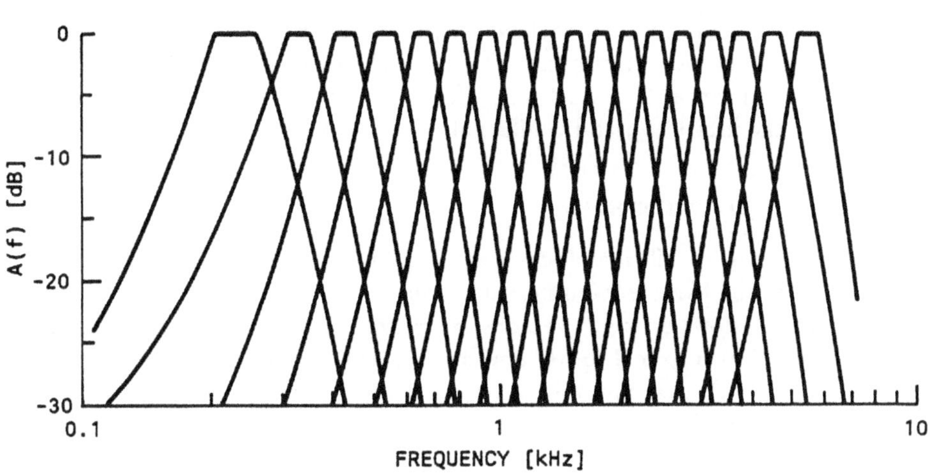

Figure 2.4: Frequency characteristics of a Mel-based set of filters

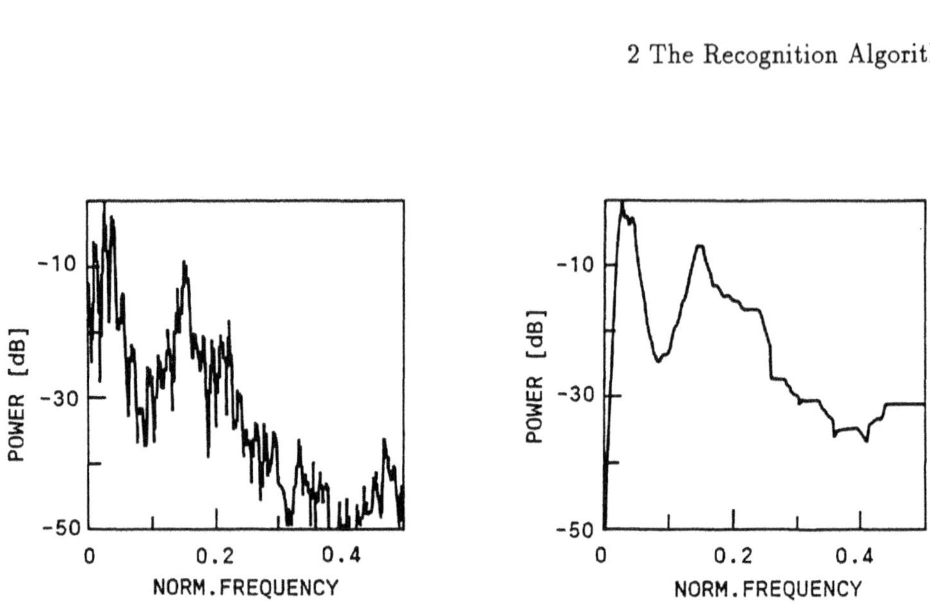

Figure 2.5: FFT and Mel-based vowel spectra

between low and high frequencies. Furthermore, cepstral coefficients C_1 to $N_F - 1$ do not depend on the frame energy. A two-dimensional histogram of the first and the second coefficients, computed over 5 minutes of speech produced by an adult male speaker, is shown in Figure 2.6.

Cepstral coefficients obtained by means of the DCT transform have the interesting property of being ordered according to their variance, with the greatest variance coming first. This allows higher-order coefficients to be discarded without affecting system performance, since low variance coefficients convey correspondingly low information about the signal. Channel and cepstral variances are compared in Figure 2.8. A plot of the cumulative variance of the cepstral coefficients (a) and of the recognition rate (b) for the difficult task of minimal pair discrimination [11] is shown in Figure 2.7. C_1 accounts for 42 % of the global variance, C_2 for 18 %, C_3 for 12 %. More than 90% of the variance is accounted for by the first 9 cepstral coefficients.

2.2.4 Vector Quantization

Vector quantization is an efficient technique of data reduction that still maintains the information needed to characterize different sounds.
Vector quantization allows an input vector $x = (x_1, x_2, \ldots, x_k)$ to be replaced by a vector $y = (y_1, y_2, \ldots, y_k)$ drawn from a finite reproduction alphabet of N elements $A = (y_i : i = 1, 2 \ldots N)$ (the codebook), that minimizes a given distortion measure.
As a matter of fact each vector of cepstral parameters is compared with a finite set of reference vectors: the nearest one, according to the distance measure, is chosen to represent the input vector. In this way the speech signal can be represented by means of a sequence of "codewords", namely labels related to the reproduction vectors. If the dimensionality of the reproduction vectors is equal to one, the technique becomes the well known scalar quantization.

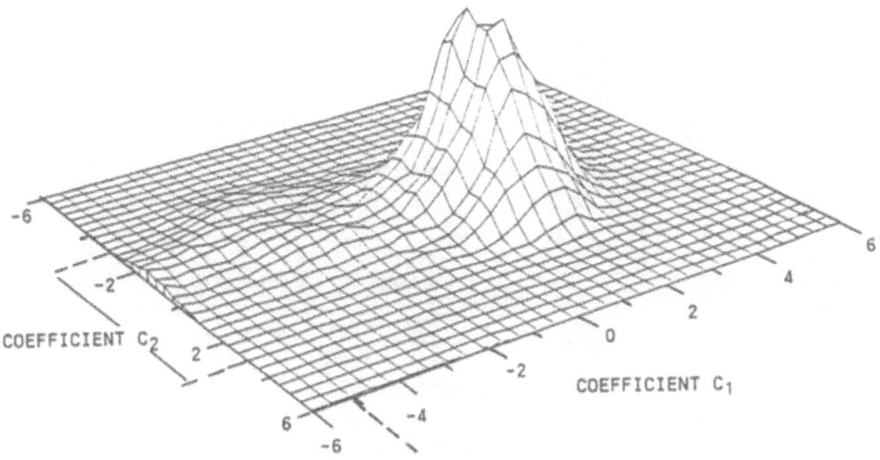

Figure 2.6: Histogram of the first and second cepstral coefficients

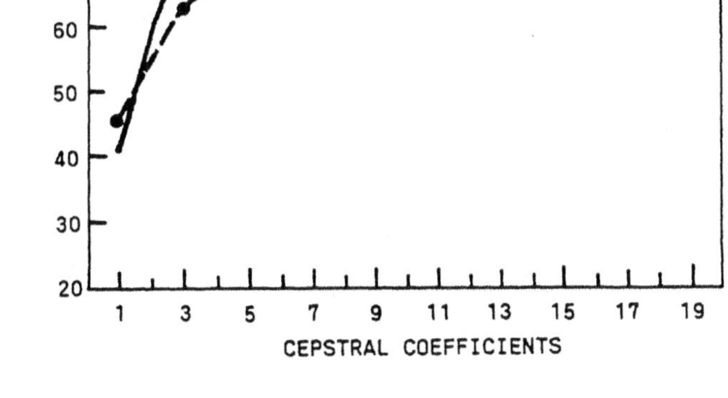

Figure 2.7: Cepstral variance and recognition rate

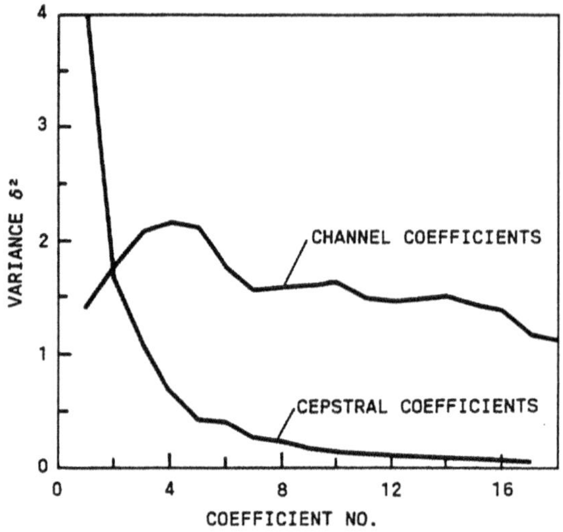

Figure 2.8: Channel and cepstral variances

Two factors affect the quantization noise, and consequently the distortion in the signal representation: the first one is the codebook design that must take into account the statistical distribution of the input vectors, while the second is the choice of the spectral distance.
Whenever detailed statistical information about the distribution of input vectors is lacking (as it happens to be in the speech signal case), the codebook can be generated by a training procedure, which requires a speech data-base usually obtained from the pronunciation of a generic text.
The LBG algorithm for codebook generation [35] is a generalization of the "K-means" algorithm, and proceeds by iterative steps: first the centroid of the whole speech data-base is computed; then a perturbation is applied to the centroid vector to obtain two new vectors, which are used to code input vectors according to the minimum distance principle. The two centroids are iteratively adjusted until a given minimum distortion threshold is reached: at this stage an optimum codebook of two (1-bit) dimensions is obtained. Each centroid is again perturbed, and a 4-level codebook is computed.
The procedure continues until a codebook of the desired dimensions is generated. Many distortion measures can be chosen: from a mathematical point of view, they must be real nonnegative functions. The mean square error measure is commonly chosen, defined as the square of the Euclidean distance between two points in the cepstral vectors space. Codebook sizes for speech applications range from 64 to 256 codewords; 128 is the size selected for the present system.
Vector quantization distortion as a function of the number of cepstral coefficients for codebooks of different sizes is shown in Figure 2.9.

Codebook generation is a rather complex and computationally expensive procedure; therefore, a speaker-independent codebook is preferable, and statistically representative material extracted from many speakers must be collected.

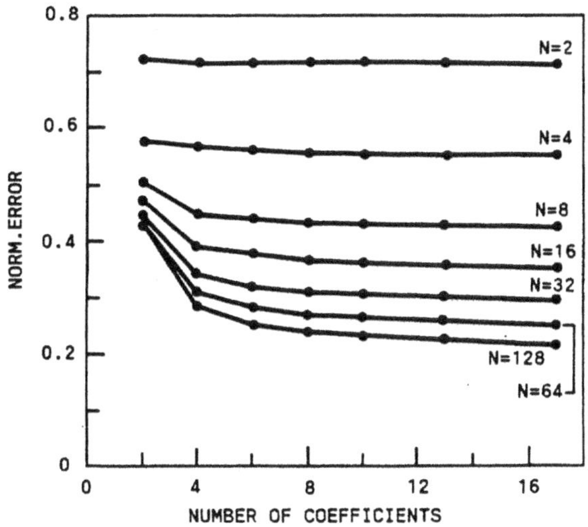

Figure 2.9: VQ distortion as a function of the number of cepstral coefficients

An efficiency measure for codebooks has been defined. Given the phonemes alphabet $P = (p_1, p_2, \ldots, p_{N_p})$ and the codewords alphabet $C = (c_1, c_2, \ldots, c_{N_c})$, an "efficiency" measure of a codebook is defined as

$$e = \frac{H(P) - H(P|C)}{H(P)} \tag{2.2}$$

where $H(P) = \sum_{k=1}^{N_p} prob(p_k) \ \log prob(p_k)$ is the "a priori" entropy of the phonemes (with respect to a given speech training data-base) and $H(P|C) = -\sum_{k=1}^{N_c} prob(c_k) \ H(P|c_k)$ is the entropy conditioned to the codebook and $H(P|c_k) = -\sum_{i=1}^{N_p} prob(p_i|c_k) \ \log prob(p_i|c_k)$ is the entropy of the phoneme alphabet given a codeword.
Codebook efficiency is 0 if $H(P|C) = H(P)$, that is if each codeword carries no information about the phoneme, while it reaches the value of 1 when each codewords univocally identifies the phoneme (actually these situations are never reached). Figure 2.10 plots the efficiency of multi-speaker codebooks as a function of the number of cepstral coefficients used for generating them.

Interestingly enough, a strong correlation can be observed between the efficiency (a) and the recognition rate (b), so that the influence of a codebook on recognition performance can be directly estimated from its efficiency score.

2.2.5 The Phonetic Representation

A vocabulary, either for recognition or training purposes, is generally available only in its orthographic form, hence the way we use to write the words using the alphabet letters. Less often the standard phonetic form is available; for instance one can have an on-line, machine-readable dictionary with the possibility of accessing the pronunciation field, but that does not solve the problem of verb inflections, proper nouns and special words that

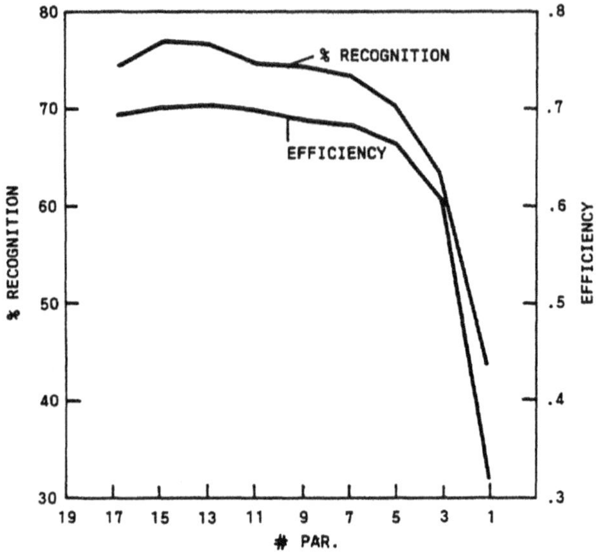

Figure 2.10: Efficiency of codebooks

do no appear in a standard dictionary (for instance jargon words). Hence it is of primary importance for doing research on large vocabulary speech recognition to have an efficient mean for handling the phonetic form of words given their orthographic description. Additionally, the phonetic form must be translated into a *recognition-unit* form that is the way a word is composed in terms of the choosen recognition units; one can simply use the phones as units, hence the last translation is trivial, but if the researcher wants to take into account more complex coarticulation phenomena and if he wants to experiment with different ways of doing that, he needs to have available a tool for defining new units starting from the phonetic description of words.

So we came to a basic representation of words lying over two levels of description. The first is the standard phonemic form of words along with additional forms accounting for inter-speaker variations due to different dialects and speaker habits. The second level is a description of each phoneme by means of smaller units called *Underlying Phonetic Structure* (UPS); they are mainly stationary segments (with a broader meaning of the word "stationary") and transitions. For instance the phoneme /k/ can be defined as a sequence of silence (the stationary portion) and transition to the following phoneme. The absence of the transition between the previous phoneme and silence means that in our unit set that event is not relevant to the recognition of /k/. Further, a set of contextual rules handles the final transcription of a word in terms of stationary and transitional units.

This development system easily allows the definition of different unit sets as phonemes, classical diphones and other.

Phonetic transcription

The first step for transforming words into recognition units is the transcription between the orthographic form to the phonetic one. For languages like Italian the transcription

keyboard	IPA	keyboard	IPA
a	*a*	u	*u*
b	*b*	v	*v*
d	*d*	z	*z*
e	*e*	-	*silence*
f	*f*	9	η
g	*g*	L	λ
i	*i*	M	η
j	*j*	N	*n*
k	*k*	&	ʃ
l	*l*	<t&>	*t*ʃ
m	*m*	<dZ>	$d\zeta$
n	*n*	<dz>	*dz*
o	*o*	<ts>	*ts*
p	*p*	w	*w*
R	*r*	(	ε
s	*s*	)	*o.*
t	*t*		

Table 2.2: The Italian phonetic alphabet

is quite straightforward and a single program, possibly including an exception table, can do the work without any manual interaction. Some problems arise with allophonic variations of phones due to regional differences in pronunciation of words. For instance in Italian, the letter *s* is sometimes pronounced as the phoneme /s/ and sometimes as the phoneme /z/ depending on the speaker provenience. In those cases we introduced multiple transcriptions of the same word; for example, the Italian word CASA (house) has two phonetic transcriptions that are */kaza/* and */kasa/* . Finally we chose an ASCII phonetic alphabet where every phone is represented by a sequence from one to 5 ASCII characters that is reported in Table 2.2 along with the IPA (International Phonetic Alphabet) symbols.

Also, a semicolon (;) following a consonant means that consonant is a geminate cluster, like /*t*/ in the word OTTO (eight, /*ot*; *o*/).

Underlying phonetic structure

As said before, the lower level of phonetic description consists in the so called Underlying Phonetic Structure (UPS); the idea is to transcribe each phoneme into a sequence of elements (*Underlying Phonetic Elements* or UPE) which, roughly speaking, show uniform acoustic characteristics. Incidentally, the alphabet used to describe UPS is the same as the phonetic one: while at the higher phonetic level each symbol represents a whole phoneme, at the lower UPS level a symbol represents a phoneme portion. The plus (+) symbol has the meaning of transition from the preceding or to the following phoneme; so writing $a = +aaa+$ means that the phoneme a (on the left of the production) can be

& = &	w = +u u u+
(= +((	z = z
) = +))	L; = L; L;+
- = -	d; = b; d;+
9 = n	n; = n;
L = L L+	N; = N N;+
M = n	f; = f;
N = N N+	g; = b; g;+
R = +R R R R+	p; = - p;+
a = +a a	b; = b; b;+
b = b b+	<dz>; = <dz>; <dz>;+
d = b d+	<ts>; = <ts>; <ts>;+
e = +e e	s; = +s; s;
f = f	k; = - k;+
g = b g+	t; = - t;+
i = +i i	<t&> = <t&> <t&>+
j = +i i i+	l; = +l; l;
k = - k+	m; = m;
l = +l l	v; = +v; v; v;+
m = m	<t&>; = <t&>; <t&>;+
n = n	&; = &;
o = +o o	<dZ>; = <dZ>; <dZ>;+
p = - p+	R; = +R R R; R+
s = +s s	<dZ> = <dZ> <dZ>+
t = - t+	<dz> = <dz> <dz>+
u = +u u	<ts> = <ts> <ts>+
v = +v v v+	

Table 2.3: UPSs for Italian

translated into a left transition ($+a$), a stationary portion (a) and a right transition ($a+$). In Table 2.3 a complete UPS for the Italian phonetic system is reported. Notice that unvoiced plosives are translated into silence (-) plus transition to the following sound while voiced plosives start with a sonorant bar (b).

The UPS of a certain phoneme is unique: a single sequence of UPE describes the phoneme. This decision leads to some considerations on the entire unit system. All the variations at the UPS level will be included in the same acoustic model. This has the effect of eventually increasing the ambiguity of the model, so degrading the performance of the recognition system. Therefore, if the variations in the phonetic structure are reasonably strong, it is better to defer to the higher level the specialization of the model, that is, to consider the variation as a different unit. Moreover, if some variations strictly depend on the context, it is easier to handle them at the higher level. The translation of a word from its phonetic form to its description in terms of recognition units starts with the translation of each phoneme into the correspondent string of UPE. According to Table 2.3 , as an example, the italian word APPARTIENE, in its basic phonetic form /ap;aRtj(ne/ can be translated into:

```
+a a - p + +a a +R R R R+ - t+ +j j+ +( ( n +e e
```

The second step consist of detecting where the transitions are, or better, of merging two consecutive transitional UPEs into one single transition unit and deleting the remaining transitional UPEs. So, following the previous example, we obtain:

```
+a a - p;a a +R R R R+ - tj j( ( n +e e
   a - p;a a    R R    - tj j( ( n    e
```

It should be noticed that defining the UPS of the generic phoneme /x/ as x = +x x+ produces the classical diphone definition.

At this point the description of the word can be handled by a set of rules to take into account the possible effects of a particular phonetic context that cannot be caught by the general UPS.

Contextual rules

Contextual rules can be expressed in the following general form:

$$U_1\ U_2 \cdots U_n \;=\; W_1\ W_2 \cdots W_m$$

where U_i and W_j are generic recognition units: the sequence of units U_i, $i = 1, 2, \ldots, n$ is translated into the sequence W_j $j = 1, 2, \ldots, n$. In our system, rules are applied sequentially in the given order to the whole word. Table 2.4 gives an example of a rule set; the symbol # is a wildcard having the meaning of a generic phoneme. In that rule set we want the phoneme /R/ to show a stationary portion only when it is not intervocalic; the UPS of /R/ is made up of two consecutive stationary portions (+R R R R+), as in Italian is impossible to utter an /R/ between two consonants and as according to the rules each vowel cuts away an R we obtain the desired transcription. The rules dealing with /v/ permits us to define only left transitions for the vowels and to have only right transitions when the vowel is followed by /v/.

The rules 1-4 cause the two vowels o and) to be represented by the same symbol o as well as the two vowels (and e; this is done because of the acoustic similarity of the sounds and due to the fact that in Italian the use of the two o's and of the two e's depends on speaker habits. Finally rule 17 transforms each geminate into the corresponding singleton as we defer the distinction between them to higher levels of knowledge.

Extending the rules to the previous example it can be easily obtained that

```
a - p;a a    R R    - tj j( ( n    e
a - pa  a   aR R    - tj je e n e
```

This formalism, developed in order to easily transcribe large lexicons into recognition units given different unit definitions (including "phonemes" and "classical diphones"), was implemented by a program (LDS Lexicon Development System) whose output is compatible both with the HMM training procedure and with word recognition and hypothesization programs. Such a program supports lexicon creation, automatic transcription from orthographic to basic phonetic forms (including main variants), unit rules compilation, transcription of words into the defined units system and relative statistics.

1	#) = #o	12	o R = o oR
2	)# = o#	13	i R = i iR
3	#(= #e	14	u R = u uR
4	(# = e#	15	R Rj = Rj
5	R Ra = Ra	16	R Rw = Rw
6	R Re = Re	17	#; = #
7	R Ri = Ri	18	a v = a av v
8	R Ro = Ro	19	e v = e ev v
9	R Ru = Ru	20	i v = i iv v
10	a R = a aR	21	u v = u uv v
11	e R = e eR	22	o v = o ov v

Table 2.4: Contextual Rules

2.3 Lexicon Structure

Knowledge representation is a central issue in the design of a large-vocabulary word recognizer. Several representations of words have been devised and experimented with that rely on different models and codes for accessing the lexicon. All models, however, describe words through a level of representation corresponding to phonemes. This assumption is also implicit in models like LAFS [28], where words are described as sequences of diphone spectral templates, and an acoustic code is the basis of the lexical access. According to most of these models, words are recognized by means of a single-step matching strategy that use all available acoustic-phonetic information. Several experiments [36, 22, 52, 32], however, pointed out that the structure of words, even partially specified, is a powerful source of constraints that is able to substantially reduce the lexicon search space. The reduction is obtained by grouping words sharing the same phonetic features into equivalence classes. According to this approach, words are described by means of a limited number of phonetic classes rather than by means of phonemes. Then, a detailed pattern matching process is performed only against the subset of candidates obtained through a less expensive selection that rules out unlikely words. A preliminary experiment, not reported here, was performed taking as its test bed a large vocabulary of Italian words [20], to evaluate the relative focusing capability of different representation schemes, with the aim of selecting a representation of words suitable to an effective lexical access. Several different classes of phonetic descriptions were considered (ranging from a very rough one to others quite close to the phonemic form), to clarify the relationship between the accuracy of the phonetic description and its selective capability. Obviously, detailed classifications result in higher selective capabilities, but a capability must be related to the complexity of achieving a detailed classification. A tradeoff consists in selecting phonetic features simple enough to be reliable and robust, but carrying sufficient information to reduce the words candidates to a reasonable size. This tradeoff must be found by taking also into account the possibility of misclassifications of a feasible acoustic-phonetic front-end.

The most important conclusions emerging from the results of that experiment are summarized in the following:

- Even a very rough description of words, as given in terms of three classes only (Sonorant, Nonsonorant and Vowel) presents a powerful discriminating ability, confirming the importance of the phonotactic shape of words.
- A detailed bottom-up classification of the Italian vowels is far less important for word discrimination than a representation that allows the separation among Front, Central and Back vowels.
- As far as consonants are concerned, the distinction between liquids and nasals is not as relevant as the distinction between fricatives and plosives.
- A good tradeoff between accuracy and selection ability is obtained by describing words in terms of six phonetic classes corresponding to plosives, fricatives, liquids/nasals, front vowels, central vowels, and back vowels.

It is worth noting that this description alphabet is very close to the classification schemes proposed for lexical access of the Italian language [30] as well as of other languages [22, 52, 47] on the basis of different analyses. Similar categories, in particular, have been proposed on a linguistic basis in the pioneering work of Shipman and Zue [49]. Even more interesting, however, is the consideration that similar broad phonetic classes are produced as a result of automatic clustering of phonemes using several different statistical methods. Consider, for example, the results of Poritz's experiment on a 5-state HMM cited in [33], and classes obtained through different optimization criteria such as the the maximization of the mutual information or transinformation [47]. Moreover, phonemes can be clustered into classes on the basis of the distance between phoneme HMMs [51], between cepstral parameters [40], or between more complex feature vectors [16, 39], confirming that the above mentioned classes can be reliably discriminated.

Lexical access is performed, therefore, through the code obtained by segmenting and classifying an utterance in terms of six broad phonetic classes.

2.3.1 Phonetic Segmentation

Phonetic segmentation is performed by two modules that work in sequence: a frame-by-frame phonetic classifier and a phonetic segmenter.

Phonetic classification

The frame-by-frame labeler estimates, by means of a hierarchical cubic polynomial classifier [1], the likelihood that a cepstral vector belongs to the phonetic classes described by the following symbols:

$k_1 = pl$: silence or plosive consonant
$k_2 = fr$: fricative consonant
$k_3 = ln$: liquid or nasal consonant
$k_4 = fv$: front vowel
$k_5 = cv$: central vowel
$k_6 = bv$: back vowel

This set of labels will be referred to in the following as "classification alphabet". It has been chosen as a result of a preliminary study on the discrimination of words in a large Italian lexicon by partial descriptions [30, 2]. A phonetic tree for the Italian and German language is represented in Figure 2.11a, in Figure 2.11b the similarity among phonemes is represented by a dendogram, and the separability into different classes by using cepstral features is illustrated in Figure 2.12.

These phonetic features are simple enough to be extracted reliably, but, at the same time, they carry sufficient information to reduce the set of words that are described by the same sequence of symbols to a reasonable size. For each 10 msec speech frame, 18 Mel based cepstral parameters $(c_0, c_1, \ldots, c_{17})$ are computed. The components of the primary pattern vector $\mathbf{x}$ used for classification are only the coefficients c_1 to c_9 and the total energy of the frame. The ideal classification of a given frame can be described by a target vector:

$$\mathbf{z} = [z_1, z_2, \ldots, z_6] \tag{2.3}$$

where

$$\begin{aligned} z_i &= 1 \quad \text{if the frame belongs to the } i\text{-th class} \\ z_j &= 0 \quad \text{if } j \neq i \end{aligned}$$

The classifier gives an estimation $\mathbf{d}$ of target vector $\mathbf{z}$

$$\mathbf{d} = [d_1, d_2, \ldots, d_6] \tag{2.4}$$

by using a cubic function of vector $\mathbf{x}$:

$$\mathbf{d} = K(\mathbf{x}) \tag{2.5}$$

The estimation is optimized for minimum mean-squared error S, defined as:

$$S = E\left[(\mathbf{z} - \mathbf{d})^T(\mathbf{z} - \mathbf{d})\right] \tag{2.6}$$

where T is the transpose operator, and $E[.]$ is the expected value. Let $\mathbf{y}$ be the secondary pattern vector obtained by appending to the vector $\mathbf{x}$ the quadratic and cubic combinations of the parameters. The relation (2.5) can be expressed by the following linear matrix equation:

$$\mathbf{d} = \mathbf{A}^T\mathbf{y} \tag{2.7}$$

and the minimization of (2.6) leads to the equation [19] :

$$E\left[\mathbf{y}\mathbf{y}^T\right]\mathbf{A} = E\left[\mathbf{y}\mathbf{z}^T\right] \tag{2.8}$$

The training procedure for the classifier estimates matrices $E\left[\mathbf{y}\mathbf{y}^T\right]$ and $E\left[\mathbf{y}\mathbf{z}^T\right]$ by using a data base of labeled speech, then it computes matrix $\mathbf{A}$ from equation (2.8) by means of a recursive procedure described in [53] and [25]. The main characteristic of this procedure is that the components of vector $\mathbf{y}$ are ordered on the basis of their significance with respect to the discrimination of the classes. The number of components of $\mathbf{y}$ used for the estimation is increased by one at each iteration which, therefore, produces a temporary result that takes into account the most important components only. The most correlated components, that are redundant for discrimination, are eliminated. On the average, a

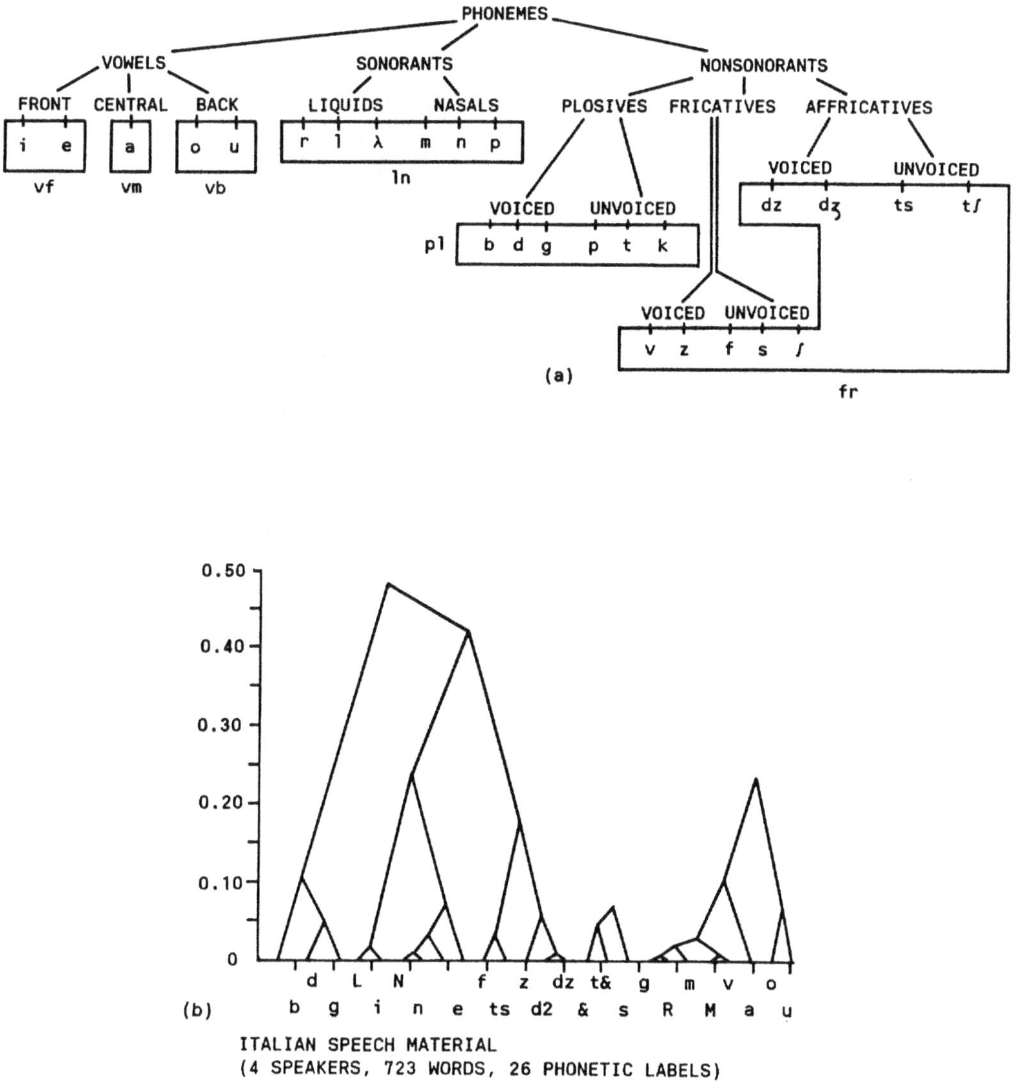

Figure 2.11: Phonetic tree and dendogram

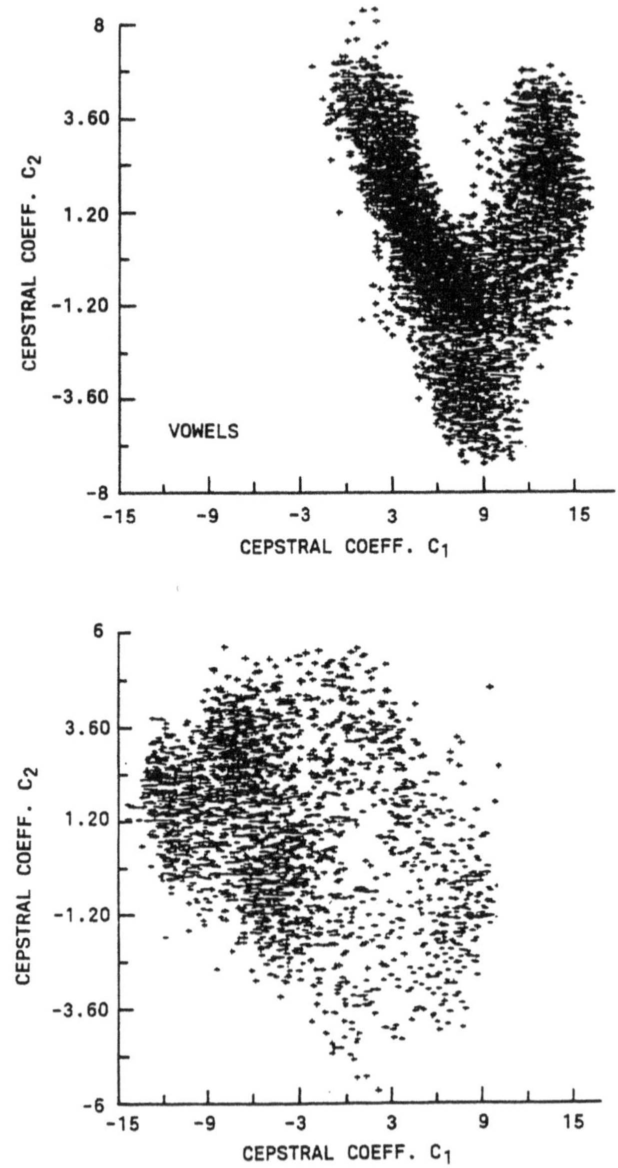

Figure 2.12: Scatter plot of C_1 vs C_2 for vowels (top) and fricatives (bottom)

test	n. of frames	pl	fr	ln	fv	cv	bv	rejection	error rate
pl	173853	86.5	4.7	3.8	1.6	1.7	1.7	0.0	13.5
fr	72875	3.2	84.3	8.1	2.6	0.4	1.4	0.0	15.7
ln	88473	1.2	1.9	83.4	7.6	2.9	3.0	0.1	16.6
fv	140065	0.6	2.0	6.4	89.9	0.8	0.2	0.1	10.1
cv	94987	0.8	1.0	3.3	1.7	91.1	2.1	0.0	8.9
bv	121844	2.1	4.4	10.2	0.4	1.7	81.1	0.1	18.9

Table 2.5: Class-to-class confusion matrix, best first decision

test	n. of frames	pl	fr	ln	fv	cv	bv	rejection	error rate
pl	173853	94.9	1.4	1.7	0.8	0.7	0.4	0.1	5.1
fr	72875	1.3	90.2	5.5	1.5	0.2	1.3	0.0	9.8
ln	88473	0.5	0.9	93.4	2.8	1.3	1.1	0.0	6.6
fv	140065	0.4	1.0	1.7	96.5	0.3	0.1	0.0	3.5
cv	94987	0.4	0.5	1.6	0.8	95.7	0.9	0.1	4.3
bv	121844	1.0	2.5	5.6	0.3	0.5	90.0	0.1	10.0

Table 2.6: Class-to-class confusion matrix, first two best decisions

reduction of the number of components, from 285 (10 linear, 55 quadratic, and 220 cubic) to 90, is observed for the secondary pattern vector.

The classifier assigns to each input frame the class k_j corresponding to the highest value component d_j of the estimation vector $\mathbf{d}$. Uncertainty and reject regions are also considered in the $\mathbf{d}$ space. If the estimated vector $\mathbf{d}$ falls in the neighborhoods of the nearest target vector, a single label is assigned to the analyzed frame, if its distance from two target vectors is within a given threshold, two labels are assigned, otherwise no decision is drawn.

A set of 1105 isolated Italian words (TRA dictionary) pronounced by 5 male and 2 female speakers was collected for training the frame-by-frame classifier. These 7735 utterances were automatically labeled in terms of phonetic units as will be described in Sect. 2.5, where training of HMMs is illustrated, and used for estimating the parameters of 7 speaker-dependent classifiers.
Another set of 1011 words, belonging to the dictionary of the geographic data base query application (GEO), was recorded by the same speakers and all the tests were performed on this set of 7077 utterances. The classifier performance, averaged among the speakers, given in terms of percentage of frames assigned to the six phonetic classes, is summarized in the class-to-class confusion matrix of Table 2.5 and 2.6. Table 2.5 shows the results considering the best first decision only, while Table 2.6 considers also the possible alternative decision. In 63% of the cases a single label is assigned to a frame.

Phonetic segmentation

The phonetic segmenter module has as input the sequence of frame-by-frame classification labels and the related component of the estimated target vector, thus for each frame n:

$$(k_{I_n}, d_{I_n}, k_{J_n}, d_{J_n}) \quad n = 1, \dots N \tag{2.9}$$

where k_i is one out of the six coarse phonetic labels and d_i is the *i-th* component of the estimation vector **d**. For those frame where only one label is generated, I_n and J_n are equal. The output of the phonetic segmenter is the so-called *micro-segmentation*, namely a structure constituted by a sequence of elements (the micro-segments) defined as:

$$M(t) = (b^t, e^t, s_1^t, a_1^t, s_2^t, a_2^t, \rho^t); \quad t = 1, \dots, T \tag{2.10}$$

where b^t and e^t are the beginning and ending frames of the micro-segment, s_1^t and s_2^t are its first and second phonetic labels, a_1^t and a_2^t are its classification reliabilities and ρ^t is a number related to the minimum segment energy. Of course, s_2^t and a_2^t are equal whenever a single hypothesis is produced.

In the following the function of the different modules constituting the phonetic segmenter will be detailed.

- PLOSIVE module
  This module give the label plosive to all frames that have 1 (silence) as energy quantization symbol. Additionally, it forces unconditionally to plosive the first and the last 3 frames of the utterance that could be inaccurate.
- REJECT module
  Since the frame-by-frame classifier can reject the classification of one or more frames, this module links unlabeled frames to labeled ones. Its rule is to give the same label as on the first left-hand frame to the left half of the unlabeled segment and the same label as on the first right-hand frame to the right half.
- MAJORITY VOTING FILTERS module
  The majority voting module processes the sequences of phonetic labels

  $$\begin{pmatrix} k_{I_n} \\ k_{J_n} \end{pmatrix} \quad n = 1, \dots N \tag{2.11}$$

  assigned to each frame n as the first and second decision respectively, and obtains a smoothed representation

  $$\begin{pmatrix} k_{I'_n} \\ k_{J'_n} \end{pmatrix} \quad n = 1, \dots N \tag{2.12}$$

  A preliminary spot-like smoothing is performed by applying the following filters:

  $$\begin{pmatrix} aba \\ *a* \end{pmatrix} \Longrightarrow \begin{pmatrix} aaa \\ *b* \end{pmatrix} \tag{2.13}$$

  $$aba \Longrightarrow aaa \tag{2.14}$$

  $$aaabaaa \Longrightarrow aaaaaaa \tag{2.15}$$

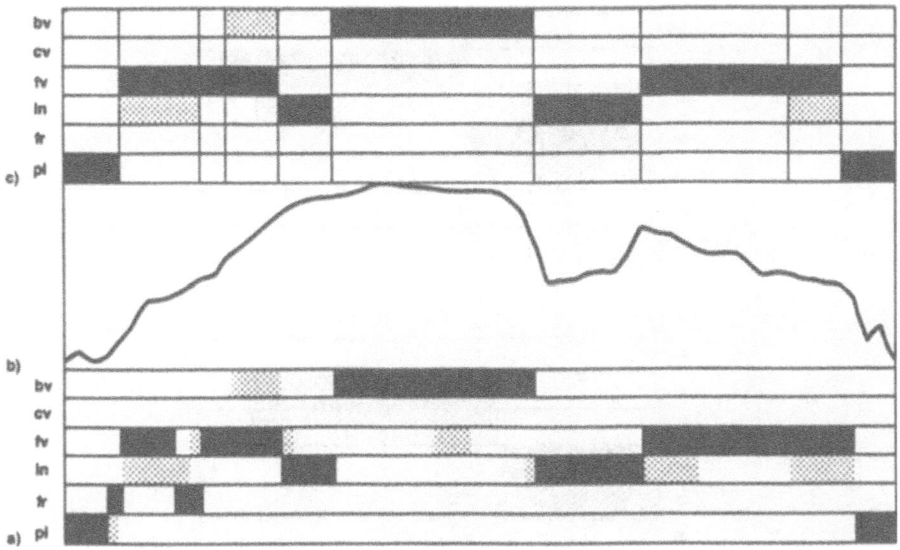

Figure 2.13: a) Frame-by-frame classification before MAJORITY VOTING FILTERS, b) energy, c) Micro-segments after MICROSEGMENTATION for the Italian word /fiume/

where a and b are specific frame-by-frame labels, $*$ is a generic label and the two rows in the formula refer to the first and second decisions respectively. Then a majority voting filter (MVF) is applied to each one of the symbol strings separately. The MVF substitutes to the central symbol of the moving window the most frequent symbol within the window. Window length is an important parameter that affects the performance of the lexical access module. As will be shown in Sect. 2.4.2, a 5-frame window length gives the minimum number of word candidates as well as the minimum computational complexity.

- MICROSEGMENTATION module
  This module defines the micro-segments as those portions of the utterance in which both the first and the second label remain unchanged. Figure 2.13 shows an example of frame-by-frame classification before the application of the majority voting filters and micro-segmentation after the application of module MICROSEGMENTATION for the Italian word *fiume* (/fjume/). The black segments represent first decision symbols, while gray ones represent alternative decision symbol frames; the phonetic class symbol corresponding to a segment can be read on the left hand side of the figure.

- ENDPOINT module
  This module detects the longest path in the micro-segmentation, from the beginning of the utterance, crossing only ***plosive*** segments. The same operation is done back-

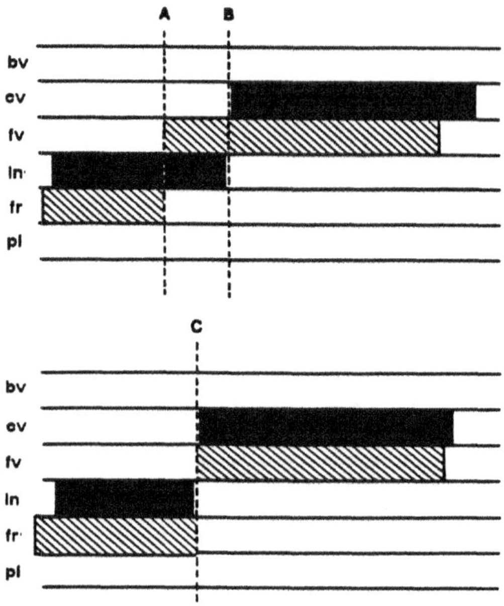

Figure 2.14: Example of how the STEP module operates

ward starting from the end of the utterance. The segments corresponding to the two paths are forced to plosive segments. This operation eliminates most of the errors due to the noise that sometimes is present at the beginning/ending of an utterance.

- ELIM module
  This module deletes from the micro-segmentation all one-frame segments.
- STEP module
  This modules eliminates the micro-segments deriving from the partial overlap of two adjacent hypotheses like the example reported in Figure 2.14. In that example the micro-segment between lines A and B is due to misalignment between the best labels and the second-best ones. There is practically no change of information in eliminating the micro-segment putting a boundary C between A and B.
- RELIABILITY module
  This module computes the reliabilities a_1^t and a_2^t for each micro-segment $M(t)$ according to the following formula:

$$a_j^t = \sum_{i=b^t}^{e^t} r_i(s_j^t) \tag{2.16}$$

where

$$r_i(s_j^t) = \begin{cases} d_{I_i} & \text{if } k_{I_i} = s_j^t \\ d_{J_i} & \text{if } k_{J_i} = s_j^t \\ 0 & \text{otherwise.} \end{cases} \tag{2.17}$$

2.4 Word Representation

Each word of the lexicon can be automatically translated, by means of a set of context-sensitive rules, from its orthographic form into a number of possible phonemic transcriptions taking into account the main speaker variations. From the phonemic forms, a set of phonetic representations of the words with different degrees of detail can be derived. The choice of a representation alphabet depends on a tradeoff between the speed-up of the lexical search due to the introduction of equivalent phonetic classes and the confusability given by a less detailed phonetic knowledge. For example, phonemes /s/ and /v/ are both fricatives, but strong fricatives like /s/ are very likely to be correctly classified as fricative consonants, while weak fricatives like /v/ are quite often classified as liquid/nasals. It is possible to better account for these classification errors by representing words in terms of more detailed classes, but this advantage must be traded with an increase of the lexical search space. A compromise has been established by evaluating the results of a set of experiments, described in Sect. 2.4.2, using three representation alphabets:

- A_1, described by the following phonetic classes:

h_1	=	Spl	:	plosive consonant
h_2	=	Lpl	:	silence or geminate plosive consonant
h_3	=	Wfr	:	weak fricative consonant
h_4	=	Sfr	:	strong fricative consonant
h_5	=	Wln	:	weak liquid or nasal consonant
h_6	=	Sln	:	strong liquid or nasal consonant
h_7	=	I	:	unstressed front vowel
h_8	=	II	:	stressed front vowel
h_9	=	A	:	unstressed central vowel
h_{10}	=	AA	:	stressed central vowel
h_{11}	=	U	:	unstressed back vowel
h_{12}	=	UU	:	stressed back vowel

  where each symbol of the classification alphabet splits into two different representation labels accounting for the difference between stressed and unstressed vowels and between strong and weak consonants,

- A_2, where the distinction between stressed and unstressed vowels has been eliminated.

- A_3, the same alphabet used for classification.

As an example, the Italian word FIUME (river), whose standard phonemic transcription is /fjume/, is represented by the following strings of symbols, depending on the description alphabet:

A_1	:	Wfr	I	UU	Wln	I
A_2	:	Wfr	I	U	Wln	I
A_3	:	fr	fv	bv	ln	fv

The representation of a word, in terms of the symbols of a description alphabet, will be referred to as:

$$W = w^1 w^2 \dots w^M \tag{2.18}$$

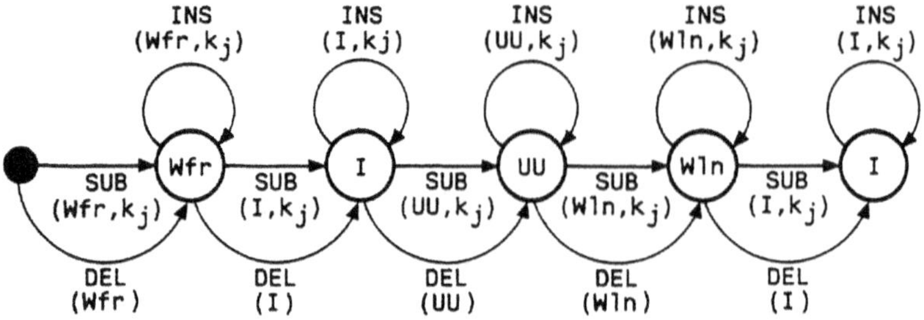

Figure 2.15: Error model of the word /fjume/

where M is the length of the representation.

2.4.1 Three-Dimensional DP Matching

A word representation which takes into account misclassifications can be modeled by a graph such as the one shown in Figure 2.15 where the symbols of alphabet A_1 are used and each link is associated with a cost $C(op(h_i, k_j))$ corresponding to the alignment operations $op(h_i, k_j)$ below:

$sub(h_i, k_j)$: substitution of test symbol k_j for reference symbol h_i

$ins(h_i, k_j)$: insertion of test symbol k_j after reference symbol h_i

$del(h_i)$: deletion of reference symbol h_i

The problem of finding the best matching of a reference word model against a test micro-segmentation can be stated as follows:

- Select one path in word description and one in micro-segmentation; each path corresponds to a string of symbols belonging to the representation and to the classification alphabet respectively.
- Compute the best alignment cost between these strings by using the costs defined in Sect. 2.4.1.
- Repeat this procedure for all path pairs.
- Select the minimum cost path pair.

Two optimizations must be performed: the innermost computes the best alignment cost between two strings, the outermost finds out the minimum cost path pair. These optimizations are carried out in a single pass by a Dynamic Programming procedure (three-dimensional DP or 3DP) that develops warping paths in the three-dimensional space illustrated in Figure 2.16. The three dimensions represent the nodes of the reference word model (dimension R), the sequence of the test micro-segments (dimension T), and the levels of the micro-segmentation lattice (dimension L). A local cost function $G(r, t, l)$ is defined in the RTL space, where r is a node of the word model associated with a symbol of the representation alphabet, t is the index of a micro-segment and l, the lattice level, assumes the values 1 or 2 referring to the best and to the second-best segmentation labels respectively. The cost function $G(r, t, l)$ can be computed, for every r, t, and l, by the

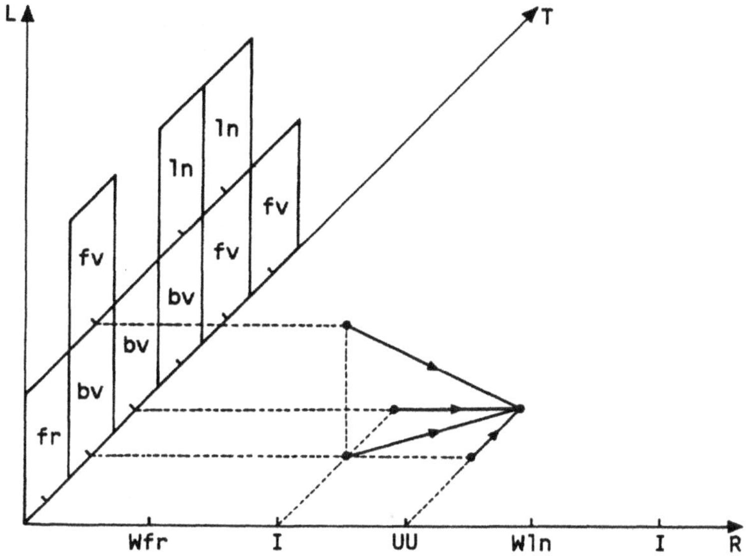

Figure 2.16: Three dimensional space

DP equations:

$$G(r,t,l) = \min_{k=1,2} \begin{cases} G(r-1,t-1,k) + C(sub(w^r, s_l^t)) & (2.19) \\ G(r,t-1,k) + IC(w^r, s_l^t) & (2.20) \\ G(r-1,t,k) + C(del(w^r)) & (2.21) \end{cases}$$

where:

$$IC(w^r, s_l^t) = \begin{cases} C(ins(w^r, s_l^t)) & if \quad s_l^t \neq s_k^{t-1} \\ CC(sub(w^r, s_l^t)) & \text{otherwise} \end{cases} \qquad (2.22)$$

where l and j assume the value 2 only if the t-th micro-segment has two classification symbols. Equations (2.19), (2.20), (2.21) account for symbol substitution, insertion, and deletion respectively. It is worth noting that this structure can lead to "false insertion" events whenever adjacent micro-segments have the same phonetic symbol. Equation 2.22 solves this case by considering a micro-segment as the continuation of the preceding one if they have the same label, CC being the "continuation cost".

For each value of r and t, 10 equations must be evaluated in the above formulation, (equation (2.21) does not depend on l). A suboptimal solution, reducing the number of equations to 4 is used instead, that, given the statistical characteristics of the segmentation process, does not substantially affect the performance of the system. In fact, the system of equations (2.19), (2.20), (2.21) carries on all locally optimal warping paths. For any given t, two optimal alignment paths exist because both the first and the alternative phonetic label of the t-th micro-segment are considered. It must be noticed, however, that if the t-th micro-segment has one label only, optimal partial paths associated with point $(r, t-1, 1)$ and with point $(r, t-1, 2)$ in the RTL space are forced to converge, in the next step of DP, to the same point $(r, t, 1)$ and the DP algorithm keeps the best one only. As a single label is associated, on the average, with 65% of the micro-segments, even if the best path selection is made at each step t, the results of the matching procedure should not be appreciably affected. In addition, as far as insertion is concerned, only the first label of an inserted micro-segment can be considered, if statistics referring to the insertion costs have been obtained accordingly in the training phase. These hypotheses have been confirmed by the experimental results given in Sect. 2.4.2. The DP equations, thus, can be modified as follows:

$$H(r,t) = \min_{k=1,2} \begin{cases} H(r-1,t-1) + C(sub(w^r, s_l^t)) & (2.23) \\ H(r,t-1) + IC(w^r, s_1^t) & (2.24) \\ H(r-1,t) + C(del(w^r)) & (2.25) \end{cases}$$

As the decision about the lattice level is made at each step, the cost function H depends only on the variables r and t.

Matching costs

A simple function for the local matching cost is:

$$C_1(op(h_i, k_j)) = -Log\left[Prob(op(h_i, k_j))\right] \tag{2.26}$$

hence:

$$\begin{aligned} C_1(sub(h_i, k_j)) &= -Log\left[Prob(substitution\ of\ k_j\ for\ h_i)\right] \qquad (2.27) \\ C_1(ins(h_i, k_j)) &= -Log\left[Prob(insertion\ of\ k_j\ after\ h_i)\right] \\ C_1(del(h_i)) &= -Log\left[Prob(deletion\ of\ h_i)\right] \end{aligned}$$

These costs are estimated in the training phase by using the same phonetically balanced vocabulary (TRA) used for training the phonetic classifier. Every uttered word is aligned to its phonetic description by means of the 3DP procedure. If a word has more than one phonetic description, the model attaining the minimum alignment cost is considered. A backtracking procedure collects, for each word, the number of substitutions, deletions, and insertions of phonetic symbols:

$$\begin{aligned} Nsub(h_i, k_j) &= Number\ of\ substitutions\ of\ k_j\ for\ h_i \\ Nins(h_i, k_j) &= Number\ of\ insertions\ of\ k_j\ after\ h_i \\ Ndel(h_i) &= Number\ of\ deletions\ of\ h_i \end{aligned}$$

When all vocabulary has been processed, the alignment costs can be estimated as follows:

$$\begin{aligned} Ntot(h_i) &= \sum_j \left[Nsub(h_i, k_j) + Nins(h_i, k_j)\right] + Ndel(h_i)] \qquad (2.28) \\ C_1(sub(h_i, k_j)) &= -Log(Nsub(h_i, k_j)/Ntot(h_i)) \\ C_1(ins(h_i, k_j)) &= -Log(Nins(h_i, k_j)/Ntot(h_i)) \\ C_1(del(h_i)) &= -Log(Ndel(h_i)/Ntot(h_i)) \end{aligned}$$

Speaker	Sex	Deletions	Incorrect Substitutions	Insertions
LA	f	8	176	2530
FR	f	25	279	2121
PD	m	18	399	2384
LF	m	38	257	2103
GM	m	44	320	1716
RP	m	34	311	2220
GP	m	15	174	2236

Table 2.7: Number of segmentation errors

These costs are re-estimated by iterating the training procedure until they do not change appreciably. Two or three iterations are generally sufficient for obtaining a stable solution. The "continuation cost" CC is null using this metric.
The initial costs are set as:

$$C_1(sub(h_i, k_j)) \begin{cases} = 0 & \text{if } h_i \text{ belongs to class } k_j \\ = 2 & \text{otherwise} \end{cases}$$

$$C_1(ins(h_i, k_j)) = 1$$

$$C_1(del(h_i)) = 1$$

This initial setting corresponds to performing a 3DP matching using a modified Levenshtein distance.
The error rates of the phonetic segmentation, computed during the estimation of the alignment costs, in terms of the number of deleted, substituted and inserted segments, are shown in Table 2.7 for 7 speakers.

The transcription in terms of 6 phonetic classes of the 1105 words of the training dictionary (TRA) generates a total of 7115 phonetic segments, i.e. 6.4 distinct phonetic segments per word, on the average. Segmentation produces, instead, 14.4 segments per word on the average. It is worth noting, however, that every segment in excess is not a spurious insertion since many of them are continuation segments as defined in equation (2.22) of Sect. 2.4.1. Table 2.7 shows that deletions and substitutions of segments are not very frequent, while more than 2 insertions per word can be expected. The highest contribution to the insertions is due to fricative and liquid/nasal consonants as shown in Table 2.8, where the percentage of substituted, deleted and inserted segments, averaged over all speakers, is detailed for each class.

An example of the 3DP matching is offered in Figure 2.17 where the best alignment path is outlined.

Duration of micro-segments

Metric C_1, defined in (2.26), does not take into account the micro-segmentation timing structure, a very important cue for word hypothesization. A straightforward way to include the duration of micro-segments in the matching cost is the following:

$$C_2(h_i, k_j, op(h_i, k_j)|len(M_j)) = -Log(Prob(op1(h_i, k_j)) * len(M_j) \qquad (2.29)$$

Class	Deletions	Substitutions	Insertions
pl	0.47	0.23	0.54
fr	0.23	0.36	5.20
ln	0.11	0.11	7.10
fv	0.17	0.34	1.74
cv	0.06	0.00	1.41
bv	0.08	0.12	1.32
Total	1.12	1.16	17.31

Table 2.8: Percentage of deleted, substituted, and inserted segments for each phonetic class

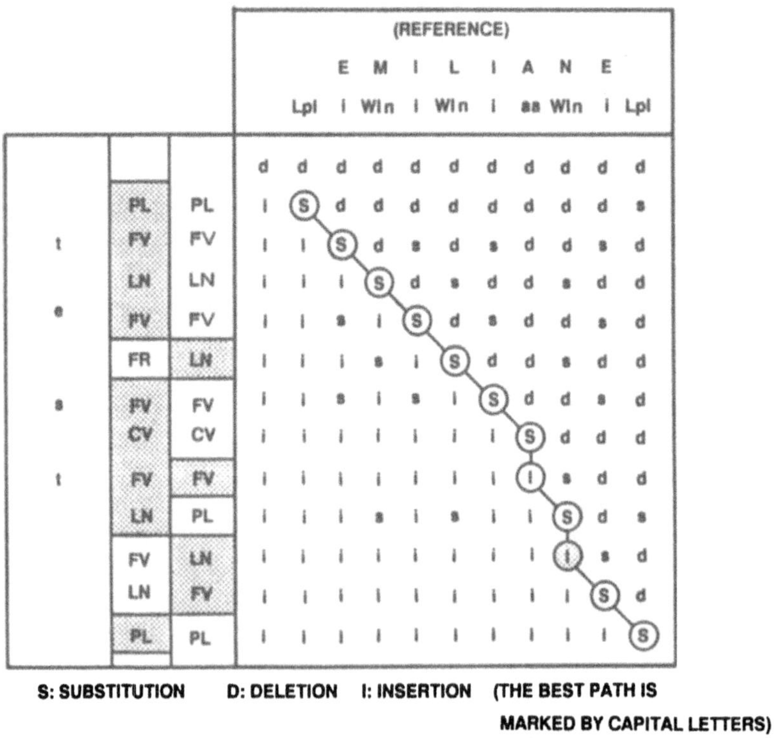

Figure 2.17: 3DP matching: best alignment path

where $op1(h_i, k_j)$ is the basic alignment operation of *one test frame*, labeled k_j, against the reference symbol h_i, and $len(M_j)$ is either the duration of micro-segment M_j, if $op1(h_i, k_j)$ is a substitution or an insertion operation, or it is the average duration of the h_i phonetic class corresponding to a deletion operation.

Reliability of micro-segments

Let the reliability $rel(k_j, M_j)$ of the k_j label of micro-segment M_j lie in the interval $[x1, x2]$; a local alignment cost can be defined in term of the conditional probability of the segment:

$$CR(op(h_i, k_j), rel(k_j, M_j), x1, x2) = -Log(Prob(op(h_i, k_j)|x1 < rel(k_j, M_j) \leq x2)) \tag{2.30}$$

By applying the Bayes relation, equation (2.30) can be rewritten as:

$$CR(op(h_i, k_j), rel(k_j, M_j), x1, x2) = \tag{2.31}$$
$$-Log\left[\frac{Prob(op(h_i, k_j)) * Prob(x1 < rel(k_j, M_j) \leq x2|op(h_i, k_j))}{Prob(x1 < rel(k_j, M_j) \leq x2)}\right]$$

Factor $Prob(x1 < rel(k_j, M_j) \leq x2)$ can be neglected because it neither depends on the reference symbol h_i nor on the $op(h_i, k_j)$. Hence, assuming the statistical independence of the alignment operations and of the micro-segment reliability, a matching cost function can be defined as the sum of two contributions, an alignment cost A and a reliability cost R as follows:

$$\begin{aligned} CR'(op(h_i, k_j), rel(k_j, M_j), x1, x2) = \\ A(op(h_i, k_j)) + \\ R(x1 < rel(k_j, M_j) \leq x2|op(h_i, k_j)) \end{aligned} \tag{2.32}$$

where

$$A(op(h_i, k_j)) = -Log(Prob(op(h_i, k_j)) \tag{2.33}$$

and

$$R(op(h_i, k_j), rel(k_j, M_j), x1, x2) = -Log(Prob(x1 < rel(k_j, M_j) \leq x2|op(h_i, k_j)) \tag{2.34}$$

thus, by taking into account the duration of micro-segments:

$$\begin{aligned} CR''(op(h_i, k_j), rel(k_j, M_j), x1, x2) = \\ C_2(h_i, k_j, op(h_i, k_j)) + \\ R(x1 < rel(k_j, M_j) \leq x2|op(h_i, k_j)) \end{aligned} \tag{2.35}$$

The reliability cost R can be estimated in the training phase by collecting statistics for each operation $op(h_i, k_j)$ into an histogram. The estimation can be simplified by considering that a micro-segment has associated with it one or two phonetic labels (s_1 and s_2) ordered according to their reliability. The histogram, thus, can be reduced to only three cells accounting for the following events:

- r_1 : the drawn symbol k_j is s_1 and $s_2 = nil$
- r_2 : the drawn symbol k_j is s_1 and $s_2 \neq nil$
- r_3 : the drawn symbol k_j is s_2

As a second simplification, the following classes of events are clustered:

- $e_1(k_j) : op(h_i, k_j) \equiv sub(h_i, k_j) \ \forall \ h_i \in k_j$ (correct matching)
- $e_2(k_j) : op(h_{i,j}) \equiv sub(h_i, k_j) \ \forall \ h_i \ni k_j$ (incorrect substitution)
- $e_3(k_j) : op(h_i, k_j) \equiv ins(h_i, k_j) \ \forall \ h_i$ (insertion)
- $e_4(k_j) : op(h_i, k_j) \equiv del(h_i) \ \forall \ h_i$ (deletion)

Hence, given p and q, the reliability cost can be computed as:

$$R(r_p \ , \ e_q(k_j)) = -Log(Prob(r_p \mid e_q(k_j)) \ ; \ p = 1, \ldots, 3 \ ; \ q = 1, \ldots, 4 \tag{2.36}$$

and the local cost as:

$$C_3(h_i, k_j, op(h_i, k_j)) = C_2(h_i, k_j, op(h_i, k_j)) + R(r_p \mid e_q(k_j)) \tag{2.37}$$

2.4.2 Lexical Access

Even if the number of phonetic micro-segments in a word is, on the average, less than the number of centisecond frames by about an order of magnitude, the complexity of matching a micro-segmentation against every vocabulary word is impractical when the lexicon size is of the order of thousands. A representation that reduces storage costs and leads to an efficient lexical access is obtained by merging the sequences of phonetic classes that describe the words in a tree in which the initial common subsequences are shared [50, 14, 30, 47]. If the nodes of the lexical tree represent phonetic classes, all words which share the same coarse phonetic description can be associated with the same node (the node representing the last phoneme) as they become a set of phonetically indistinguishable lexical items. An example of a simple 12-word lexical tree is shown in Figure 2.18, where all leaves and some (terminal) nodes are associated with the set of lexical items having the same phonetic structure. A tree is best suited to the lexical access task, rather than a more compact graph structure, because the former allows the N best word candidates to be easily obtained. The 3DP algorithm, in fact, can evaluate the alignment costs of all vocabulary words in parallel. This operation would be more complex and expensive if performed on a graph.

Given the micro-segmentation of an uttered word belonging to a lexicon represented by a tree TN, lexical access is performed by detecting the sequences of phonetic nodes TN(i), and hence the corresponding words, whose costs computed by means of the 3DP lie within a fixed range of the best one.

Each node TN(i) of the lexical tree is characterized by the following (static) information:

PHON_ID	:	phonetic identifier
FIRST_SON	:	pointer to its first son
BROTHERS	:	pointer to the list of its brothers
LEQW	:	list of phonetically equivalent words

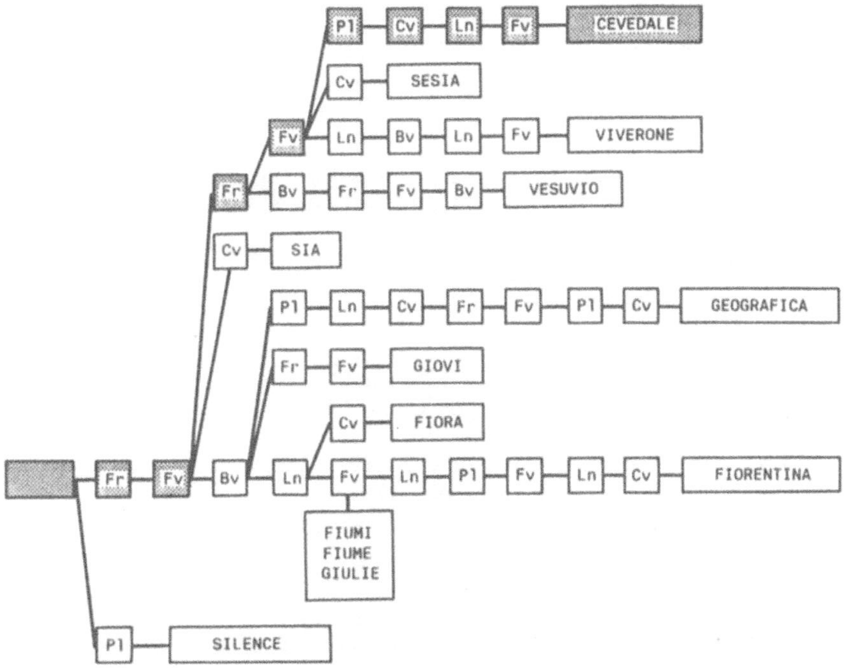

Figure 2.18: A lexical tree

where the phonetic identifier PHON_ID is a symbol of the representation alphabet, the FIRST_SON pointer identifies the first son of node TN(i), while BROTHERS is the pointer to the list of the brothers of TN(i), i.e. the nodes sharing the same father, and LEQW is the (possible empty) list of words that share the same path from the root node TN(0) to TN(i).

As lexical access is based on the expansion of the TN tree, a set of two-element arrays is added to the static information of the nodes in order to save the values needed for carrying on the 3DP procedure:

COST	:	optimal alignment cost array
DECS	:	decision symbol array
LINK	:	next active node pointer array

For each $t = 1, \ldots, T$, the COST array stores the cost values of the current best alignment paths ending in node TN(i) after micro-segments $M(t)$ and $M(t+1)$ have been observed. The DECS array stores the labels of micro-segments $M(t)$ and $M(t+1)$ that are drawn by the optimization process; this information is used for dealing with "false insertion" events. Links to nodes that must be expanded are held into the LINK array. The expansion of the TN tree is controlled by a beam search strategy, working on two lists of active nodes, according to the basic steps described in Pascal-like language in Table 2.9.

When the last micro-segment M(T) is observed, the active nodes in list L(*current*) are processed only for possible extension of the best paths through deletions. The set of

Initialize variables *current* and *next* (referring to the list of active nodes to be expanded in the current and next cycle of the search) to 0 and 1 respectively;
Initialize t, and COST(0) of the root node TN(0) to 0; The COST elements of all other nodes are set to ∞;
Create empty lists L(*current*) and L(*next*), append TN(0) to L(*current*).
repeat

  Reset list L(*next*);
  repeat
    Examine TN(i), next node in L(*current*).
    If TN(i).COST(*current*) exceeds the cost of the optimal path up to the micro-segment M(t-1) of a fixed threshold, **then**
      the best path associated with node TN(i) is extended no further, according to the beam search strategy.
    Else
      If the insertion cost obtained by using equation (2.24) of the 3DP procedure is less then TN(i).COST(*next*), the cost of the current best path up to micro-segment M(t+1) and node TN(i), **then**
      Set TN(i).COST(*next*) to the new value and append node TN(i) to list L(*next*) using TN(i).LINK(*next*) unless it is already there.
      For every son TN(j) of TN(i) **do**
        If the deletion cost obtained by using equation (2.25) of the 3DP procedure is less then TN(j).COST(*current*), the cost of the current best path up to micro-segment M(t) and node TN(j), **then**
        Set TN(j).COST(*current*) to the new value and add node TN(j) just after node TN(i) into list L(*current*).
        If the substitution cost obtained by using equation (2.23) of the 3DP procedure is less then TN(j).COST(*current*), the cost of the current best path up to micro-segment M(t) and node TN(j), **then**
        Set TN(j).COST(*current*) to the new value and append node TN(j) to list L(*next*).
      Reset TN(i).COST(*current*) to ∞ and TN(i).LINK(*current*) to *nil.*
  until every node in L(*current*) has been processed;
  swap(C,S);
  Set t to t+1;
until micro-segment M(T-1) has been observed.

Table 2.9: Lexical access algorithm

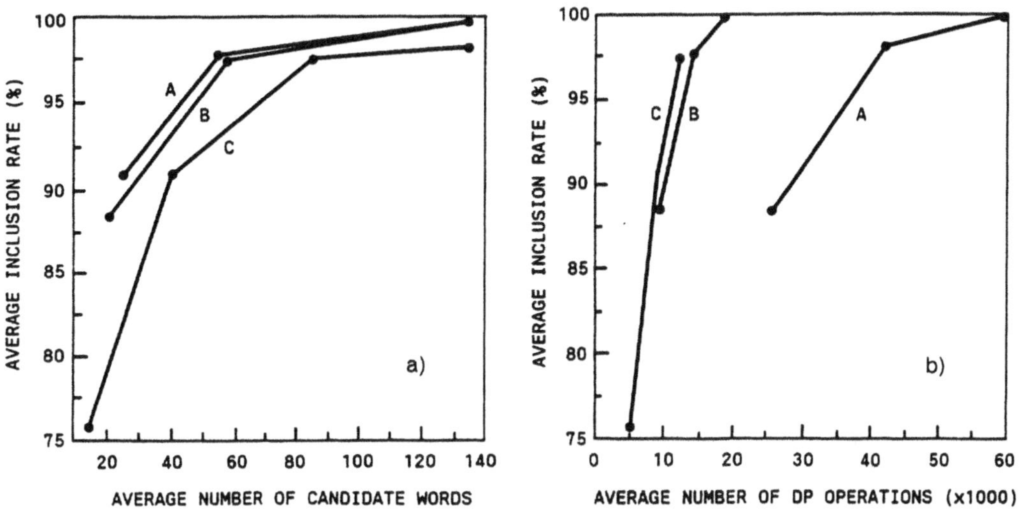

Figure 2.19: DP matching procedure comparison: a) Average inclusion rate vs. average number of candidate words, b) Average inclusion rate vs. average number of DP operations

candidate words can be, then, easily retrieved by means of the LEQW of nodes contained in the list L(*current*). The average dimension of this set is controlled by the value of the beam search threshold.

Experimental results

A first set of experiments was devoted to the assessment of the 3DP method. The complete set of 1011 words of the GEO vocabulary pronounced by a male speaker was used as test.

Figure 2.19a shows the rate of inclusion of the correct word in the candidate list versus the average number of candidate words for three different matching procedures, namely optimal 3DP (curve A), sub-optimal 3DP (curve B), and DP matching of the best first segmentation hypotheses only (curve C). Word models were represented by means of the symbols of alphabet A_1, and the C_1 metric was used for the evaluation of the costs. The curves were obtained as a function of the beam search threshold.

The 3DP procedure performs considerably better than classical DP: fewer candidate words and higher inclusion rates are obtained. The optimal and the sub-optimal procedure give very close results but the complexity of the sub-optimal procedure is comparable with the complexity of the classical DP (see Figure 2.19b), in fact, for each reference node and for each micro-segment, four equations rather than three must be evaluated. Sub-optimal 3DP has been, therefore, used in all remaining experiments.

A second set of experiments was carried out for selecting the best representation alphabet. The same test was performed by representing the GEO vocabulary words through the symbols of the alphabets A_1, A_2 and A_3 introduced in Sect. 2.3.1. Table 2.10 shows the number of nodes (N), the number of leaves (L), the terminal nodes (T), and the average branching factor of the obtained lexical trees.

Alphabet	N. of nodes	N. of leaves	N. of terminal nodes	Branching factor
A_1	2656	801	894	1.431
A_2	2178	704	875	1.477
A_3	1739	569	749	1.485

Table 2.10: Number of nodes, leaves, terminal nodes, and branching factor of the 1011 word GEO lexical trees using three representation alphabets

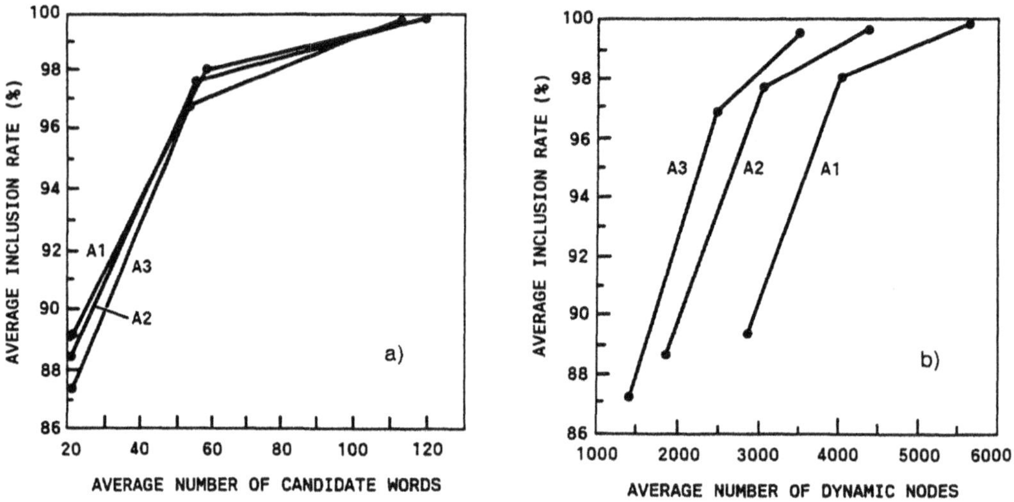

Figure 2.20: Representation alphabets comparison: a) Average inclusion rate vs. average number of candidate words, b) Average inclusion rate vs. average number of expanded nodes

The curves of Figure 2.20a, that present the inclusion rate versus the average number of candidates obtained by varying the beam search threshold, suggest that a more detailed specification of the lexical tree, such as that offered by alphabets A_1 and A_2, does not substantially reduce the candidate average size at inclusion rates greater than 99%. Better performance of alphabets A_1 and A_2, with respect to alphabet A_3, for more constraining beam search thresholds, is not surprising because more information is conveyed by their alignment cost matrices. However, due to the scarce redundancy of the micro-segmentation code, large values of the beam search threshold must be used for obtaining acceptable high performance. Thus, coarseness of matching renders the accuracy of the model unhelpful. Furthermore, the computational load increases when more detailed representation alphabets are used, as shown in Figure 2.20b, where the inclusion rate is plotted versus the average number of nodes expanded during the search. A_3 has been, therefore, used as the representation alphabet in all successive experiments.

The third experiment has been carried out to assess system performance as a function of the above described metrics C_1, C_2 and C_3. Its results are summarized in Figure 2.21a

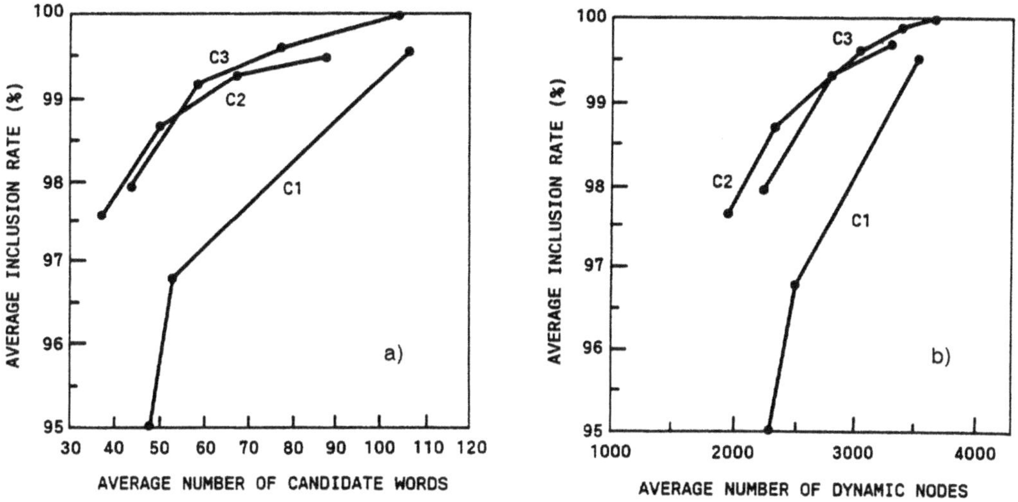

Figure 2.21: Comparison of different metrics: a) Average inclusion rate vs. average number of candidate words, b) Average inclusion rate vs. average number of expanded nodes

and Figure 2.21b.

Timing information (metric C_2) gives substantial improvements, and further improvements are obtained by using the reliability of the phonetic labels (metric C_3).

The next set of experiments was performed for seven speakers, in the best conditions suggested by the previous experiments: sub-optimal 3DP, A_3 representation alphabet and C_3 metric. Figure 2.22a shows, for various beam search thresholds, the inclusion rates and candidate list size for all speakers, while Figure 2.22b presents the averaged results. The difference of the average inclusion rate among speakers is within 1% for the same beam search threshold value. Larger values of the threshold do not affect appreciably the accuracy of the hypotheses, while they considerably increase the average number of candidate words, and the computational load. On the average, only about 10% of the items in the lexicon must be verified, and substantial improvement can be obtained by taking into account the heuristics introduced in Section 2.4.2. Figure 2.23 shows the average number of word candidates as a function of the number of syllables in a word; superimposed, as a bar graph, is the distribution of words in the GEO vocabulary as a function of their number of syllables. Short words generate a large number of candidates because the shorter the uttered word is, the easier it is to find, in a large vocabulary, similar or slightly different words in terms of a phonetic description into coarse classes. Errors are uniformly distributed among words composed of 2, 3 and 4 syllables. No errors were observed for monosyllabic or very long words. Monosyllabic words are generally well segmented and classified, when pronounced in isolation, because they are pronounced slowly with respect to the syllables of polysyllabic words, as can be observed in Figure 2.24 where the average syllable duration is shown as a function of the number of syllables in a word.

Figure 2.25a shows the inclusion rate as a function of the position of the correct word

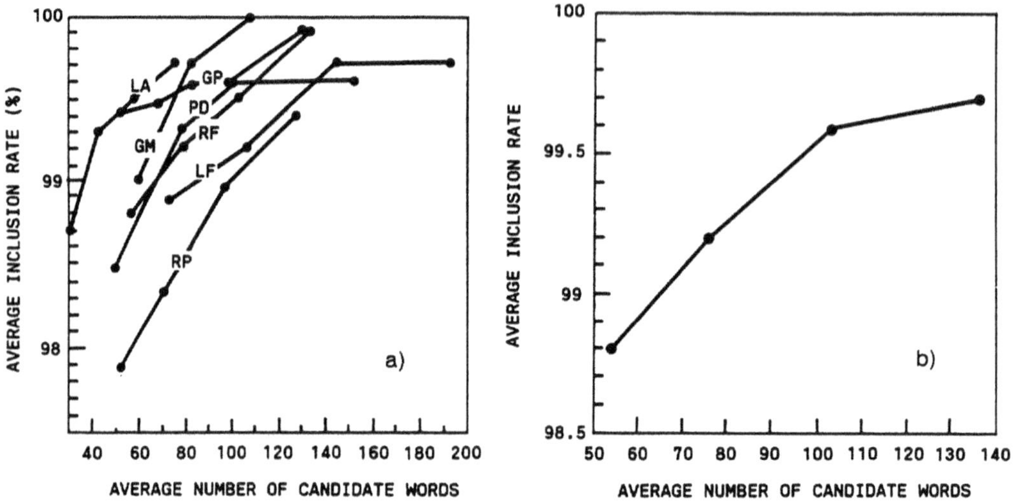

Figure 2.22: Results as a function of the beam search threshold: a) for 7 speakers (5 male, 2 female), b) averaged results

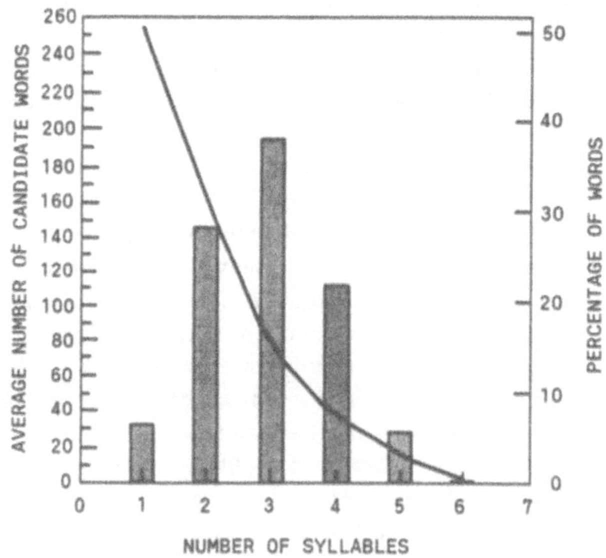

Figure 2.23: Average number of hypothesized words and distribution of words in the GEO vocabulary vs. their number of syllables

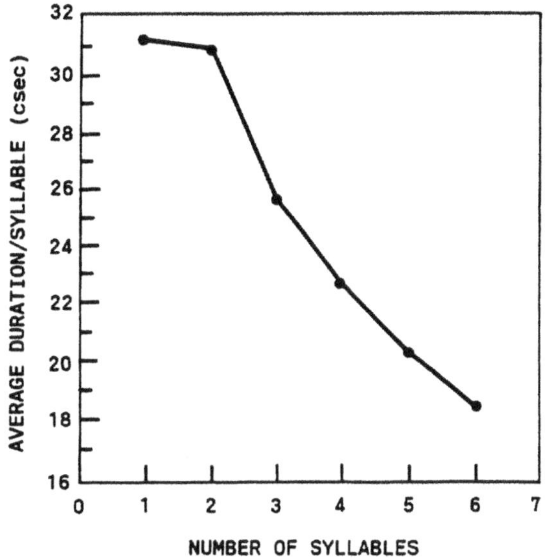

Figure 2.24: Syllable duration

in the list, ordered by cost, of candidates generated by the lexical hypothesizer; in 62% of the cases the correct word is in the set of the best-scored phonetically equivalent words and only one word is hypothesized in 13% of the cases, as shown in Figure 2.26a where a histogram representing the distribution of the size of the candidate word list is reported.

Use of heuristics

Robust heuristics can be introduced in the lexical access procedure to reduce the average number of hypothesized words and to speed up computation.

The first one (H_1) smoothes out the strings of phonetic labels produced by the frame-by-frame classifier through a majority voting filter. Two strings of symbols are considered: one corresponding to the best first classification and the other one corresponding to the sequence of alternative decision labels. The second decision symbol is set to the value of the best one whenever the classifier has taken a single decision. The majority voting filter, applied to a shifting window of N (odd) frames, associates to the central frame of the window the phonetic labels that most frequently appear as the best first and the alternative decision respectively. Fewer micro-segments are obtained because many spurious segments are eliminated. This reduction of the number of micro-segments reduces the number of operations needed for matching as well. Unfortunately, by increasing the window length, some correct segment disappears. Therefore, the number of spurious insertions decreases, but the number of deleted segments increases. The optimal window length depends on the speaking rate. Several experiments were performed for all 7 speakers varying the beam search threshold in order to achieve, for a given length of the majority voting filter window, an average inclusion rate of 99.7%, the same obtained excluding any filtering (window length equal to 1). The results are shown in Figure 2.27 where the

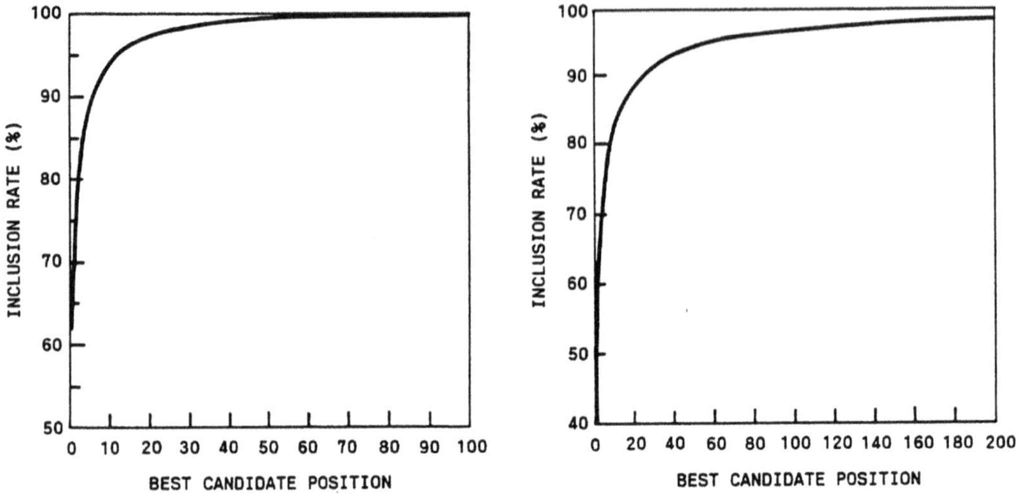

(a) 1011 words (b) 18388 words

Figure 2.25: Cumulative inclusion rate as a function of N-best candidate words

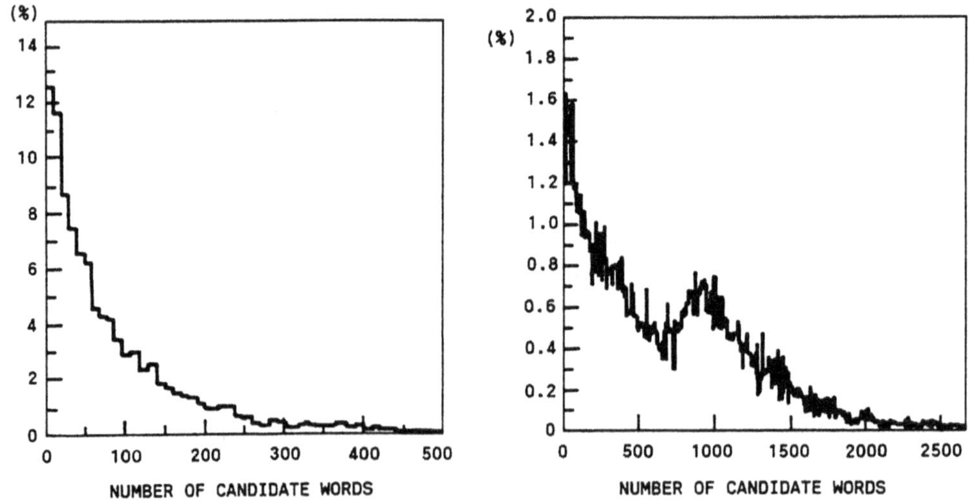

(a) 1011 words (b) 18388 words

Figure 2.26: Number of candidate words histogram

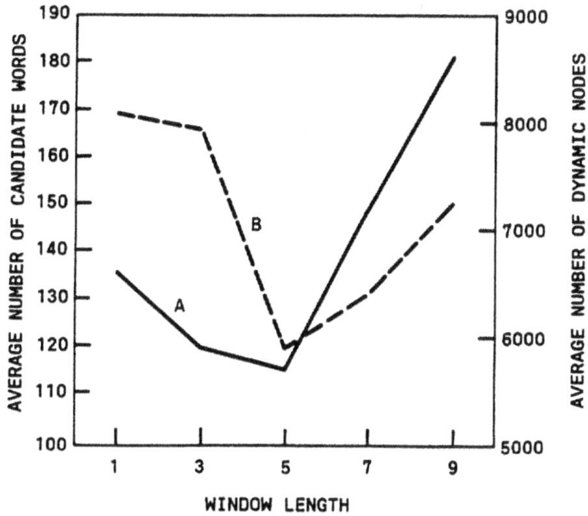

Figure 2.27: Average number of candidate words (A) and average number of expanded nodes (B) per word vs. filter window size

average number of candidate words (curve A), and the average number of nodes expanded per word (curve B) are plotted as a function of the filter window size. A window size value of 5 frames gives the minimum number of word candidates as well as the minimum computational complexity. A second heuristic (H_2) refers to reliable segments. Let R (s_i^t) be a function that associates a number r_i^t to the label s_i^t of a micro-segment $M(t)$ and let R (s_i^t) be monotonically increasing with the probability that s_i^t is a correct classification of the micro-segment $M(t)$. If such a function exists, and if it is continuous, a threshold **z** and a value **v** can be found such that:

$$r_i^t > \mathbf{z} \quad \Longrightarrow \quad Prob\left[s_i^t \text{ } is \text{ } a \text{ } correct \text{ } classification \mid M(t)\right] > \mathbf{v} \qquad (2.38)$$

Thus, in principle, a threshold **z** can be chosen such that it is possible to detect segments whose probability of being misclassified is below a fixed value, or, in other words, segments that can be considered correctly classified with a given confidence value. The reliability measure associated with micro-segments can be chosen as function R according to the results shown in Figure 2.28, where an estimation of the probability that a micro-segment label is correct, given its reliability, is presented for each phonetic class. A value of the threshold $\mathbf{z_m}$ (shown by an arrow in the figures) was fixed for each class k_m, so that all training set segments with reliability greater than $\mathbf{z_m}$ were correctly classified:

$$a_i^t \; > \; \mathbf{z_m^t} \quad \Longrightarrow \quad s_i^t \; \in \; k_m^t \qquad (2.39)$$

During lexical access, a segment satisfying the above mentioned conditions is considered correctly classified. Hence, it cannot be inserted or substituted for a reference symbol that does not belong to the same phonetic class. This further local path constraint in the

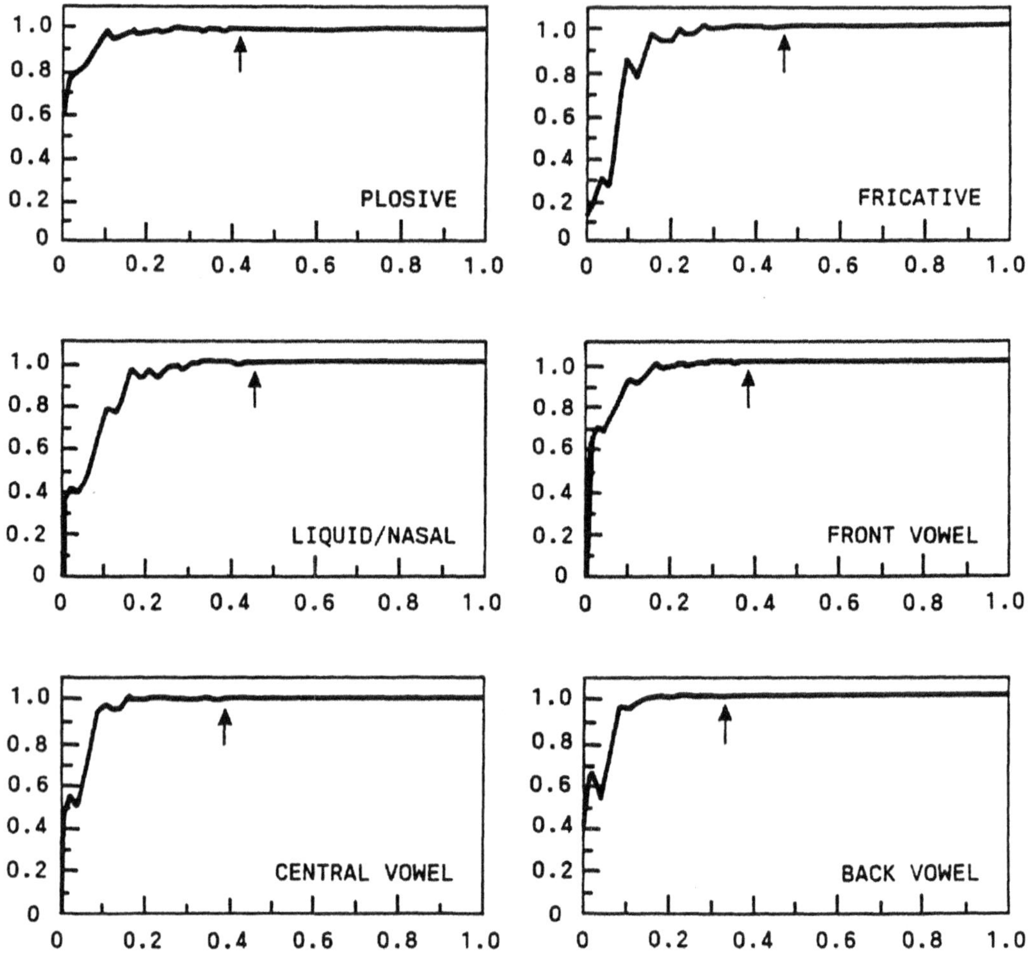

Figure 2.28: Probability of correct classification of a micro-segment vs. its reliability

3DP procedure has two beneficial effects: an appreciable reduction of the computational load and of the average number of word candidates, for the same inclusion rate. As can be observed in Table 2.11, a small reduction of the inclusion rate is traded for a significant reduction of the average candidate word number and of the computational load expressed in terms of average number of expanded nodes.

Similar considerations lead to a third heuristic (H_3) that exploits robust cues for deciding that a particular phonetic class cannot be hypothesized for a given segment. If phonetic class k_n cannot definitely be assigned to a micro-segment, it cannot be substituted in the 3DP matching for a symbol of the representation alphabet belonging to class k_n. Frame energy, for example, has been used as a cue for deciding that a high energy micro-segment cannot be substituted for a plosive sound. Table 2.11 shows the performance obtained by using the H_2 and H_3 heuristics, and a 5-frame window majority voting filter (H_1).

As mentioned in the preceding subsection, the largest set of word candidates is generated by short words which, however, are generally well segmented. Hence, it is likely that the correct word is in the first few positions in the candidate list. candidate word list, but the list is generally very short. The fourth heuristic (H_4) introduces, therefore, a constraint on the maximum number of active *nodes* of the lexical tree that are considered for word retrieval at the end of the search: only the N best nodes are allowed to generate word hypotheses. This constraint is not used during the search because it would be too expensive to order the best partial paths according to their cost, rather than performing a simple beam search.

Figure 2.25 shows that more than 99% of inclusion rate can be obtained keeping only the first 60 best candidate *nodes*. Recall that the number of candidate *nodes* is different from the number of candidate *words*, since more than one word can be associated with a candidate node. This result is also illustrated in Figure 2.29, that shows the inclusion rate and the average number of word candidates obtained by varying the value of the maximum number (j) of best candidate *nodes* (Mj, $j = 40, \ldots, \infty$) kept by the hypothesizer. The performance of the system using all these heuristics, constraining the maximum number of final active nodes to 140 is detailed in last column of Table 2.11

In Figure 2.30a the average inclusion rate is shown as a function of the vocabulary size. Refer also to Figure 2.25b and Figure 2.26b for statistics about experiments made with the 18388-word vocabulary. By increasing the vocabulary size from 1011 words to 18388, and using the same beam search threshold, a slight reduction (0.7%) of the average inclusion rate is observed. The increase of the average number of word candidates is presented in Figure 2.30b. It is worth noting that the percentage of the vocabulary words that must be verified decreases as vocabulary size increases: the bold right lines in the figure represents 10% and 2% of the vocabulary size respectively.

2.5 Verification Module

This module applies a more detailed phonetic knowledge than the phonetic classification.

A Word Translator generates, from the orthographic form of the words, one or more phonetic transcriptions through a set of rules. Multiple transcriptions are due, for example, to the ambiguity introduced by affricates and by intervocalic /s/ that, in Italian, can be voiced or unvoiced depending on the speaker's regional attitude. Moreover, diphthongs

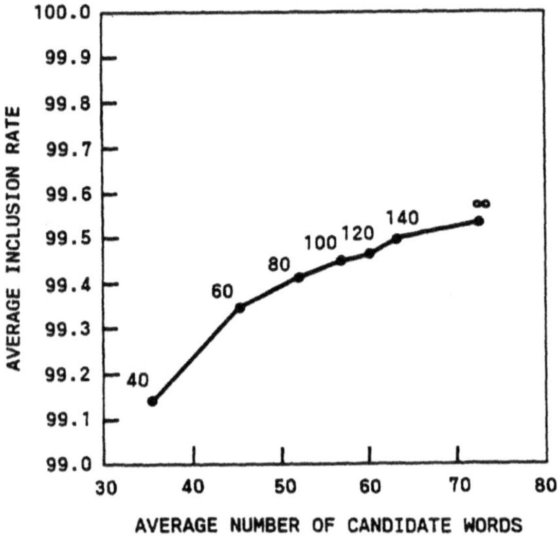

Figure 2.29: Average inclusion rate and average number of candidate words as a function of the number of final active nodes

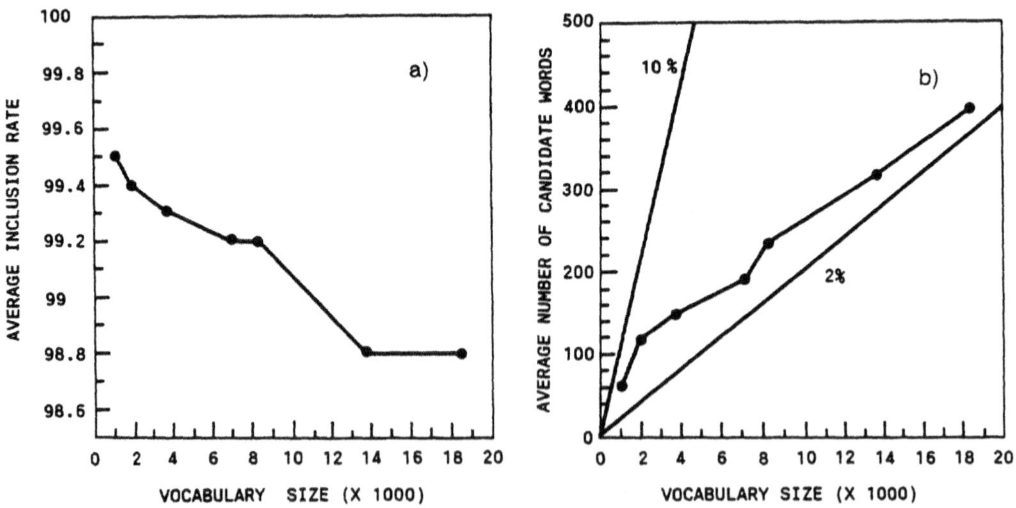

Figure 2.30: a) Average inclusion rate and b) average number of candidate words as a function of the vocabulary size

Heuristics	H_1	$H_1 + H_2$	$H_1 + H_2 + H_3$	$H_1 + H_2 + H_3 + H_4$
Insertion rate	99.7	99.5	99.5	99.5
Average number of candidates	115.2	79.1	72.4	62.8
Average number of expanded nodes	5874	4546	4280	4280
Average number of operations	14570	9828	8771	8771

Table 2.11: Word hypothesization module performance

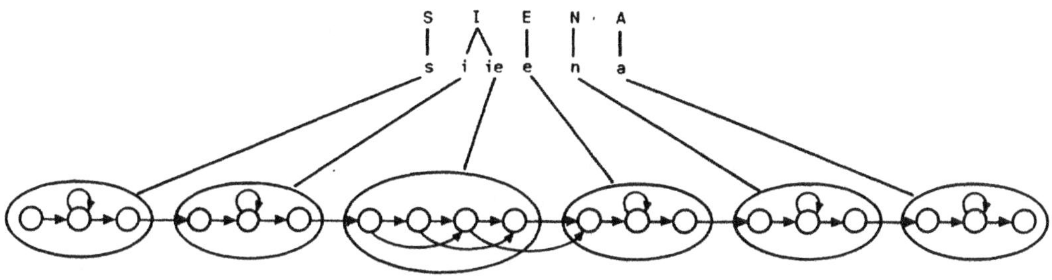

Figure 2.31: HMM of the word /SIENA/

and hiatuses are not discriminated by the Word Translator, which always includes both these forms in the translation. The current set of rules produces 1.44 transcriptions per word, on the average. Every word phonetic transcription is then represented by a sequence of phonetic units, taking into account coarticulation phenomena occurring in the transitions between different sounds. Phonetic units are modeled by left-to-right HMMs with different number of states [11, 42, 1]. An example of an HMM for the Italian word /SIENA/ is given in Figure 2.31.

The verification module accepts as input the list of word candidates produced by the lexical access module. The HMMs sequences corresponding to this set of words are organized into a tree structure, where transcriptions with common initial parts share the same branches. Then, a beam search Viterbi procedure is performed on the tree to evaluate the most likely words.

2.5.1 The Recognition Units

Sub-word recognition units offer several advantages over whole word models. If the vocabulary of the system is very large or if the application needs frequent updates, a whole word approach is not appropriate because it requires new training sessions. Furthermore, a substantial saving in storage is obtained since the same unit appears in different words but its parameters are stored only once. Another advantage of sub-word units is that properly designed units can perform better than whole words models in discriminating words that include similar parts (e.g. minimal pairs) [45]. In fact, the difference often observed for the emission densities of the common part of two slightly different words is generally due to the limited size of the training set. The consequent difference in the partial likelihoods computed during recognition can exceed the differences in the phonetically different part. Since in the sub-word approach phonetic portions that are equal are represented by the same model, the differences in the discriminant parts are enhanced. This consideration suggests the representation of steady parts of the phonemes (whenever they can be defined) by the same model, and to account for transitions by means of additional models only if they carry significant discriminant information [11]. This definition of the sub-word units was first proposed for template-based systems [46], leading to satisfactory results both for Italian [9] and for English [44].

These recognition units can be considered as a tradeoff between diphones and phonemes.

Diphones, defined as the speech portion between two consecutive phonemes, take coarticulation effects into account since the transitional effects are included in the units. However, as in a language there is a large inventory of diphones, it is difficult to obtain good estimates of the parameters of their HMMs using a limited training set. Moreover, the model of the steady part of phonemes is included, with relevant differences, into a number of different models leading to poor estimation. Finally, the information included in the transition part of many diphones is not significant for the discrimination of adjacent phonemes. For instance, the transition between an unvoiced fricative and a vowel (like /fa/ in the Italian word /fare/) does not carry necessary information for the classification of the phonemes /f/ and /a/.

On the other hand, the number of phonemes is very small, thus they are suitable for accurate statistical modeling [5] but their performance [11] is poor when the discriminating information among different sounds depends on the transition towards adjacent phonemes. For instance, unvoiced plosives, such as /k/, are acoustically realized by a short interval of silence followed by a burst. It is impossible to classify a plosive only by means of the burst since most of the discriminating information is in the transition.

In order to define an appropriate set of recognition units, a special language and the corresponding compiler [12] was designed. It allows any set of recognition units to be easily defined and words to be automatically translated from their orthography directly into the sequence of defined units. 26 phonetic units were considered for the Italian language. Of the 650 ($26 * 25$) possible transition units, 101 only were selected according to phonetic knowledge and observing the results of recognition experiments carried out with difficult vocabularies such as minimal word pairs [11]. These 101 transition units include all plosive/vowel, affricate/vowel and some sonorant/vowel transitions in addition to some consonant clusters. Transitions from vowel to sonorant are considered only for consonants

orthographic form	phonetic form	translation into units										
SETTE	sette	s	e	-	te	e						
APPARTIENE	appartjene	a	-	pa	ar	r	-	ti	ie	e	n	e
AVERE	avere	a	av	ve	e	er	re	e				
AVREBBE	avrebbe	a	av	v	vr	r	re	e	b	be	e	
ANDARE	andare	a	n	b	da	a	ar	re	e			

Table 2.12: Some examples of word translation from orthographic form to phonetic form and to their corresponding sequence of recognition units. The recognition units '-' and 'b' are the silence and the voicebar respectively

/r/ and /v/. As duration is not modeled very well by Markov models [41], geminate consonant are represented as singletons, so that words SETE (thirst) and SETTE (seven), for instance, are phonetically indistinguishable; their disambiguation can be deferred to the higher level linguistic processing. 22 steady units complete the unit inventory: 5 vowels, 6 sonorants, 5 fricatives, 4 affricates, silence and voicebar. Some examples of the translation from the orthographic form to the phonetic one and finally to the corresponding sequence of units is given in Table 2.12. Details about the context-sensitive translation rules can be found in [12, 11]. Similar approaches have been proposed in the past: using left and right contextual phonemic units [48] obtained as a weighted estimation from context-free and context-dependent units, where the weights depend on the number of occurrences of each unit in the training set, or manually choosing which context could improve the recognition rate by means of an accurate error analysis [15].

2.5.2 Model Estimation

Hidden Markov modeling of the sub-word units allows model training to be performed automatically performed.

As it is impossible in most of the cases, and always impractical to pronounce an isolated sub-word unit (such as a phoneme or a diphone), the training procedure relies on the observation of larger events such as words or sentences. The required observation subsequences could be made available by hand segmentation and labeling of the utterances, a time-consuming and error-prone process. It is possible, on the contrary, to estimate the model parameters of a set of sub-word units, without human interaction. From the orthographic form of the training vocabulary words, different phonemic transcriptions are generated according to their possible pronunciations and taking care also of the ambiguity arising in the translation process. These alternatives are automatically converted into the unit sequence. This operation is performed for all training words. The training set is composed of one or more utterances of the training vocabulary represented as sequences of Vector Quantization codewords. For each utterance, a forward and a backward matrix is computed bootstrapping the system from untrained HMMs (uniform transition and emission matrices). For every sub-word unit appearing in the training data base the transition and emission probabilities are estimated by using a generalization to multiple observations of the classical re-estimation formula [34, 1]. This procedure is repeated until convergence is reached. An important problem arises when a state is assigned zero

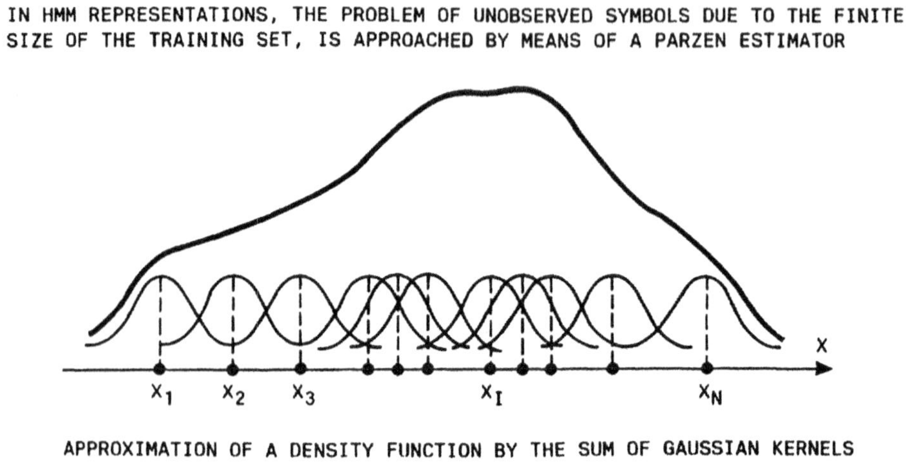

Figure 2.32: Approximation of a density function by the sum of Gaussian kernels

probability for a given symbol because it has never been observed in that state during the training session. This problem, generated by a poor estimation of probability densities, due to the limited number of training examples, is generally solved by interpolation [5] or by setting to a small constant value the probabilities that are null [34].

In our system, a Parzen estimator with normal kernel [19] (see Figure 2.32) smoothes the emission probability estimates $b_i(k)$ of codeword k being in state i after the Forward-Backward iterations [10, 11] according to this formula:

$$\tilde{b}_i(k) = C_i \sum_{m=1}^{nc} b_i(m) * \exp \frac{-d^2(k,m)}{D} \tag{2.40}$$

where $d(k,m)$ is the Euclidean distance between the k-th and the m-th vector quantizer codewords, nc is the number of codewords in the codebook, D is a fixed parameter (the Parzen radius) and C_i is a normalization factor such that

$$\sum_{k=1}^{nc} \tilde{b}_i(k) = 1 \tag{2.41}$$

2.5.3 Experimental Results

Each speaker trained a set of 126 unit models by pronouncing once the words in the TRA dictionary. Each training set consists of about 20 minutes of speech. These utterances were then coded by means of a speaker-dependent 7-bit vector quantizer. Five iterations of the Forward-Backward algorithm were sufficient for obtaining stable estimates of the parameters of the models.

The confusion matrix among phonemes (steady units) is shown in Figure 2.33.

Curve B in Figure 2.34 shows the recognition rate, averaged over all speakers, as a function of the best candidate position for the two-pass approach (hypothesis generation

	b	d	g	t	k	p	ʃ	f	v	s	z	l	r	λ	m	n	η	dz	ts	dʒ	tʃ	j	w	a	e	i	o	u
b	87								3							9												
d		77	19													3												
g			90													9												
t				70	19	9																						
k					96														3									
p					3	96																						
ʃ							96	3																				
f						3	9	83																				
v	19		3						41							22	3						9					
s										100																		
z											93	6																
l												96						3										
r	1		3			3			3	1		3	77				3		3									
λ						6		3				3		83			3											
m	3								9		9	9			25	29	12											
n	3	3										3			9	67	12											
η														6		22	70											
dz											3							96										
ts																			100									
dʒ							48				3									41		6						
tʃ							32														67							
j														36								63						
w																							100					
a																								100				
e																									93	6		
i																										100		
o																											90	9
u																												100

Figure 2.33: Confusion matrix among phonemes

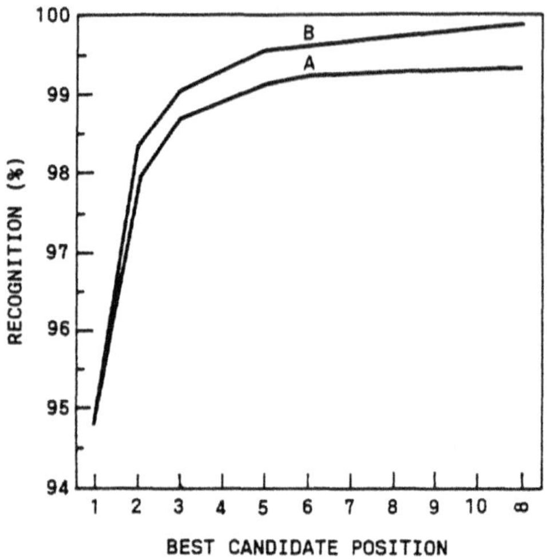

Figure 2.34: Recognition rate vs. N-best word scores for the 1011 word vocabulary

by partial phonetic description and successive detailed verification by stochastic decoding). Every word hypothesized by lexical access is represented by the set of its transcriptions into recognition units. All these representations are then compiled into a tree whose branches are the sequences of states of the HMM recognition units, and whose leaves identify words. A beam search Viterbi procedure operates on a tree to evaluate the best state sequences. The paths that are still active at the end of the search generate a set of word hypotheses ordered according to their likelihood. 95% of words attain the best first likelihood, while 99.3% of the uttered words are correctly included in the final set of hypotheses, whose average size is 4.4. Curve A in Figure 2.34 refers, instead, to the results obtained in the direct approach, excluding the lexical access module, hence by applying the same beam search Viterbi procedure to the tree representing all vocabulary words. Obviously, slightly worse results are obtained in the former approach because the lexical access module propagates its errors (correct words missing in the candidate list) to the verification module. It is worth noting, however, that there is no difference in the recognition rate for the best first hypothesis. This means that a missing word in the candidate list produced by lexical access is also missed as the best scored one by the direct approach. By increasing the rank of the accepted hypotheses, the difference between the two curves keeps constant and depends only on the error of lexical access (0.5%). In Table 2.13 a comparison of the performance of the two approaches can be found. As far as complexity is concerned, the phonetic segmentation and the generation of the hypothesis tree for verification are negligible in comparison to the matching. Matching requires a basic computation both for lexical access and for verification: the dynamic expansion of a trellis node. It consists in the evaluation of the cost of expanding a partial path from an origin node to a destination node, and in its comparison with the cost of the current best path reaching the destination node. As the complexity of cost computation is approximately equal for lexical access and for verification, a good approximation of the

	Direct approach	Two pass approach
Best first recognition rate	95.0 %	94.9%
Inclusion rate	99.9 %	99.3%
Number of hypotheses	5.5	4.4
Number of operation/word in lexical access	-	9342
Number of operation/word in the verification module	99795	17200
Total number of operation/word	99795	26542

Table 2.13: Comparison of the one and two pass lexical access strategies

computational complexity of the two approaches can be given in terms of the average number of expansion operations. A complexity reduction of about 82% is achieved.

Figure 2.35 shows the recognition rate as a function of the best candidate position for the two-pass approach for the 18388-word vocabulary. The best first recognition rate is 84.7%. Relevant improvements, similar to those in Figure 2.34, can be observed for the best two candidates, reaching more than 91% of accuracy. About 99.2% of the words are included in the final set of hypotheses whose average size is 21.3.

2.5.4 Conclusions

The main suggestions deriving from the hypothesize and test approach can be summarized as follows:

- It is very easy to reach about 90% of inclusion rate in the set of candidate words, but the real problem in large-vocabulary lexical access is to obtain almost 100% of inclusion rate and a reasonably small number of word candidates.
- A coarse phonetic segmentation can be more accurate than a detailed one, but few misclassifications can dramatically reduce the performance of a lexical access due to the small redundancy of the code.
- Robust phonetic segmentation can be achieved by generating, rather than a sequence of segments, a lattice of phonetic hypotheses to be matched against the vocabulary words which can be represented by a graph model including statistics about possible segmentation errors.
- the lexicon can be effectively represented as a tree, of phonetic nodes in the hypothesize step, and of HMM sub-word units in the verification step.
- A three-dimensional DP matching algorithm has been introduced that performs better than other conventional algorithms.
- A suboptimal version of the matching procedure can be used without appreciable performance degradation.

The experimental results show the capability of the statistical models and of the lexical constraints to cope with the errors of the segmentation module. The accuracy of the HMMs of the sub-word phonetic units in the verification phase has also been assessed.

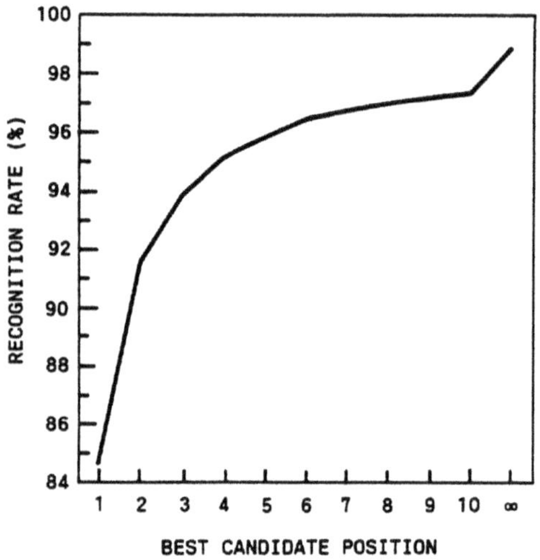

Figure 2.35: Recognition rate vs. N-best word scores for the 18388-word vocabulary

Over 99% of the correct words are within the first 5 best candidates for a 1011-word vocabulary; this accuracy reduces to about 96% for a 18388-word vocabulary.

2.6 Continuous Speech

The hypothesize and test approach has also been applied to the continuous speech recognition task (see Figure 2.36). It is mainly suggested by efficiency issues. A continuous speech hypothesizer for a large vocabulary requires that its task is constrained by lower and higher level knowledge. While higher level constraints generally increase accuracy, the same cannot be said for bottom-up constraints. As shown in Sect. 2.4.2, word preselection reduces computational complexity at the expense of a small increase in error rate. The *Hypothesize process* (Hp) generates a lattice of word candidates spanning the whole sentence, and the *Test process* (Tp) verifies each hypothesized word by computing scores of acoustic matching (see Figure 2.37). Each hypothesis consists of word identifier, log-likelihood or probabilistic score and time boundaries. This lattice is then passed to the linguistic module for syntactic/semantic parsing [17]. This straightforward strategy has a number of drawbacks.

The first one is that quite often the boundaries of candidate words detected by the hypothesization stage are incorrect. This inaccuracy has been observed to be the cause of very bad scores for the true hypotheses when the Tp is based on Hidden Markov Models. Hence the Tp should not rely on the boundaries of the word hypotheses, rather only on the *region* in which the word could be observed.

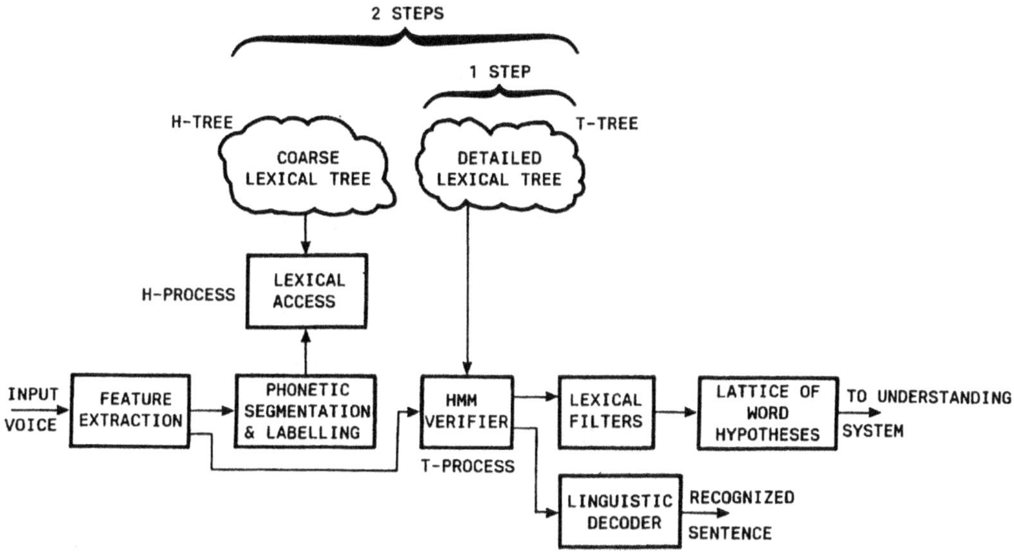

Figure 2.36: Block diagram of the continuous speech system

	H-PROCESS	T-PROCESS
INPUT:	COARSE PHONETIC LATTICE	VECTOR QUANTIZATION SYMBOLS
KNOWLEDGE:	TREE OF COARSE PHONETIC WORD DESCRIPTIONS	TREE OF HMM SUB-WORD UNITS
ALGORITHM:	THREE DIMENSIONAL DYNAMIC PROGRAMMING (3DP)	VITERBI DECODING OR FORWARD DECODING
TIME UNIT	1 MICRO-SEGMENT $\sim$ 80 ms	1 FRAME = 10 ms

Figure 2.37: Processes involved in the two-step strategy

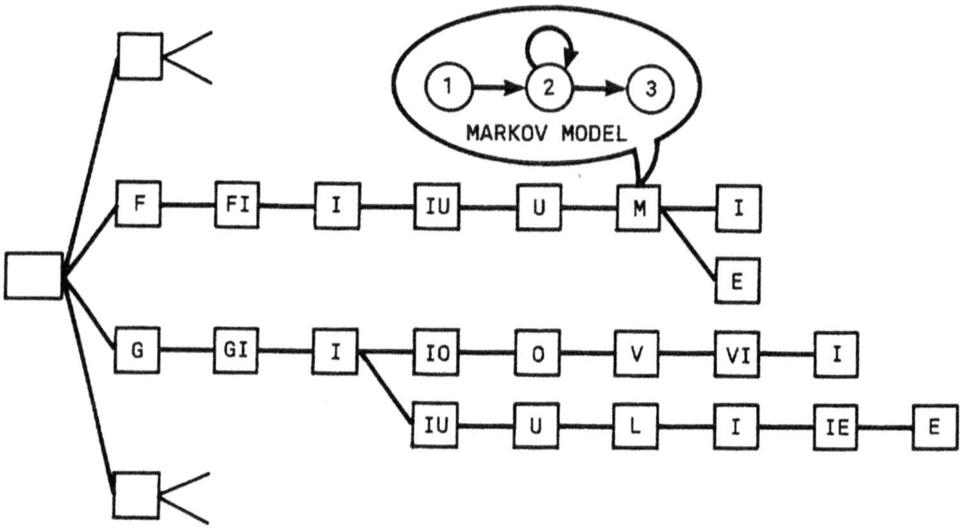

Figure 2.38: Lexical tree for the hypothesisation level

The second disadvantage is a lack of efficiency for the computationally expensive task Tp: it is not easy to organize the lattice into a structure like a tree or a graph, that would allow an effective search strategy to be implemented.

Finally, the above described approach is strictly sequential: the verification process cannot begin before the ending of the hypothesization process. This is in contrast with a real-time implementation, hence a more tight interaction between the two modules is suggested.

In the following, the extension to continuous speech of the basic modules implemented for isolated utterance recognition will be decribed, and two control strategies will be outlined for a continuous speech word hypothesizer.

2.6.1 Control Strategies

At the hypothesization level, word phonetic transcriptions are organized into a tree (Ht) whose nodes represent coarse classes (Figure 2.18); terminal nodes have a link to the set of words they represent.

At the Test (verification) level, the whole lexicon is organized into a tree (Tt) whose nodes represent Hidden Markov Model states (HMMs) of sub-word units (Figure 2.38). Hp and Tp algorithms are similar except for the score computation and for the kind of transitions among the nodes of the trees. As syntactic constraints are not presently used at this level, a word can be followed by every word of the vocabulary. Hence, whenever a node in the Ht (as well as in the Tt) has to expand beyond a terminal node, a word is hypothesized and the root node is considered again as a possible expansion. Each active node must keep the information about the score and the time corresponding to the

path expansion to the root node for generating the beginning and ending time of a word hypothesis along with its likelihood score.
This algorithm is an extension of the syntax-driven recognition algorithm described in [10] where the syntax was described by a graph. While a graph allows only the best-matching model to be detected, a tree permits all matching paths to be kept. The number of paths can be limited at each step by the beam-search threshold.

Unless other constraints are introduced, the extension of the Tp (as well as the Hp) to continuous speech produces, for each active terminal node, a complete word hypothesis at each input frame (or micro-segment). A terminal node is generally active in a large range of frames in the neighborhood of the correct word ending frame; all terminal nodes would be *always* active unless beam search is used. Thus a decision process (Dp) keeps only the best scored among all the hypotheses concerning the same word and starting at the same frame.

Cascade integration

The first integration consists in the sequential application of Hp and Tp as depicted in Figure 2.39

The Hp generates a coarse lattice of word hypotheses through the above described procedure. The basic information shared among the Hp and the Tp is a mapping between each vocabulary word and its corresponding nodes in the Tt. As the same word can be hypothesized more than once in a given region, a decision process (DH) selects as boundaries for a word candidate hypothesis the beginning and ending point of the largest segment over which it appears. Thus, when the whole sentence has been processed by the Hp, the information about the nodes of the Tt that could be active in each detected segment is sent to the Tp. Thus the Tp estimates word likelihoods on the Tt. A node of the Tt is expanded only towards the nodes that the Hp has marked as active at that particular time interval. The decision process (DT) finally finds, for each detected word, the best ending point.

The cascade integration solves the problem of inaccuracy of time boundaries, since only regions in which word hypotheses can be found are given to the Tp, which uses this information only to reduce the number of nodes that it will expand at each frame. As the Tp cannot start before the Hp has processed the whole sentence, this kind of interaction is not particularly suited to a real-time implementation. This problem has been solved by devising a second scheme that exploits a tighter integration.

Full integration

In the full integration scheme of Figure 2.40 the Tp performs the main hypothesization activity whose task is dynamically constrained by the Hp. After a micro-segment has been processed by the Hp, the current Ht active nodes constrain the Tp node expansion. To that purpose a mapping between the Ht and the Tt has been established. The Tp expands a Tt node only if its corresponding node in the Ht is active at that particular time frame.

For mapping the Tt on the Hp, each sub-word unit is associated to a string of coarse classes, for instance: /ta/ = (pl, cv), /ts/ = (pl, fr).
Then the Tt, representing the whole lexicon, is built. Each node represents the HMM of

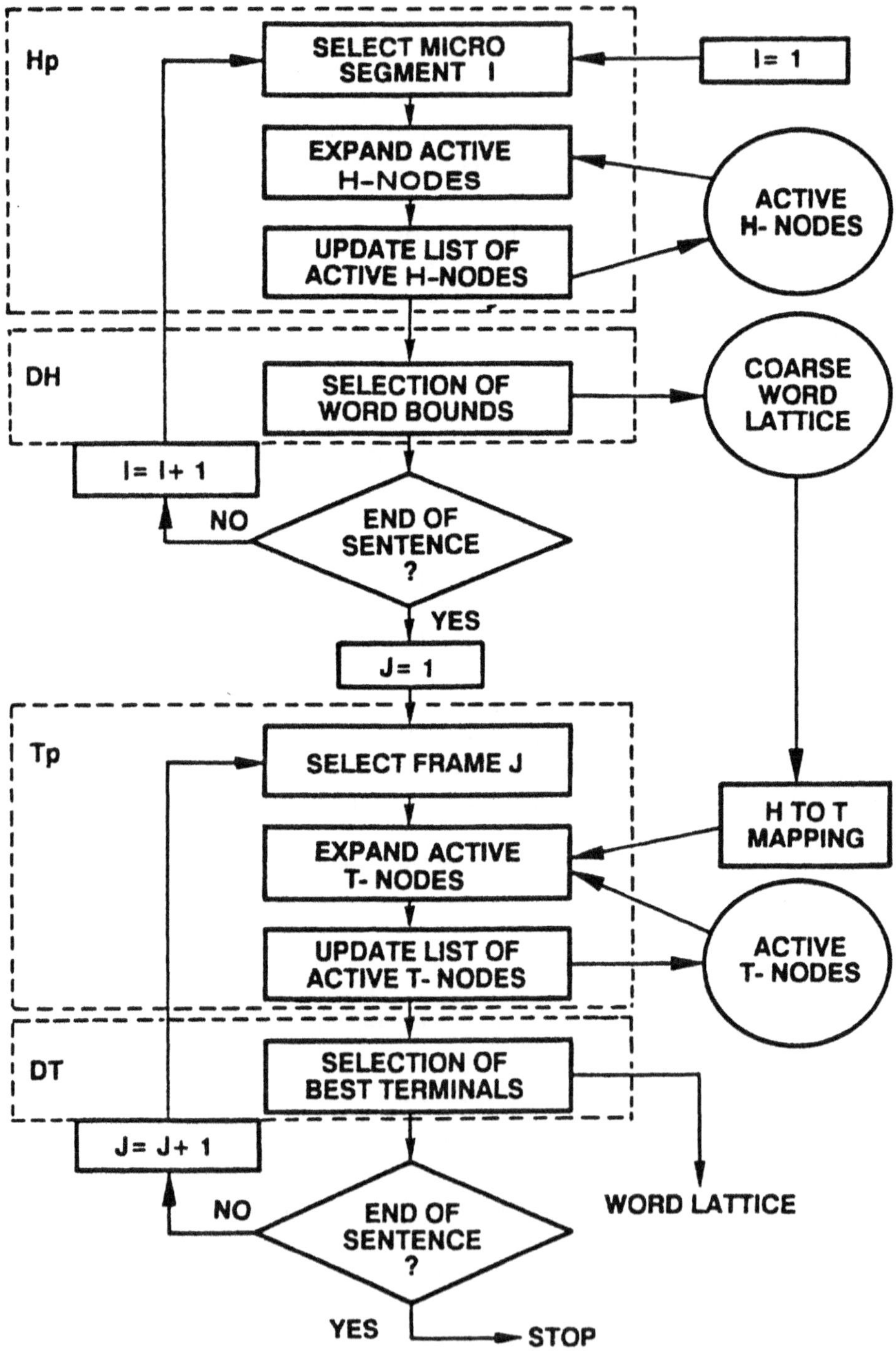
Hp
SELECT MICRO SEGMENT I
I= 1
EXPAND ACTIVE H-NODES
ACTIVE H- NODES
UPDATE LIST OF ACTIVE H-NODES
DH
SELECTION OF WORD BOUNDS
COARSE WORD LATTICE
I= I+ 1
NO
END OF SENTENCE ?
YES
J= 1
Tp
SELECT FRAME J
H TO T MAPPING
EXPAND ACTIVE T- NODES
UPDATE LIST OF ACTIVE T- NODES
ACTIVE T- NODES
DT
SELECTION OF BEST TERMINALS
J= J+ 1
NO
END OF SENTENCE ?
WORD LATTICE
YES
STOP

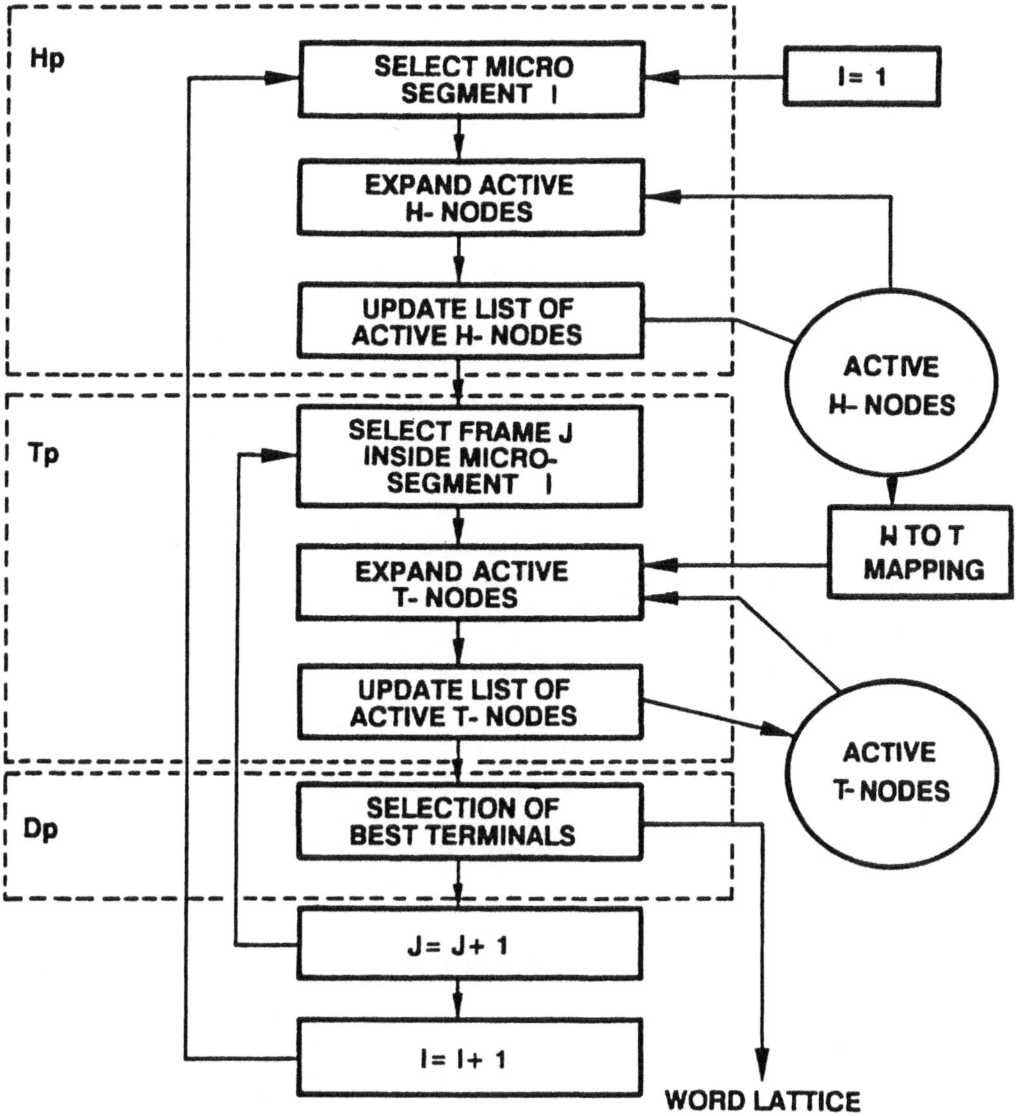

Figure 2.40: Full integration

a recognition unit. The Ht is then built by substituting the corresponding sequence of coarse classes to each node of the Tt. Of course, each Ht node subsequence showing the repetition of the same symbol is collapsed into a unique node. The mapping from the Tt to the Ht is obtained through a pointer from a Tt node to the identifier of the Ht nodes it has generated. Figure 2.41 illustrates a simple example of a Tt and its corresponding Ht for a three-word lexicon. The numbers within the Tt nodes are the pointers to the corresponding Ht nodes.

The control strategy can be summarized as follows:

- For a given input micro-segment, the paths corresponding to active Ht nodes are expanded according to the constraints given by the SID model. Node expansion, controlled by a beam search threshold, generates a new list of active nodes.
- Each path corresponding to an active Tt node could be expanded according to the transition defined for the HMMs, but the expansion is allowed only if the destination node has a pointer to an active Ht node. The node expansion, controlled by a beam search threshold, generates a new list of active Tt nodes. All active terminal nodes are given as input to the Dp. This step is iterated for each input frame belonging to the currently analyzed micro-segment.
- The Dp receives as input all active Tt terminal nodes for each processed frame. The corresponding word hypotheses are then compared with the list of those already collected at previous frames. If no hypothesis exist in the list having the same word identifier and the same beginning frame, the new hypothesis is inserted, otherwise the new one updates the old one only if it has a better score.

This tighter integration scheme is suitable for real-time implentation as the three steps outlined above can be easily pipelined. Its drawback is that the Hp constraints are looser than those imposed by the cascade integration. In fact, as the cascade integration allows only complete word hypotheses to be propagated from the Hp to the Tp, the full integration activates also the nodes corresponding to partial word hypotheses, that could be later pruned away.

2.6.2 Word Hypothesis Normalization

The score of a word hypothesis is computed by subtracting the score of the tree path at the root node to the score that the path attains at the terminal node. The log-likelihood score $L(W)$ of a word hypothesis obtained by the Viterbi algorithm is defined as the logarithm of the best path probability given the observed sequence of codewords $\mathbf{O}$, $L(W) = \log \max_i P(S_i, \mathbf{O} \mid M)$, where S_i is a state sequence within the word model M. Hence the likelihood of word hypotheses generated in different regions of the sentence cannot be compared as they refer to different observations. As the scores activate an island driven parser at the syntactic-semantic level of the system [18], they must be normalized to correctly represent the evidence of the hypotheses.

Rather than using the Viterbi likelihood, the observation probability defined as:

$$P(M|\mathbf{O}) = P(\mathbf{O}|M)P(M)/P(\mathbf{O})$$

can be used. $P(\mathbf{O}|M)$ is computed by the Forward algorithm, while $P(M)$ probabilities, in the present implementation, are uniformly distributed. Finally $P(\mathbf{O})$ can be derived

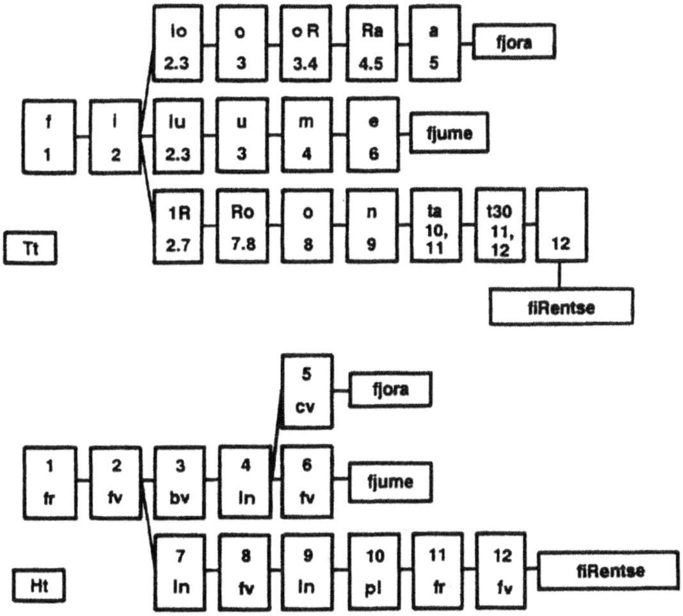

Figure 2.41: Example of Ht/Tt mapping

from the sum of the probabilities that each node of the tree attains at the ending frame of the considered word hypothesis. An efficient procedure for the computation of $P(M|\mathbf{O})$ has been implemented that relies on the scaling coefficients introduced in [34] for the computation of the forward probabilities in the HMM training. The above probability can be computed in the Tp, by using the forward probability estimation within a word model, while the Viterbi decoding selects the best terminal node for the path expansion to the tree root.

2.6.3 Lattice Filters

Word lattices produced by the HMM verifier are of the order of 500-2000 words, with reference to sentences of 4-10 words. Two heuristic lexical filters, F1 and F2 (equation 2.42), are applied to the output of the verification process for pruning out unlikely hypotheses, so as to reduce the lattice size to the order of 200-500 words. Parameters of the lexical filters have been tuned to reduce as far as possible the lattice size without affecting correct word hypothesization rate.

The following rules are applied, with reference to two conflicting hypotheses, w_a and w_p, with log-score s_a and s_p, and time limits t_a^1, t_a^2 and t_p^1, t_p^2 respectively.
In F1 w_a filters out w_p if:
a) $t_a^1 < t_p^1$ and $t_a^2 > t_p^2$
b) $s_p < s_a - d_r$
In F2 w_a filters out w_p if:
c) $t_p^1 > t_a^1 + d_s^1$ and $t_p^2 < t_a^2 - d_s^1$

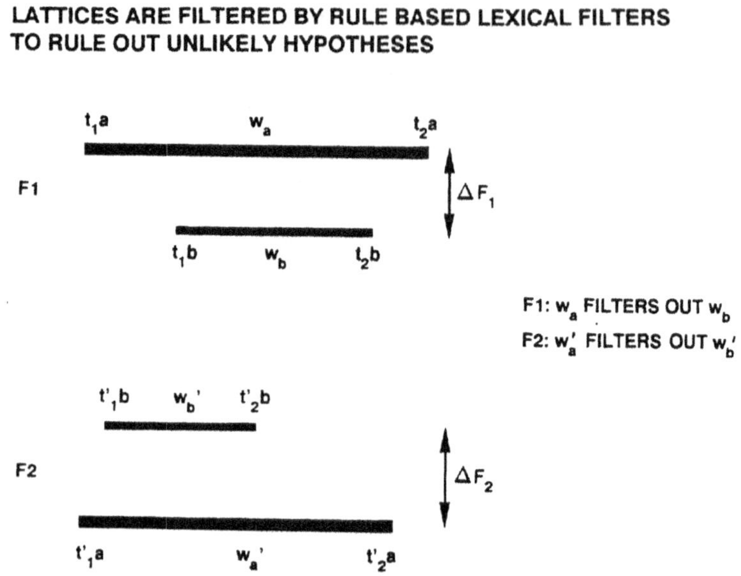

Figure 2.42: Lexical filters

d) signal energy in $\left[t_a^1, t_p^1\right]$ and $\left[t_p^2, t_a^2\right] > e_0$
e) no other word in the lattice ends in $\left[t_a^1, t_p^1\right]$ or starts in $\left[t_p^2, t_a^2\right]$
f) $s_a < s_p < s_a + d_s^2$

d_r, d_s^1, d_s^2, and e_0 are properly chosen thresholds.

2.6.4 Efficiency Measures

A desirable property of a good word hypothesizer for a continuous speech recognition and understanding system is that all uttered words appear in the lattice, they are correctly aligned and that the lattice is as small as possible. Furthermore, correct hypotheses should have better acoustic scores than wrong ones, so that their retrieval during a score-driven syntactic and semantic analysis is not delayed; this property directly affects the linguistic analysis performance, both in terms of successful understanding rate and of computing time.

These considerations lead us to choose the "efficiency" measure introduced by Smith and Erman [50] for assessing the "quality" of a lattice. This quality can be measured by the level of appearance of correct hypotheses among all the others. So the RANK R_i of a correct hypothesis H_i can be defined as the number of wrong hypotheses H which compete with H_i and have a better score:

$$R_i = [H : s(H) > s(H_i), Comp(H, H_i)] \tag{2.42}$$

$Comp(H, H_i)$ is a predicate with value *true* if the overlapping in time of H and H_i is greater than the half of the shorter of the two.

The *efficiency* of the i-th correct hypothesis E_i is defined as $E_i = \frac{1}{R_i}$
The *average efficiency* of a lattice is defined as

$$E = \frac{1}{n} * \sum_{k=1}^{n} E_i$$

where n is the number of words of the uttered sentence. E_i can be thought of as the amount of work that has to be done to retrieve the correct hypothesis H_i, in terms of number of wrong hypotheses that must be evaluated before H_i.

The possible values for E range from 0, when no correct hypothesis has been generated, to 1, when all uttered words are the best in the lattice. One consideration can be made: it is questionable that only wrong conflicting hypotheses can delay the retrieval of a correct one: in fact, if a region R_1 of the lattice has better scores, on average, than a different region R_2, then wrong hypotheses in R_1 can delay the retrieval of correct hypotheses in R_2, even if they are not competing. Furthermore, the definition of competition is rather arbitrary. So a new measure has been defined, based on the concept of Degree of Appearance (DoA) of a correct word H_i, defined as the rank of H_i but relaxing the constraint of competition:

$$DoA_i = [H : s(H) > s(H_i)] \tag{2.43}$$

By taking the average on the whole lattice, we get the degree of appearance of the lattice

$$DoA = \frac{1}{n} * \sum_{k=1}^{n} DoA_i$$

The minimum degree of appearance is defined as $DoAmin = \min_i DoA_i$, which measures the depth of the correct word having the worst score in the lattice.

To find the sequence of correct words in the lattice, a forced parser has been implemented, which is based on the concept of adjacency in time and on the scores of the hypotheses. The forced parsing problem has been approached as a search in a directed and weighted graph, whose nodes are the lexical hypotheses in the lattice. Two nodes are joined by an arc when they are instances of consecutive words in the uttered sentence, and satisfy adjacency constraints in time, properly defined; the arc is weighted according to the amount of gap and overlap intervals between the corresponding lexical items. This problem of minimum distance search in a graph has been solved by Dynamic Programming and by the A* algorithm.

2.6.5 Experimental Results

In the following, experimental results computed over a corpus of 150 continuously uttered sentences are discussed. A speaker-dependent voice-activated data retrieval system is simulated, applied to a geographical semantic domain; the size of the lexicon is 1016 words.

The number of lexical hypotheses in the lattice affects the complexity of the understanding task, thus it is important to obtain small and reliably scored lattices, possibly avoiding missing semantically meaningful words.

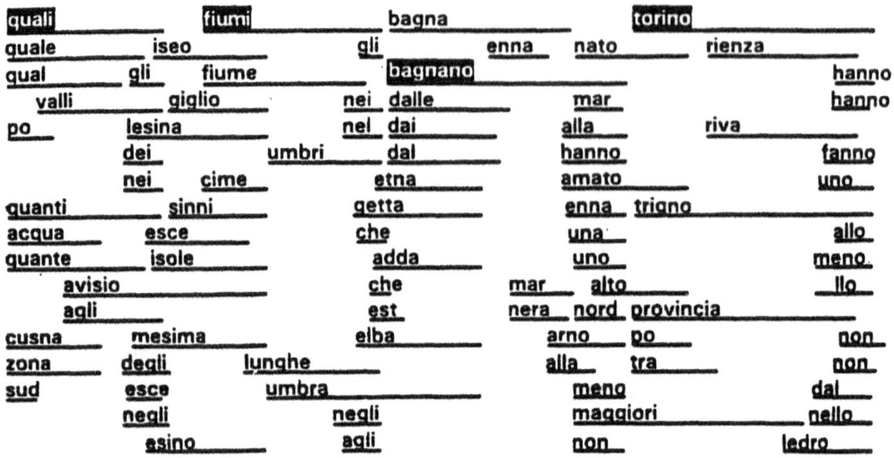

Figure 2.43: Example of a lattice

In Figure 2.43 an example lattice is presented corresponding to the question "Quali fiumi bagnano Torino ?" ("Which rivers wash Torino ?"). Words are ordered according to descending scores, starting from the top of the figure; correct words are enhanced while short connective words are not displayed.

Two approaches have been compared: in the first one (1 step), the decoding algorithm is applied to the whole lexicon, represented by a tree of speaker-dependent HMMs, and lattice growth is limited only by a beam search control strategy. In the second approach (2 step), the HMM decoding process is driven by a word preselection process based on coarse phonetic segmentation of speech into 6 rough classes. The *cascade integration* scheme is used for controlling the two processes, allowing an improvement of system efficiency at the expense of accuracy. In fact errors made in the preselection stage cannot be recovered during HMM decoding.

Two different decoding algorithms have also been tested (Figure 2.44), namely the Viterbi (VIT) algorithm with log-likelihood score normalized according to the hypothesis duration and the Forward (FORW) algorithm with score normalization as outlined in the preceding subsection.

Statistics collected from the 150 sentences are shown in Table 2.14.

The top of the table shows statistics referring to computational load. The first and the second row represent the average number of DP operations computed at each 10 msec frame by the hypothesization and by the verification process respectively. The total DP activity is shown in the third row. In the fourth and fifth row the average number of active nodes for the two processes is displayed. The bottom part of the table refers to lattice measures.

The Forward algorithm with score normalization yields a remarkable performance improvement on the number of missing words, on the lattice efficiency and on the degree of appearance (DoA). In particular, a few lattices in the one-step Forward case reach $E =$ 1: the n words of the sentence are the n best hypotheses, and they are correctly aligned in time.

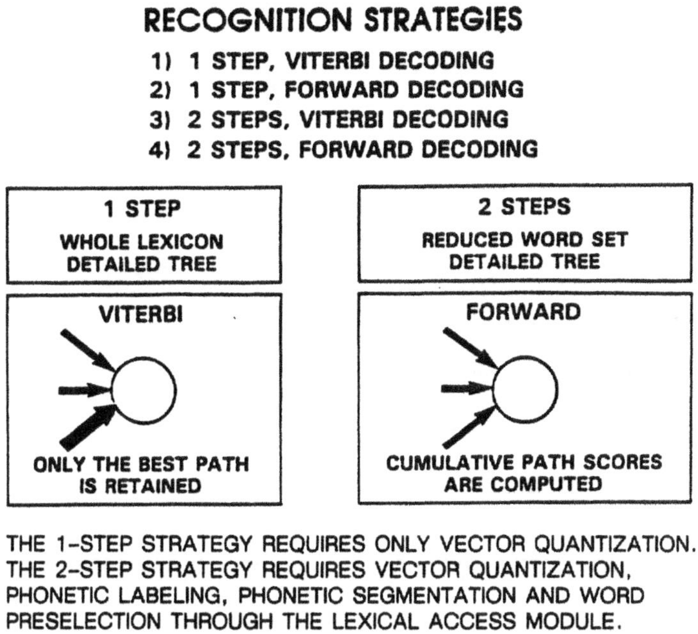

Figure 2.44: Decoding algorithms

	VITERBI		FORWARD	
	One pass	Two pass	One pass	Two pass
Hp DP operations/frame	0	448	0	448
Tp DP operations	4133	3379	2901	2457
Total DP operations/frame	4133	3630	2901	2905
Active Hp nodes	0	87	0	134
Active Tp nodes	2370	1980	1619	1396
Lattice size	310	275	433	369.5
Missing words (%)	2.2	3.3	0.7	2.0
Lattice efficiency	0.465	0.465	0.697	0.690
Degree of appearance	0.149	0.151	0.479	0.464
Successful forced parsing	89.3 %	83.3 %	96 %	89.3 %

Table 2.14: Comparison of the one- and two-pass lexical access strategies

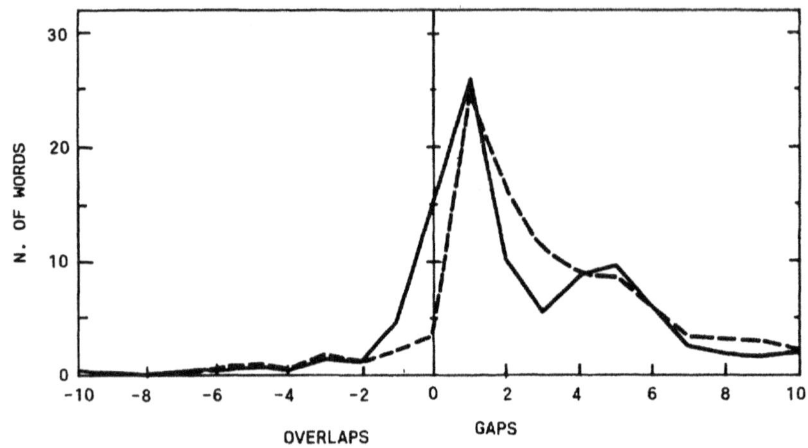

Figure 2.45: Histogram of gaps and overlaps along the best path in the lattice

The number of missing words increases, in the two-step approach, due to two kinds of problems:

- bad phonetic segmentation, which results in a mismatching at the hypothesization level, with consequent pruning of the correct path during the search along the Ht tree;
- bad starting or ending point detection of the H process, which reflects in a bad acoustic matching at the verification level, with consequent pruning of the correct path during the search in the Tt tree.

The Viterbi verifier gives slightly worse results compared to the Forward one. The latter, besides, better aligns in time the correct words. A histogram of the gaps and of the overlaps (positive and negative side of the abscissa respectively) computed through the forced parser is plotted in Figure 2.45.

Viterbi-scored hypotheses (dotted line) are shifted toward the increasing gap side with respect to the Forward-scored ones (continuous line). It is worth noting that a better alignment in time reflects in the possibility to adopt tighter adjacency constraints at the linguistic parser level, therefore reducing the number of sub-sentence hypotheses to be generated before obtaining the parse of the whole sentence.

2.7 Conclusions

A high-performance speaker-dependent continuous-speech word hypothesizer for large vocabularies is the final result of the activities carried out in subtask 2.1 of project P26. Most of the algorithms and architectures developed in this framework are applicable also to high performance, large and very large isolated-word recognition tasks. The major results of this sub-task can be summarized as follows:

- Use of state-of-the art signal processing techniques for speech analysis.

- Novel techniques for phonetic classification, phonetic matching, and lexical access to large vocabularies. A phonetic classifier that segments and labels speech in terms of six broad phonetic classes attains 86.2 % and 93.7 % correct classification rates when the first best choice and the two best choices are taken into account respectively. Lexical access is performed by means of an original extension to 3 dimensions of the classical Dynamic Programming algorithm.
- Sub-word speech units optimization. 24 stationary units and 101 transitory units have been carefully selected and modeled.
- Efficient control strategies for interaction between lexical access and detailed Hidden Markov Model verification. 82% of computational complexity reduction can be achieved by means of the lexical access module which is integrated with the verification module for real-time oriented implementations.
- Vocabulary flexibility. Changing the lexicon is a simple operation that needs only the new list of words in their orthogaphic form, because every word in a given language can be synthesized as a HMM chain of sub-words units;
- Automatic training. No hand-labeling of speech data is required; automatic labeling is obtained as a side-effect of the Forward-Backward algorithm used for statistical training of Hidden Markov Models.
- Adaptability to different languages by proper definition of new sub-word units and by replacement of orthographic-to-phonetic rules.
- Capability of real-time performance with a multi-DSP architecture and of parallel implementability of the algorithms.
- Efficient continuous speech word hypothesization and connection with linguistic modules.

A multi-speaker speech data base has been collected, and graphic packages have been developed for interactive monitoring of each level of processing, from analog waveforms to lattices of lexical hypotheses. Demonstrators have been defined and implemented for the final technical meeting scheduled in October 1988. Extensive experiments have been carried out for tuning system parameters, and system performance has been assessed at each stage of processing.

System performance complies with the specified project target, and can be summarized as follows:

- *isolated word task*: Several lexicons, ranging from 1K words to 18K words have been used in the tests. Average recognition rates range from 95.3% for the 1K lexicon to 85% for the 18K lexicon with respect to the best-scored word; corresponding figures with respect to the 5 best-scored words are 99.2% and 95.3% respectively.
- *continuous speech task*: correct words are hypothesized and properly aligned with a 96.4% success rate. This results in an 80% successful understanding rate when linguistic processing is applied. Average lattice size, with respect to uttered sentences of 5.7 words on average, is less than 400 words.

Recognition tests were performed in a typical office environment by using a close-talk microphone. Continuous sentences were produced at normal speed and with naturalness.

Algorithms and architectures are real-time oriented, and most computationally expensive procedures are suitable for a parallel implementation.

Bibliography

1. "Annex to 1st six-month report of ESPRIT Project P26." Technical Report, ESPRIT P26, 1985
2. "Annex to 2nd half-year report of ESPRIT project P26." Technical Report, ESPRIT P26, 1986
3. "Deliverable 11a: preliminary report on studied algorithms and first evaluation experiment of speech data reduction stage." Technical Report, ESPRIT P26, 1986
4. "Preliminary report on algorithms for speech data reduction." Technical Report, ESPRIT P26, 1984
5. L.R. Bahl, F. Jelinek, R. Mercer: "A maximum likelihood approach to continuous speech recognition." *IEEE Trans. on Pattern Analysis and Machine Intelligence*, vol. 5, pp. 179-190, March 1983
6. J.M. Baker: "State of the art speech recognition, U.S. research and business update." *Proc. of the European Conf. on Speech Technology*, pp. 440-447, Edinburgh (UK), Sept. 1987
7. R. Billi, G. Massia, F. Nesti: "Word preselection for large vocabulary speech recognition." *Proc. of the ICASSP '86*, pp. 65-68, Tokyo, Japan, Apr. 1986
8. D.M. Carter: "The use of speech knowledge in automatic speech recognition." *Computer Speech and Language*, vol. 2, pp. 1-11, March 1987
9. A.M. Colla, D. Sciarra: "Automatic diphone bootstrapping for speaker adaptive continuous speech recognition." *Proc. of the ICASSP '84*, pp. 35.2.1-35.2.4, San Diego, Ca., March 1984
10. M. Cravero, L. Fissore, R. Pieraccini, C. Scagliola: "Syntax driven recognition of connected words by Markov models." *Proc. of the ICASSP '84*, pp. 35.5.1-35.5.4, San Diego, Ca., March 1984
11. M. Cravero, R. Pieraccini, F. Raineri: "Definition and evaluation of phonetic units for speech recognition by hidden Markov models." *Proc. of the ICASSP '86*, pp. 2235-2238, Tokyo, Japan, Apr. 1986
12. M. Cravero, R. Pieraccini, F. Raineri: "Definition of recognition units through two levels of phonemic description.": *Proc. of the Montreal Symposium on Speech Technology*, pp. 53–54, Montreal, Canada, July 1986
13. K.H. Davis, P. Mermelstein: "Comparison of parametric representation for monosyllabic word recognition in continously spoken sentences." *IEEE Trans. Acoust., Speech and Signal Processing*, vol.28, pp 357-366, Aug. 1981

14. P. Demichelis, P. Laface, E. Piccolo, G. Micca, R. Pieraccini: "Recognition of words in a large vocabulary." *Int. Workshop on Recent Advances and Applications of Speech Recognition*, pp. 115-123, Rome, Italy, May 1986

15. A.-M. Derouault: "Context-dependent phonetic Markov models for large vocabulary speech recognition." *Proc. of the ICASSP '87*, pp. 360-363, Dallas, Tex., Apr. 1987

16. P. D'Orta, M. Ferretti, S. Scarci: "Phoneme classification for real-time speech recognition of Italian." *Proc. of the ICASSP '87*, pp 81-84, Dallas, Tex., Apr. 1987

17. L. Fissore, E. Giachin, P. Laface, G. Micca, R. Pieraccini, C. Rullent: "Experimental results on large vocabulary continuous speech recognition and understanding." *Proc. of the ICASSP '88*, pp. 414-417, New York, NY, Apr. 1988

18. L. Fissore, P. Laface, G. Micca, R. Pieraccini: "Interaction between fast lexical access and word verification in large vocabulary continuous speech recognition." *Proc. of the ICASSP '88*, pp. 279-282, New York, NY, Apr. 1988

19. K. Fukunaga: *Introduction to Statistical Pattern Recognition*. Academic Press, 1972

20. A. Giordana, P. Laface, L. Saitta: "Discrimination of words in a large vocabulary using phonetic descriptions." *Int. Journal of Man-Machine Studies*, vol.24, pp. 453-473, May 1986

21. V.N. Gupta, M. Lenning, P. Mermelstein: "Integration of acoustic information in a large vocabulary word recognizer." *Proc. of the ICASSP '87*, pp. 697-700, Dallas, Tex., Apr. 1987

22. D.P. Huttenlocher, V.W. Zue: "A model of lexical access from partial phonetic information." *Proc. of the ICASSP '84*, pp. 26.4.1-26.4.4, San Diego, Ca., March 1984

23. F. Jelinek: "Continuous speech recognition by statistical methods." *IEEE Proc.*, vol.64, pp. 532-556, Apr. 1976

24. F. Jelinek: "The development of an experimental discrete dictation recognizer." *IEEE Proc.*, vol.73, pp. 1616-1624, Nov. 1985

25. A. Kaltenmeier: "Acoustic/phonetic transcription using a polynomial classifier and hidden Markov models." *Proc. of the Montreal Symposium on Speech Technology*, pp. 95-96, Montreal, Canada, July 1986

26. T. Kaneko, N.R. Dixon: "A Hierarchical decision approach to large-vocabulary discrete utterance recognition." *IEEE Trans. Acoust., Speech, Signal Processing*, vol.31, pp. 1061-1066, May 1983

27. D.H. Klatt: "Overview of the ARPA speech understanding project." In: W.A. Lea (ed.) *Trends in Speech Recognition*, pp. 249-271. Prentice Hall, 1979

28. D.H. Klatt: "SCRIBER and LAFS: two new approaches to speech analysis." In: W.A. Lea (ed.) *Trends in Speech Recognition*, pp. 529-555. Prentice Hall, 1979

29. T. Kohonen, H. Rittinen, E. Reuhkala, S. Haltsonen: "On-line recognition of spoken words from a large vocabulary." *Information Sciences*, vol.22, pp. 3-30, July-Aug. 1984

30. P. Laface, G. Micca, R. Pieraccini: "Experimental results on a large lexicon access task." *Proc. of the ICASSP '87*, pp. 809-812, Dallas, Tex., Apr. 1987

31. H. Lagger, A. Waibel: "A coarse phonetic knowledge source for template independent large vocabulary word recognition." *Proc. of the ICASSP '85*, pp. 862-865, Tampa, Fla., March 1985

32. J.N. Larar: "Lexical access using broad acoustic-phonetic classification." *Computer Speech and Language*, vol.1, pp. 47-59, March 1986

33. S. Levinson: "Structural methods in automatic speech recognition." *Proceedings of the IEEE*, vol.73, pp. 1625-1649, Nov. 1985

34. S.E. Levinson, L.R. Rabiner, M.M. Sondhi: "Introduction to the application of the theory of probabilistic functions of a Markov process to automatic speech recognition." *Bell System Technical Journal*, vol.62, pp. 1035-1074, April 1983

35. Y. Linde, A. Buzo, R.M. Gray: "An algorithm for vector quantizer design." *IEEE Trans. on Communications*, vol.28, pp. 88-95, Jan. 1980

36. S.M. Marcus: "Associative models and the time course of speech." *Bibliotheca Phonetica*, vol.12, pp. 36-52, 1985

37. J.J. Mariani: "Speech technology in Europe." *Proc. of the European Conf. on Speech Technology*, pp. 431-439, Edinburgh (UK), Sept. 1987

38. W.D. Marslen-Wilson: "Speech understanding as a psychological process." In: J.C. Simon (ed.) *Spoken Language Generation and Understanding*, pp. 39-67. D.Reidel , 1980

39. B. Merialdo, A.-M. Derouault, S. Soudoplatoff: "Phoneme classification using Markov Models." *Proc. of the ICASSP '86*, pp. 2759-2762, Tokyo, Japan, Apr. 1986

40. G. Micca, R. Pieraccini, P. Laface, L. Saitta, A. Kaltenmeier: "Word hypothesization and verification in a large vocabulary." *Proc. of the 3rd Esprit Technical Week*, pp. 845-853, Brussels, Belgium, Sept. 1986

41. R.K. Moore, M.J. Russel, M.J. Tomlinson: "The discriminative network: a mechanism for focusing recognition in whole word pattern matching." *Proc. of the ICASSP '83*, pp. 1041-1044, Boston, Mass., Apr. 1983

42. R. Pieraccini, F. Raineri, A. Giordana, P. Laface, A. Kaltenmeier, H. Mangold: "Algorithms for speech data reduction and recognition." *2nd Esprit Technical Week*, Brussels, Belgium, Sept. 1985

43. D.B. Pisoni, H.C. Nusbaum, P.A. Luce, L.M. Slowiaczek: "Speech perception, word recognition and the structure of the lexicon." , *Speech Communication*, Vol.4, pp. 75-96, Aug. 1985

44. A.E. Rosenberg, A.M. Colla: "A connected speech recognition system based on spotting diphone-like segments - preliminary results." *Proc. of the ICASSP '87*, pp. 85-88, Dallas, Tex., Apr. 1987

45. M.J. Russel, R.K. Moore: "Explicit modeling of state occupancy in Hidden Markov Models for automatic speech recognition." *Proc. of the ICASSP '85*, pp. 5-8, Tampa, Fla., March 1985

46. C. Scagliola: "Language models and search algorithms for real time speech recognition." *Int. Journ. Man-Machine Studies*, Vol.22, pp. 523-547, May 1985

47. G. Schukat-Talamazzini, H. Niemann: "Generating Word Hypotheses in Continuous Speech." *Proc. of the ICASSP '86*, pp. 1565-1568, Tokyo, Japan, Apr. 1986

48. R. Schwartz, Y. Chow, S. Roucos, M. Krasner, J. Makhoul: "Improved Hidden Markov Modeling of phonemes for continuous speech recognition." *Proc. of the ICASSP '84*, pp. 35.6.1-35.6.4, San Diego, Ca., March 1984

49. D.W. Shipman, V. Zue: "Properties of large lexicons: Implications for advanced isolated word recognition systems." *Proc. of the ICASSP '82*, pp. 546-549, Paris, France, May 1982

50. A.R. Smith, L.D. Erman: "Noah - A bottom up word hypothesizer for large vocabulary speech understanding systems." *IEEE Trans. on Pattern Analysis and Machine Intelligence*, vol.3, pp. 41-51, Jan. 1981

51. S. Soudoplatoff: "Markov modeling of continuous parameters in speech recognition." *Proc. of the ICASSP '86*, pp. 45-48, Tokyo, Japan, Apr. 1986

52. A. Waibel: "Prosodic knowledge sources for word hypothesization in a continuous speech recognition system." *Proc. of the ICASSP '87*, pp. 856-859, Dallas, Tex., Apr. 1987

53. R. Zelinsky, F. Class: "A segmentation algorithm for connected word recognition based on estimation principles." *IEEE Trans. on Acoust., Speech, Signal Processing*, vol.31, pp. 818-827, Aug. 1983

54. V. Zue: "The use of speech knowledge in automatic speech recognition." *IEEE Proceedings*, vol.73, pp. 1602-1615, Nov. 1985

Chapter 3

The Real Time Implementation of the Recognition Stage

Robert Breitschaedel (Daimler Benz), Alberto Ciaramella (CSELT), Davide Clementino (CSELT), Roberto Pacifici (CSELT), Jean Pierre Riviere (Thomson-CSF), Giovanni Venuti (CSELT)

3.1 Introduction

Subtasks 2.2 and 2.3 of the P26 project have been devoted to the design of a hardware architecture and to the implementation on it, in real time, of recognition algorithms already developed and experimented within Subtask 2.1.: this real time implementation of the recognition stage will be called RICO in the following. Table 3.1 summarizes the key points we considered when we started our work, i.e. algorithmic requirements, project development constraints, hardware and software technology limits; they contributed to the definition of RICO main characteristics, summarized in Table 3.2: in the following of this paragraph we will detail these considerations. We started with the consideration that recognition algorithms can be distinguished into two principal blocks, a first "feature extraction" block till vector quantization and phonetic classification of frames, and a following "search" block extracting the lattice of most likely words using dynamic programming: this system "cut" corresponds to the minimal flow of data and besides separates blocks with different computational characteristics. The first block in fact is characterized by predictable execution times, cyclic computations, vector data structures, not-too-large data addressing requirements: this block in fact implements "traditional" DSP algorithms, for which the DSP chips fit well. Instead memory and computational requirements of the second block heavily depend on the recognition vocabulary size and on the speaking style (continuous speech of course is more demanding than isolated words) and also exhibit a time dependency for the same utterance; in each case, for the real time recognition of continuous speech with a 1K words vocabulary, the computational requirements are quite demanding, although were not clearly defined at the beginning of the project. This block could have been implemented by using chips customized [1, 2, 3, 4, 5] or optimized [6] for the dynamic programming algorithm: we had also the possibility of using an internally developed chip of this kind [7, 8]. We preferred however to retain the maximum of the flexibility allowed by a DSP implementation, although in this case the computation throughput was smaller and some of the DSP capability remained unused (typically the fast multiplier), whilst other features would be welcome, as wider data memory addressing. Hence we chose the DSP [10, 11] with the widest data addressing

Algorithmic requirements	Features extraction	Predictable execution time
	Lattice extraction	Large data addressing
		Execution time both task and time dependent
Project development constraints	Fast hardware and firmware prototyping	
	Ease of expandibility	
Hardware limits of DSP technology in P26 time frame	Reduced addressing space	
	Fixed point computations	
Firmware limits of DSP technology in P26 time frame	DSP assembler programming	
	Lack of multiprocessor operating system	

Table 3.1: Key points affecting RICO architecture

Hardware architecture and implementation	Common bus multiprocessor
	Local plus global distributed biport memory structure
	Fast common bus (VME)
	Asymmetric multiprocessor equipped with: - general purpose master CPU (68020 based) - custom DSP slave boards (TMS32020 based)
Software architecture and implementation	Large grain task partition
	Task synchronization through busy waiting
	Assembly programming for the slave CPUs
	Pascal programmming for the master CPU

Table 3.2: RICO main characteristics

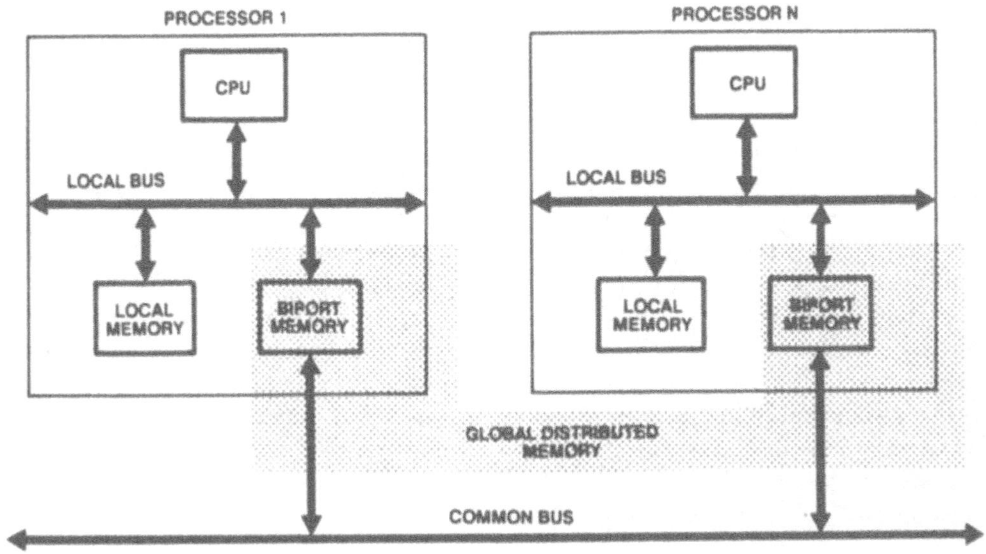

Figure 3.1: Common bus multiprocessor, with local plus global distributed biport memory

space available at the beginning of the project (64K words), and we further increased its data addressing capability by an external page register [9]. As a result we defined as a common iterable block a DSP board capable of supporting in real time all the computations foreseen at the beginning of the project for the feature extraction stage and with the data addressing capability required by the dynamic programming stage: for achieving real time this last computation however can be distributed and parallelized in different DSP boards, as required by the specific application task. This tasks partition could be mapped to different multiprocessor hardware architectures proposed in the literature, as for example tree connected [12], multiple bus connected [13], processor clusters [14]. We chose the simplest single bus architecture, and to be more specific a single bus architecture with local memories and a distributed global memory, where distributed memories allocated in different boards are biported to the common bus.This kind of architecture is the most efficient between single bus architectures [29], since no contention arises either on local buses or in biport memory accesses, but in the common bus access only. Figure 3.1 exemplifies such a kind of architecture. We further reduced common bus contentions:

- by using secondary buses for fast local transfers [18],
- by using a fast primary bus, the VME bus [15, 16, 17], which could carry heavier traffic before saturating it.

The implemented multiprocessor is asymmetric, with a general purpose master for handling i/o and mass memory accesses with standard hardware and software facilities, and some slaves DSP for speeding up more computational intensive algorithms: DSPs in fact are an order of magnitude faster than general purpose microprocessors. Since the VME bus is a widely accepted standard, we could use as far as possible commercially available boards for the master, the input-output and the memory, whilst we developed only two kinds of project-specific boards, i.e. the DSP and the converter. In addition to the hardware we developed both the system control firmware, and the algorithm firmware. The algorithm firmware was split between the Motorola master (in Pascal language [28]) for less time-critical sections and the DSP (in assembler language) for more time-critical ones: we point out that integer arithmetic and assembler language programming were a typical limitation of DSPs, whereas today DSPs allow a faster firmware development cycle since they support floating point arithmetic and C language [19, 20, 21, 22]. As a final result we demonstrated in real time and with a good recognition accuracy the intended recognition task of 1K words continuous speech using a not-so-expansive implementation. This is a clear indication that present technology is already sufficient to build a realistic speech recognition system for large-vocabulary continuous speech, and that technological advances will simplify this task more and more: some of us in fact are now experimenting this trend in new ESPRIT projects [24]. In the following we will present an overview of the system implemented, then we will detail both the hardware and the firmware blocks, finally we will show the system throughput.

3.2 System Overview

3.2.1 Functions Overview

RICO performs several functions, summarized in Table 3.3: first of all in recognition mode it hypothesizes the uttered words, forwarding to the following understanding stage the most likely ones in the application vocabulary; these are organized as a lattice for the continuous speech and as a list for the isolated words case. Three different real time recognition algorithms have been implemented:

- the single-step recognition for isolated words,
- the two-step recognition for isolated words,
- the single-step recognition for continuous speech.

All these tasks have been demonstrated in speaker-dependent mode with a high quality head-mounted microphone input. The input utterance is presently limited by an initial keystroke for the isolated words task and by an initial and a final keystroke for the continuous speech task. As detailed before, in the single-step recognition algorithm the whole application vocabulary is verified. In the two-step recognition algorithm on the other hand a subset of the whole vocabulary is selected first using a coarse preselection, then a more detailed verification is performed on this subset: this last strategy requires more computations for small vocabularies, but becomes more and more efficient as the vocabulary increases. The vocabulary size for which the single-step approach becomes

Real time recognition	- Single step for isolated words -Two steps for isolated words - Single step for connected words
Off-line recognition	- Single step for isolated words - Two steps for isolated words - Single step for connected words
Acquisition and training	
System testing	
MultiDSP program loading facility	

Table 3.3: Summary of implemented functions

more efficient than the two-step approach is larger in the continuous speech case than in the isolated words case: hence for continuous speech we have implemented only the simpler one-step strategy, given that in this project we aimed at vocabularies of the order of 1 K words and in this case the single-step strategy is both easier to implement and faster in execution time. Besides real time recognition, RICO performs other functions: first it can perform off-line recognition, using prerecorded speech parameters as input: this is useful in order to characterize the system both in accuracy and in speed and to tune some system parameters. Then RICO can be used to drive the speaker acquisition session, displaying the words to be uttered, synchronizing the utterance with start and stop keystrokes, and extracting the parameters of the uttered words: at the end of the session this parameter database is sent to a host VAX, which performs the parameter training; these parameters are finally transmitted back to RICO for the following recognition phase. Figure 3.2 summarizes the system behaviour in these three cases: the connections are enabled only if marked by the number corresponding to the case (i.e. 1 for on-line recognition, 2 for off-line recognition, 3 for parameter training). Finally we have implemented a system test for checking the system and a multiDSP loading facility, for transferring object files from the RICO mass memory to a specific DSP.

3.2.2 Architecture Overview

As anticipated, RICO is centered around a VME bus and is composed of general purpose boards (CPU, central memory and peripheral interfaces) and of boards explicitly developed in this project, which are the digital signal processor and the acquisition boards [9]. The general purpose boards are:

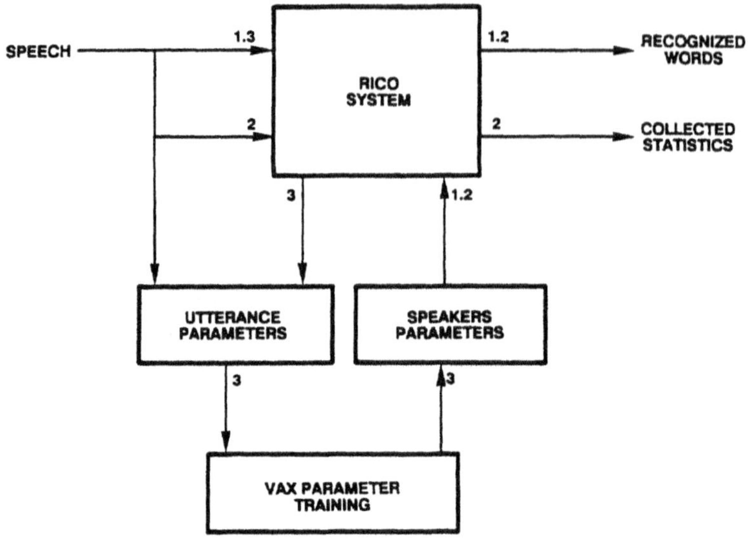

Figure 3.2: Block diagram sketching system functions

- a master CPU board, using the Motorola 68020 [25] and 1 Mbyte of internal RAM, biported also to the VME bus,
- a mass memory controller, which controls a hard disk and a floppy disk,
- an input/output board, for data transfer to the host and other functions,
- a VME/VMX biport 1 Mbyte RAM board.

The special purpose boards are:

- the acquisition board, whose functions can be expanded for debugging purposes by a piggy-back board for buffering speech samples,
- three copies of digital signal processor boards, of which the first is used for features extraction and the remaining two are used for DHMM scoring.

Local transfers are done through two VMX buses: the first allows the transfer of samples from the acquisition to the feature extraction board, the second allows the extension of the memory of the other two DSP boards used for verification with the VME/VMX biport RAM. As anticipated, memories can be distinguished into local, accessed by a specific processor, and global, addressed by all processors through the VME bus; the global area is distributed to different boards, where they behave as biport memories. Local memories and biport memories have been in fact implemented in DSP boards, whilst the VME/VMX biport 1 Mbyte board is used as an expansion of DSP boards used for DHMM scoring. The system can be expanded and configured differently for different tasks; Fig. 3.3 shows the final hardware configuration. Figure 3.4 shows instead the principal functional blocks required in the two-step verification: the gray area characterizes blocks implemented in the

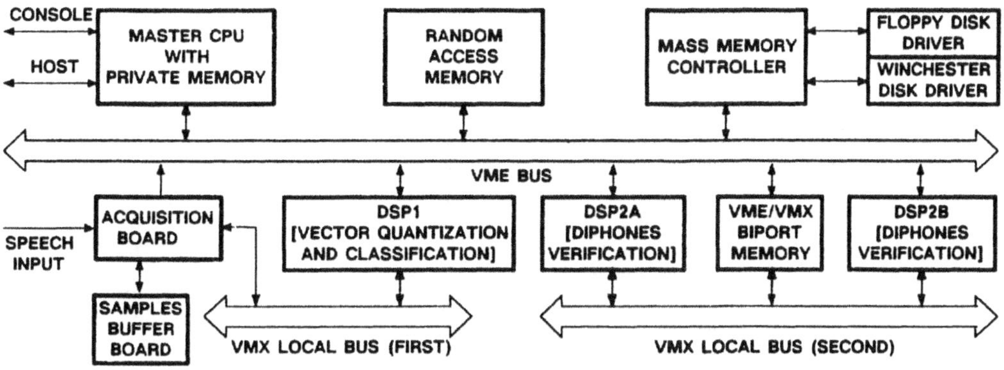

Figure 3.3: Final hardware configuration

master CPU. Feature extraction is performed by the DSP1 and is composed by DCTs and energy computation, vector quantization and classification; the hypothesization stage is performed by the master and is composed by frame segmentation, cohort hypothesization and diphones lattice building; discrete hidden Markov models (DHMM) verification is composed by the diphone tree scanner, implemented in the master, and by the diphones verification, implemented in the DSP2. In order to speed up this computation some DSP2 can be put in parallel: in our final implementation we experimented two DSP2 in parallel and we called them DSP2-A and DSP2-B[1]. This decomposition of the DHMM verification into two levels, the less computation-intensive one on the master, and the more computation-intensive parallelized in some fast slaves, is well suited to a real time multiprocessor implementation. Two-level decomposition of the verification stage has been used in all three cases for the real time implementation and is not to be confused with the one-step and the two-step approaches. In fact the one-step approach can be obtained from the two-step approach by completely discarding the hypothesization stage and by scanning only the precompiled diphone tree describing all the vocabulary, instead of traversing the on-line compiled diphone tree describing the subset of words evaluated by the hypothesization stage. Table 3.4 summarizes the blocks used for the three real time recognition applications implemented, with the specification of the board hosting them; of course the system can evolve and demonstrate a further speed-up by porting other functions from the system master to DSP boards. Off-line recognition differs from on-line recognition in the system control that in this case activates the DSP2 only by using prerecorded acquisition results; the acquisition function instead requires only the parameters extraction section driven by a suitable system control, so in this case the DSP1 only is used.

[1] Just not to confuse readers, we point out that in some earlier P26 documents and papers we referred to these DSP as DSP3-A and DSP3-B or even DSP3 and DSP4 for historical reasons. The notation used in this document however seems more appropriate.

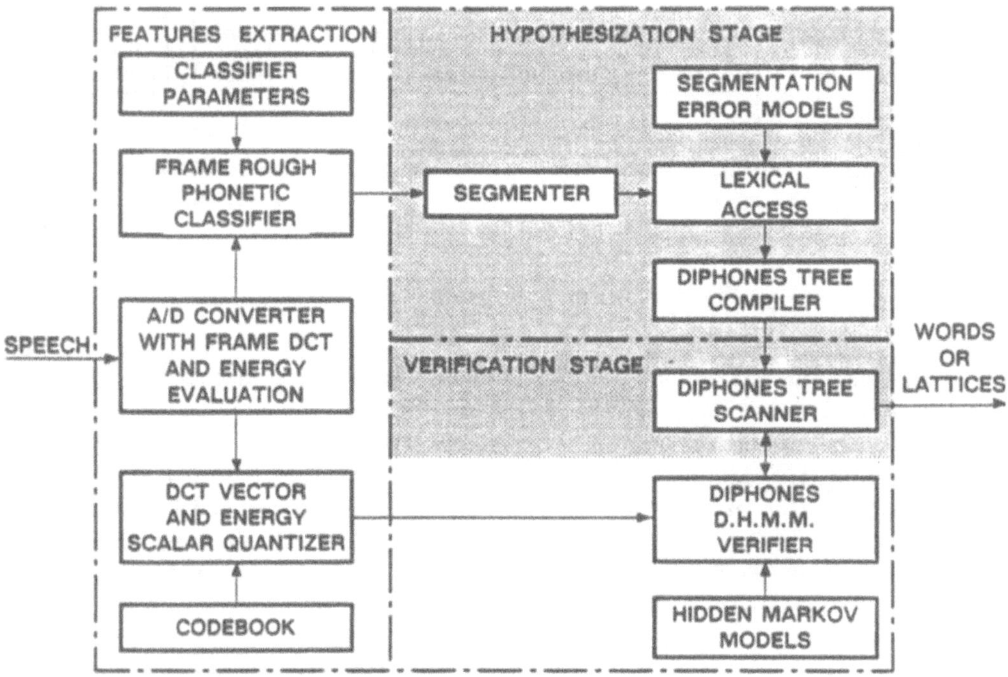

Figure 3.4: Principal function blocks for the two steps verification

MAIN FUNCTIONS	SINGLE STEP ISOLATED WORDS	TWO STEPS ISOLATED WORDS	SINGLE STEP CONT. SPEECH	IMPLEMENTED IN BOARD
SYSTEM CONTROL	Y (1)	Y (2)	Y (3)	MASTER
DCT AND ENERGY EXTRACTION (*)	Y	Y	Y	DSP 1
VECTOR QUANTIZATION (*)	Y	Y	Y	DSP 1
FRAME CLASSIFICATION (*)	N	Y	N	DSP 1
FRAME SEGMENTATION (**)	N	Y	N	MASTER
COHORS HYPOTHESIZ. (**)	N	Y	N	MASTER
ON LINE DIPHONES LATTICE BUILDING (**)	N	Y	N	MASTER
DIPHONE TREE SCANNER (***)	Y	Y	Y	MASTER
DIPHONE HIDDEN MARKOV M. VERIFICATION (***)	Y	Y	Y	DSP 2
LATTICE FILTERING	N	N	Y	MASTER

(*) Features extraction stage

(**) Hypothesization stage

(***) Verification stage

Table 3.4: Comparison of blocks used in different subcases

3.2.3 System Control and Synchronization Methods

The system is controlled by the 68020 master CPU, running the real time multiuser multitasking operating system VERSADOS [26] and a ROM resident monitor; the DSP boards and the acquisition system behave as intelligent system peripherals, which exchange data and synchronizations by reading and writing VME addressable locations. The acquisition board is controlled by an internal ROM resident monitor, which decodes commands received through the VME bus; on the other hand the DSP board exploits a PROM resident kernel with selftest and loader functions only: in this way the DSP's program and data configuration is is totally controlled by the master CPU software. For each DSP we have a command and status area, allocated in a predefined area of the VME biport memory; the specific format of this area depends on the application program and in fact it is different for DSP1 and for DSP2. In general the command and status area controls these functions:

- the DSP program bootstrap, from a VME addressable buffer to the DSP internal program area;
- the data bootstrap from a VME addressable buffer to a DSP internal data area, used for loading the codebook for the DSP1 and the DHMM for the DSP2;
- the input and output buffers configuration and enabling; these buffers can be in fact enabled or disabled by suitably writing the corresponding command area; enabled buffers can be allocated everywhere in the VME addressable area;

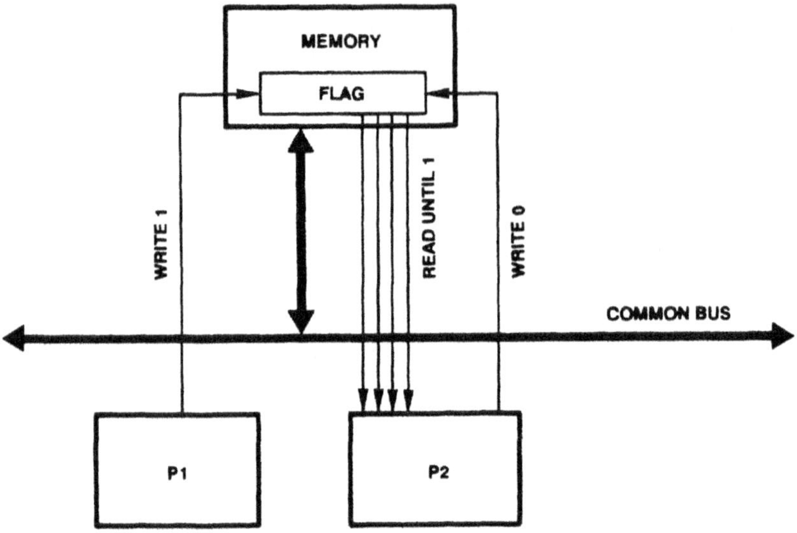

Figure 3.5: Synchronisation traffic for flag addressed through a common bus only

- the DSP program control: when the programs and data are loaded and the buffers are configured, we can start up the system by suitably writing in the command area; also in this case we can have some different possibilities, as for example "run forever until stopped" (used in the real time application), or "run for a definite number of frames" (used in the debugging).

Different boards are synchronized through flags by the busy waiting mechanism [23], in which the processor P2 to be synchronized continuously reads the flag until it is in reset state and then starts the following computations when it reads that the synchronizing processor P1 has set the flag; at the same time P2 resets the flag. This is the simplest synchronization mechanism to implement, but if the flags are not properly allocated it could generate excessive bus traffic for synchronization purpose only, due to the burst of flag readings: this happens for example if the flag is allocated in a memory which is addressed through the common bus by both processors (see Fig. 3.5). In order to avoid this problem we used VME biport memories and whenever possible we allocated the synchronization flags in the biport memory housed in the same board of the processor to be synchronized: in such a way the flag synchronization reading burst remains internal to the board and does not affect the bus traffic (Fig. 3.6). Synchronization flags are used in our system both to start up and to stop in an orderly way the different blocks, and to validate messages between different blocks.

3.2.4 System Run-Time Evolution

As far the run-time evolution is concerned, the system can be distinguished into two sections (Fig. 3.7):

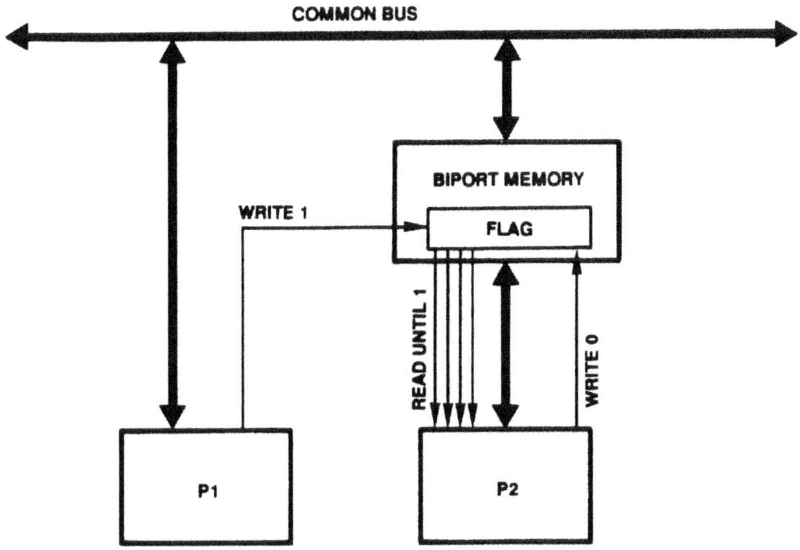

Figure 3.6: Synchronisation traffic for flag allocated in the P2 biport memory

- a first section, composed by the DSP1, which is frame synchronous, since it is assured that it extracts frame parameters each 10 ms,
- a second section, composed by the remaining computations, and centered on the frame verification, which is frame asynchronous since its timing is not predictable; as a rule of thumb beginning and ending frames of the utterance spend less time in the verification stage than central ones.

The verification stage computational time is frame dependent since not all the diphones are verified at each frame, but only the active ones: active diphones are those which have at least one state above a beam search probability threshold. The verification stage is composed by the diphone tree scanner and by the more computational intensive discrete hidden Markov model (DHMM) verifier (Fig. 3.7). On a frame basis the diphone tree scanner sends to the verifier through the broadcast area both the frame code vector and the beam search threshold, then identifies the diphones whose verification has to start and puts them in a list, called the "push list", transmitted to the diphone verifier. This last block performs dynamic programming on diphones in the push list and on the diphones already active at the end of the previous frame; then updates the active diphones by discarding those from which all the state probabilities are below the best path by a given threshold and finally identifies the diphones reaching the final state: these are organized in a "pop" list and sent to the diphone scanner, which from the pop list evaluates the push list of the next frame (Fig. 3.7). We implemented two kind of sequencing between the synchronous and the asynchronous sections: first a simpler, but less time efficient "serialized" implementation, then a more efficient "interleaved" one, which is the final one released: Figure 3.8 summarizes the frame timing in both cases. In the serialized case the asynchronous computation on the first frame starts when the synchronous computation on the last frame ends, whilst in the interleaved case the asynchronous computation on

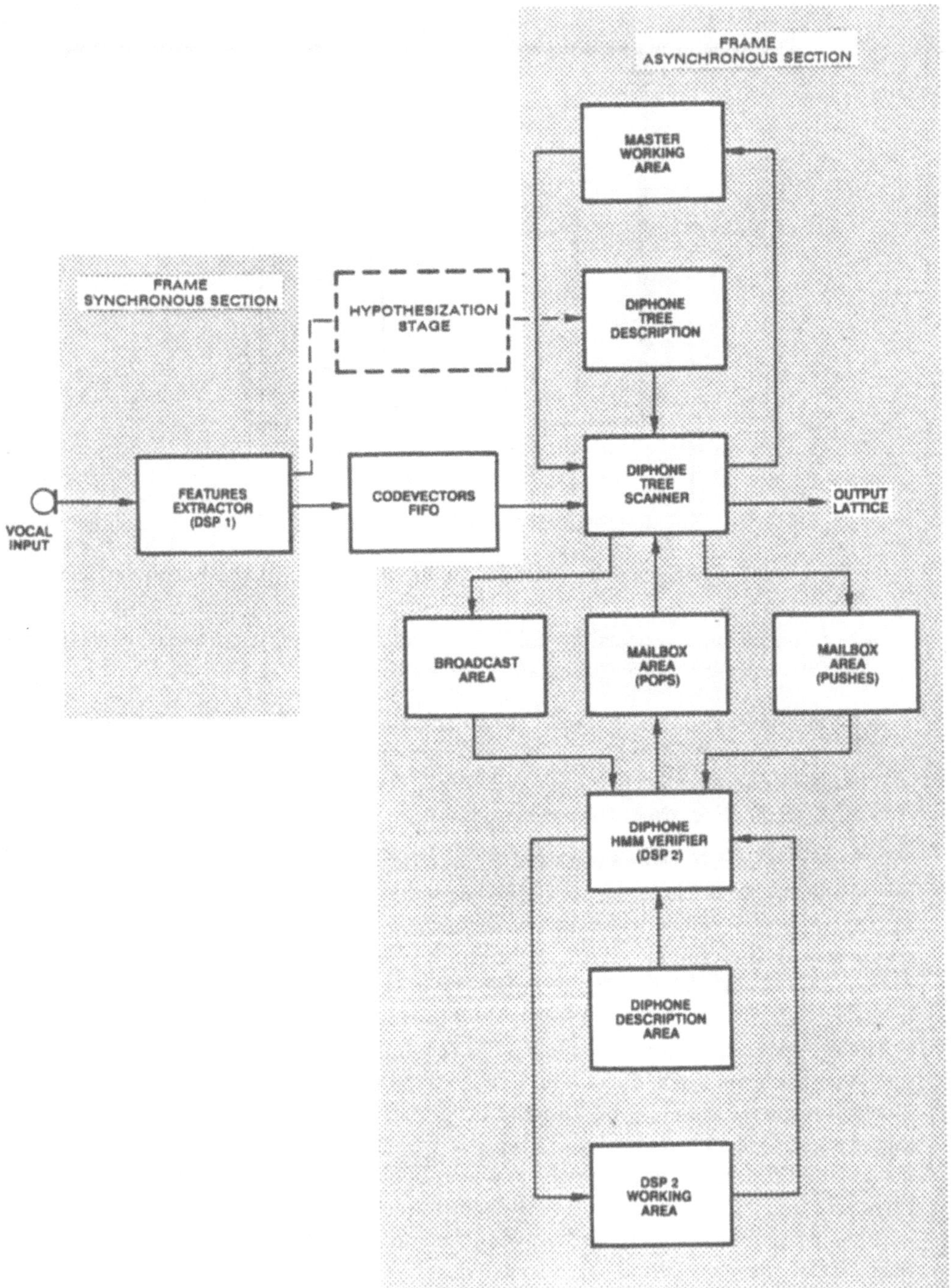

Figure 3.7: System interplay

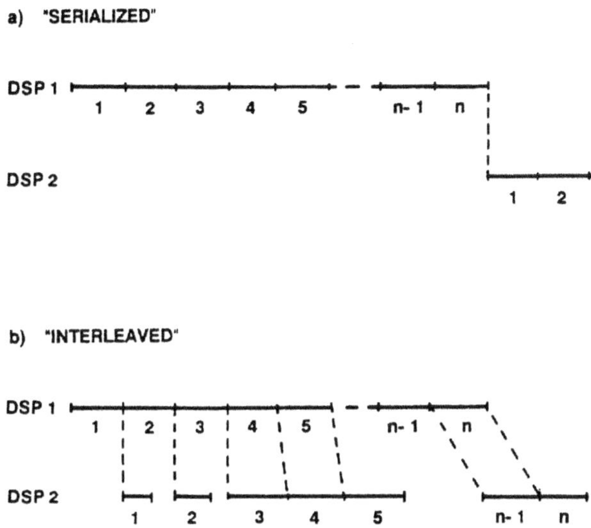

Figure 3.8: System timing

the first frame starts when the synchronous computation on the first frame ends and so on. We can see also that in the serialized case all DSP2 frame computations are performed consecutively since the input code vectors are all available from the beginning, whilst in the interleaved case some idle time can appear in DSP2 computations, and this happens when DSP1 has not yet finished the code vector evaluation of the next frame (see Fig. 3.8). In every case it is obvious that the recognition delay of the system is reduced in the interleaved case and this situation is even better when there is a balance between the frame computational time of the DSP1 and the (average) frame computational time of DSP2: for the continuous speech application with 1000 words we approximate this balance by putting two DSP2 in parallel. All these considerations can be extended from the single-step to the more involved two-step approach: in this last case we have also to take into account the hypothesization stage (dashed block in Fig. 3.7) which changes the diphone tree description on a phonetic segment basis[2]: in this last case however we cannot oversee master computational times due to the hypothesization stage.

[2]A phonetic segment typically lasts some frames (e.g. 5-10 frames).

3.2.5 Details on the Asynchronous Stage Activity

We will detail now a model of the asynchronous stage activity in order to explain both its computational load frame behavior and the criteria used in parallelizing the verification task. Figure 3.9 shows the model of the asynchronous stage activity: each frame t is characterized by a number of "push" diphones $Npush(t)$, by a number of diphones active already active $Nactb(t)$ and by a total number of diphones $Nver(t)$ which are verified by the dynamic programming. Some of these, with the probality $Pdis$, becomes inactive, and their number is $Ndis(t)$, the remaining instead remain active at the end of the frame, and their number is $Nacte(t)$. Some of these, with probability Pop , originates the final pop diphones, with probability $Ppop$. Hence in the same frame we have:

$$Npush(t) \cap Nactb(t) = Nver(t) \quad (3.1)$$

$$Nver(t) \times Pdis = Ndis(t) \quad (3.2)$$

$$Nver(t) \times (1 - Pdis) = Nacte(t) \quad (3.3)$$

$$Nacte(t) \times Ppop = Npop(t) \quad (3.4)$$

whilst the frame t activity is related to the frame t+ 1 activity by these relationships:

$$Nacte(t) = Nactb(t+1) \quad (3.5)$$

$$Npop(t) \times bm = Npush(t+1) \quad (3.6)$$

bm being an average factor which summarizes the activity of the diphone scanner in generating pushes from pops. The heavier computation of the asynchronous stage is the dynamic programming, which is iterated $Nver(t)$ times each frame: hence in a first approximation we can say that $Nver(t)$ measures the frame variable computational load of the asynchronous stage. At frame $t = 1$ we have that $Nactb(1) = Nacte(0) = 0$, hence $Nver(1) = Npush(1)$, which are the initially pushed diphones: hence the first frame computational load is quite reduced. Then $Nver(t)$ grows frame by frame because new pushed diphones $Npush(t)$ are greather than Ndis(t), i.e. disabled diphones. At the end of the utterance instead $Nver(t)$ decreases again, since in this phase there are many disabled diphones $Ndis(t)$. Figure 3.10 summarizes the typical frame-by-frame verification time of an utterance: its behaviour is similar to other search problems, i.e. the computational load increases first and then decreases. In the case that the verifier is too slow for a specific task, it can be parallelized to N different DSP2 in such a way that the computational load is balanced between these processors: this could be achieved if each of the N parallel verifiers performs dynamic programming on $Nver(t)/$ N diphones. An exact balance would require, however, the unacceptable overhead of redistributing the active nodes on a frame basis between the different processors: we found that an acceptable policy is to balance the number of pushes provided by the scanner by distributing $Npush(t)/$ N pushes to each verifier; hence in our implementation with two DSP2 in parallel new pushes are alternatively sent to DSP2-A and DSP2-B. Figure 3.11 details the cooperation of the scanner with two verifiers in parallel. Other than this we have to point out that the scanner and the verifier run completely in parallel; this is due to the fact that the push list entering the verifier is organized as a double buffer, hence while the scanner computes pushes for frame $t + 1$ as soon as pops of frame t are available, the verifier reads last pushes of the frame t and computes last pops.

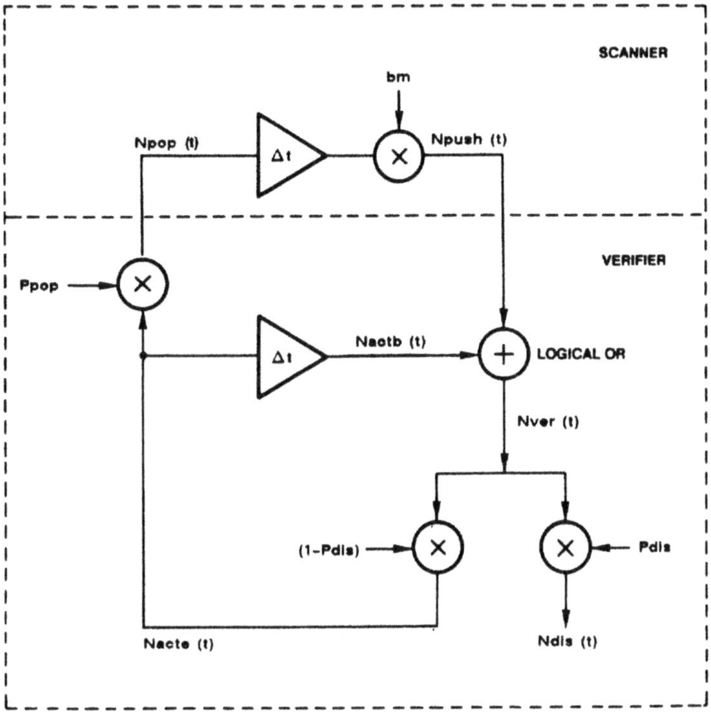

Figure 3.9: Model of subwords activation

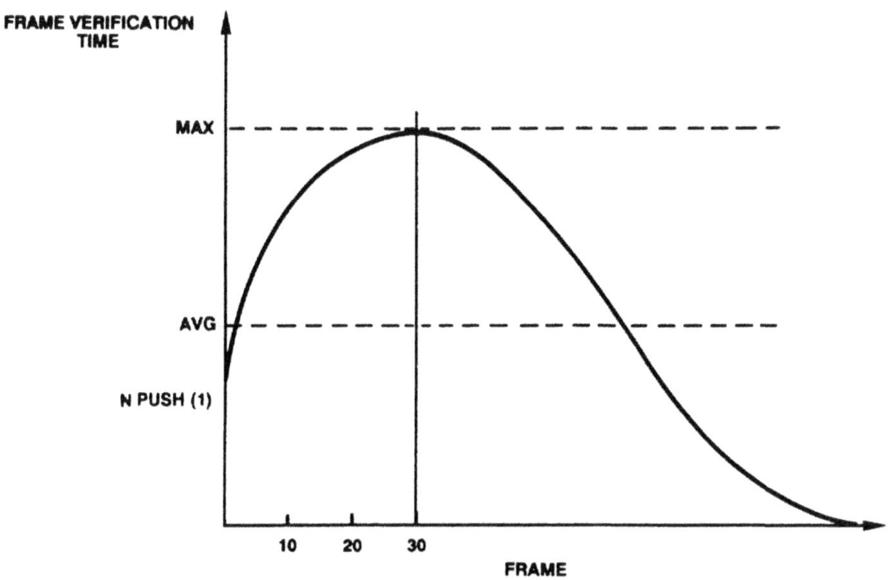

Figure 3.10: Typical verification times of an utterance

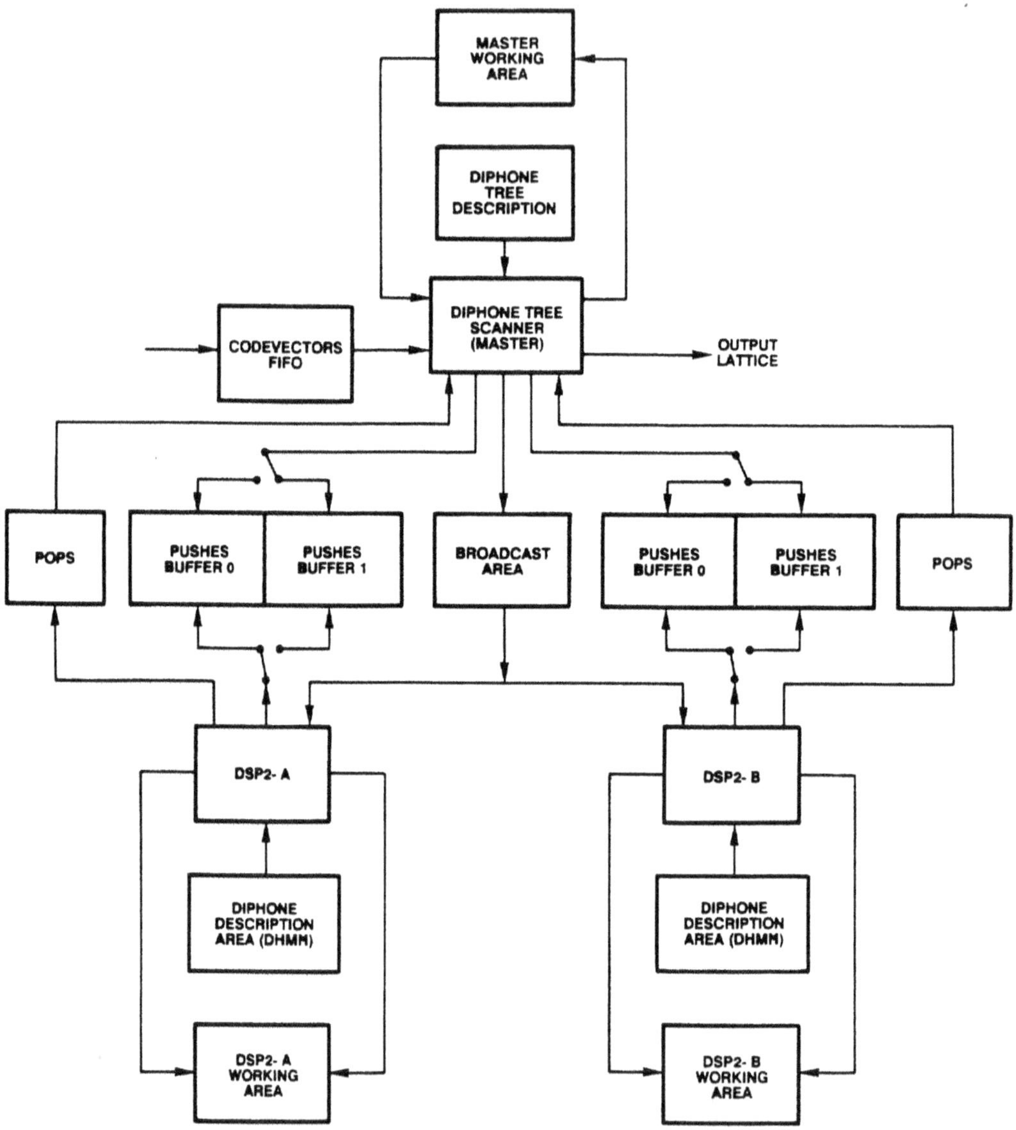

Figure 3.11: System interplay of the final system with two verifiers in parallel

3.3 Hardware Details

3.3.1 DSP Board Description

DSP board architecture requirements

The DSP board is centered around a TMS32020 digital signal processor, driven by a 20 MHz quartz clock: this was in fact the most advanced DSP available at the beginning of the project and moreover the DSP with the widest address range available: in fact our application was memory intensive, especially for the algorithm of Viterbi decoding using DHMM. Nevertheless the native addressing capability of the DSP was not enough for us, since we wanted to completely address two 24-bit buses (VME and VMX): hence the first problem that we faced was the extension of the DSP native address space through data area paging. The second problem that we had to solve was to render the board easy to reprogram, and this was achieved in the most easy and flexible way by providing the board with a bootstrap ROM. Finally we provided the board with biport memories in order to allow busy waiting synchronization through flags without producing bus overhead, as already explained.

DSP board architecture details

As anticipated, the DSP used is a TMS32020 [10]; it is characterized by an address range quite large for the category of signal processors, i.e. 64K of program and 64K of data, of which 544 words are internal to the chip and therefore faster; these are structured in three banks, B2 (of 32 words), B0 (of 256 words) and B1 (of 256 words too): these areas must be used for more frequently accessed data in order to obtain the maximum algorithm speed-up: we will detail this point later for specific algorithms. The TMS32020 interfaces to a 16-bit data bus and a 16-bit address bus, which are organized into three different spaces, i.e. a program space, a data space and an I/O space; the data space can be further split into private and general. In fact, according to the configuration of the GREG [10] register internal to the DSP chip, data addresses below a threshold address are private and above are general (in our case the threshold is 32K): when the data general area is addressed a synchronization dialogue is also activated, in such a way that this area can be shared with other processors too. Our DSP board is equipped with 4 banks of 8K words of memory, each of which can be configured as program memory or internal data memory: the start address of each bank, its nature (whether RAM or ROM, whether program or data) is jumper selectable or PLA programmable. To be more precise, the DSP1 is configured for 16K RAM words of programs and 16K RAM words of internal data, whilst the DSP2 is configured for 8K RAM words of programs and 24K RAM words of internal data: these differences reflect different requirements of the implemented algorithms. In order to be reprogrammable, the DSP board is equipped with a kernel PROM, with selftesting and bootstrap functions, summarized in Sect. 3.1.3: this 512-word PROM starts from address 0 of the TMS32020 and is in overlay with the first bank of program RAM; it is automatically addressed for reading after a power-on or a system reset signal and deselected by issuing a suitable DSP output: from this point on only the corresponding address of the overlay program memory bank will be accessed and the corresponding DSP application program executed. Using the 32K words of the general data area, we can access to the following memories:

- the VMX 2-ports RAM, of 8Kwords, allocated on the same DSP board,
- a DSP external address allocated on the whole VMX bus,
- the VME 2-ports RAM, of 8Kwords, allocated on the same DSP board,
- a DSP external address allocated on the whole VME bus.

These 4 cases are distinguished by specifying the 2 most significant bits in the EXFR register (Fig. 3.14), allocated in the TMS32020 I/O space; other 14 bits are used to specify the kind of VME or VMX transfer and the number of the 32K words page accessed on the VME or VMX bus: in our system in fact we use 24 bits for addressing the VME and VMX buses, hence we have to expand the native addressing capability of the TMS32020. Hence, to summarize, the data RAM of a DSP board is organized in a three-level hierarchy:

- the RAM internal to the chip: this is the fastest, but of 544 words only,
- the RAM internal to the board, but external to the chip: this is of intermediate speed and size,
- the RAM external to the board, addressable through the VME bus: this is the slowest, but the widest in the address range.

Other than this, the DSP board contains also two specialized VLSI circuits to generate and handle interrupts on the VME bus: the registers used for programming the functions of these VLSI are also mapped into the DSP I/O space. In this way the DSP board can generate and receive the seven different interrupt lines specified by the VME protocol; however, only 3 out of the 7 received interrupts can be forwarded to a single DSP, given that the TMS32020 accepts a maximum of 3 external interrupts: a set of jumpers defines for a specific DSP board which of the 7 input interrupts are really handled. We point out that we did not use interrupts in our application for simplifying the implementation. Figure 3.12 shows the DSP board general block diagram, while Fig. 3.13 summarizes the DSP board address map in hexadecimal notation, as always in this description.

DSP kernel

Each DSP is equipped with an identical PROMmed program kernel, with selftest and program bootstrap functions: it is mandatory to install this PROM in order that the DSP board work properly. Commands, results and synchronization words are exchanged between the master and the DSP through some predefined locations of the VME biport RAM of the corresponding DSP, allocated in the first 8 addresses of this area: of these, the first 6 words specify parameters, while the last 2 synchronize the master and the DSP. At system reset the kernel program automatically starts for each DSP board of the system, since this command sets the DSP's program counter to zero and automatically selects the selftest and bootstrap PROM in the first program bank. The subsequent program evolution is summarized here in pseudo-Pascal code.

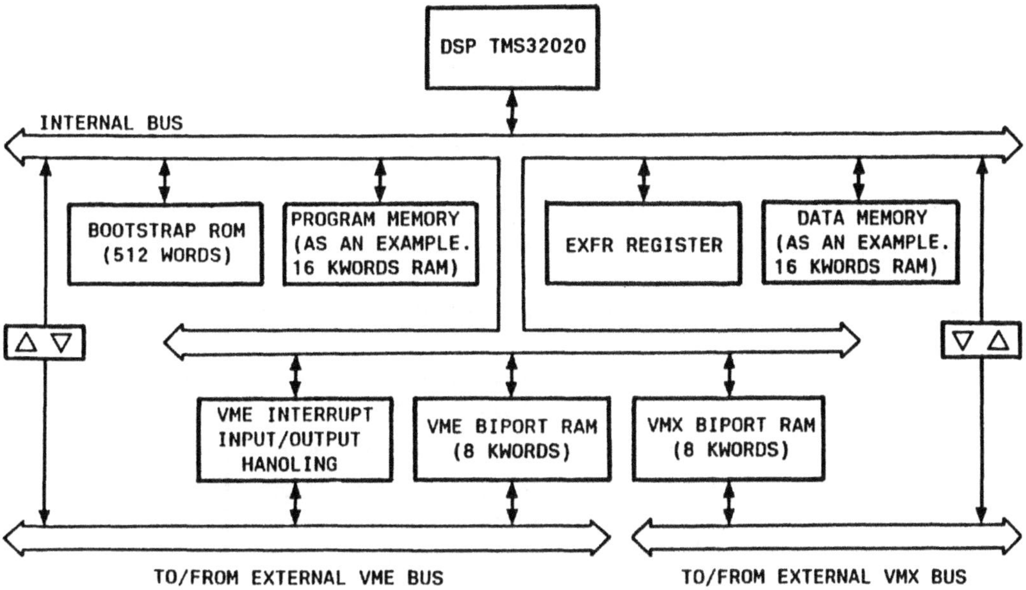

Figure 3.12: DSP board general block diagram

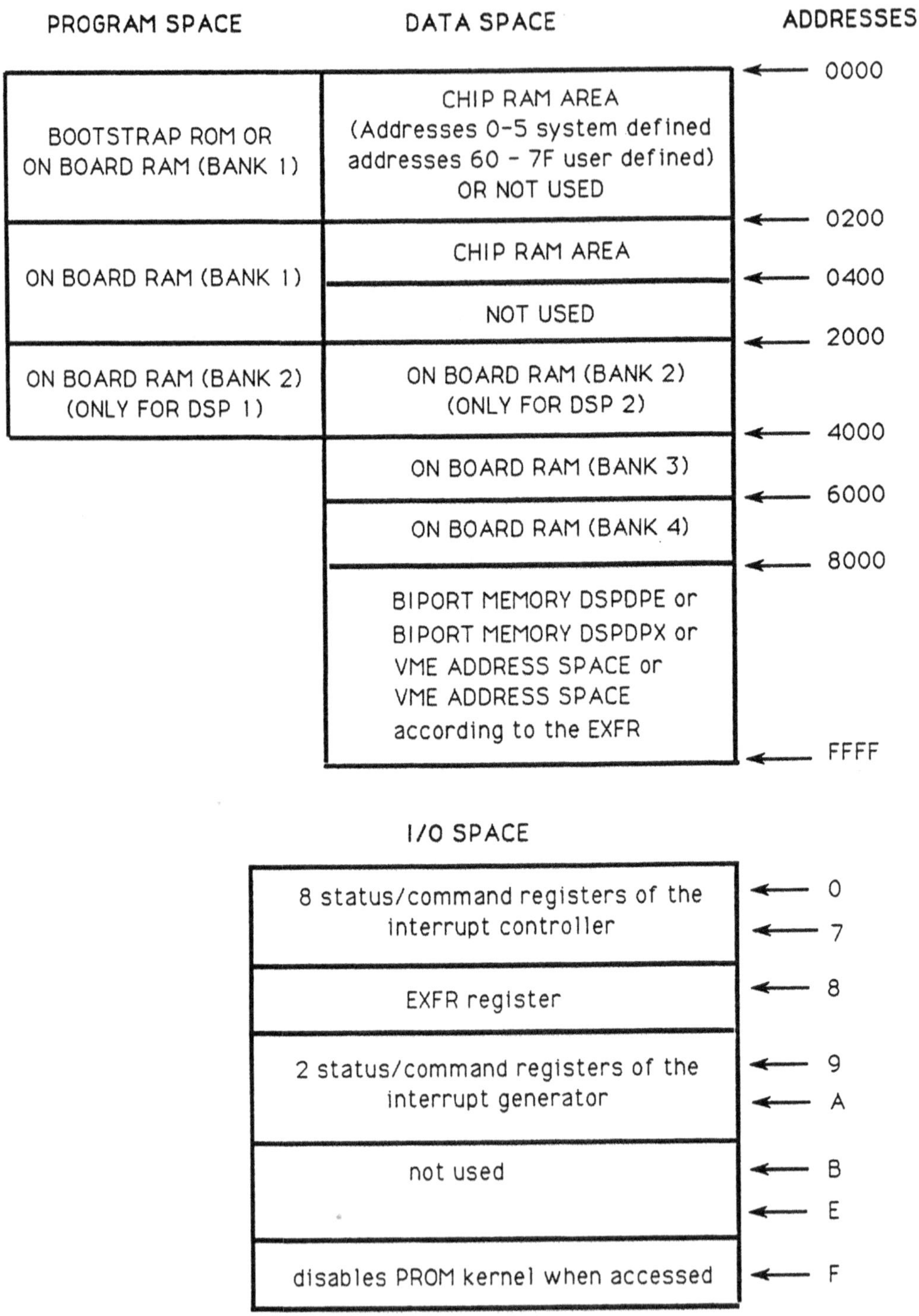

Figure 3.13: Address map of a DSP (hexadecimal addresses)

VXM biport area

15															0
0	0	X	X	X	X	X	X	X	X	X	X	X	X	X	X

VXM bus

15															0
0	1	AX1	AX0	X	X	X	X	A23	A22	A21	A20	A19	A18	A17	A16

VME biport area

15															0
1	0	X	X	X	X	X	X	X	X	X	X	X	X	X	X

VME bus

15															0
1	1	AM5	AM4	AM3	AM2	AM1	AM0	A23	A22	A21	A20	A19	A18	A17	A16

Figure 3.14: Configurations of the EXFR register

```
DSP board initializations;
first master/dsp synchronization;
if tests are required then
    repeat
        master/dsp synchronization
        to validate test parameters;
            case test of
                data test: perform data buffer area test
                                    and obtain diagnostics;
                program test: perform program data buffer area test
                                    and obtain diagnostics;
                program to data: transfer a buffer from a
                                    program to a data memory;
                data to program: transfer a buffer from a
                                    data to a program memory;
        if the test is not ok then stop;
    until no new test is required;
endif;
master/dsp synchronization to validate program bootstrap parameters;
if program boostrap is required do it;
switch the first program memory bank from PROM to RAM and
jump to the program address 200.
```

From this description, we see that the kernel loops in the first synchronization phase until an appropriate data interchange occurs with the master; it is hence possible to skip completely the test phase or to enter a test loop. Four test functions are available: a test on a data buffer area, a test on a program buffer area, a buffer transfer from a program to a data area or vice versa: these two last functions are not really test programs, but can be useful in a system test environment. Before each test the master writes in the VME biport area of the DSP the buffer starting address and the buffer length: it is hence possible to test every VME or VMX addressable area, since in the starting address we define also the EXFR register configuration. In case of fault, the program is stopped automatically and the DSP writes diagnostic information in the VME biport area arranged in two words, of which the first codes the first memory address in fault and the second codes the bit(s) in fault. The following program bootstrap phase transfers a buffer area from a global data area addressable through the VME bus to the DSP program space; in a previous phase the object programs have been transferred from the mass memory to the VME addressable data space. At the end the kernel program jumps to the fixed program location 200, after having switched the first program bank from this PROM to the RAM by issuing an output at address F; in order to be compatible with the kernel the DSP application programs must start from location 200, with addresses from 200 to 220 filled with a jump table. Since the PROM content cannot be accessed any more, if we want use again the selftest and bootstrap functions after the PROM deselection, we must have a kernel copy in the corresponding program RAM area, which is accessed in the case of a DSP local reset command.

3.3.2 Acquisition Board Description

Acquisition board requirements

In the implementation of the acquisition board we faced with two problems: the former is the simplification of the samples transfer to the following DSP board, the second one is the simplification of running experiments from prerecorded sample files. The easier way to transfer samples from the converter to the DSP board is to use a private bus (the VMX one) and to equip the converter board with a FIFO of adequate size. In this architecture the synchronization between the converter and the DSP board is obtained implicitly by using the VMX protocol characteristics, if the processing time of a frame in the DSP is faster than a time frame. For running experiments from prerecorded sample files the acquisition system has been equipped with a large samples memory (1 Mword) whose samples can feed the DSP board as an alternative to being supplied by the converter: this samples memory is mounted on a piggy-back board; hence the complete acquisition system consists really of two boards.

Acquisition boards architecture details

The acquisition section block diagram is shown in Fig. 3.15; it has been implemented in two boards, although only one board is enough for a minimal functionality excluding the use of samples memory. The data conversion path begins with the microphone, followed by an amplifier whose gain can be programmed through a command issued by the VME bus; the amplifier output is sent out to a 6 kHz low pass filter, which limits the input signal bandwidth; the following stage then digitizes the analog signal with a 12 bits accuracy and a 12 kHz sampling rate, which is enough for our application; however higher sampling rates (24 kHz and 48 kHz) can be chosen by a suitable jumper setting. Several functional modes can be set by suitable commands on the VME bus: they are read by a monitor program resident on this board, which is under control of a local CPU. Among these, the most frequently used is the normal acquisition: samples are sent to a pair of parallel FIFOs, of 1K words each, which can be independently read at two different addresses of the VMX bus: this gives the possibility of decomposing parameters extraction algorithms in two boards running in parallel, achieving real time also for more demanding computations than presently implemented [3]. A FIFO overrun can be notified through interrupts and flags, and this happens if DSP1 frame computational time is slower than 10 ms: a DSP1 program implementation of this kind is incorrect from the point of view of real time. A correct DSP1 program in fact performs frame computations faster than the 10 ms frame time, reading the FIFO at a faster speed than A/D writes it; hence for some read instructions it happens that no new sample is immediately available: in this case the VMX bus access is frozen while the read instruction of the DSP idles until a new sample is available: this is the "trick" used by the DSP1 application program for synchronizing with the acquisition system. As mentioned before, the converter board provides also other functional modes; generally these modes use the piggy-back samples memory, which has a maximum size of 1 Mword and can store up to 80 seconds of speech with a sampling

[3] In this implementation only one DSP board for feature extraction was enough, but the possibility of using two DSP boards in parallel can be taken into account for future system expansions, as for example larger codebook or multicodebook implementations.

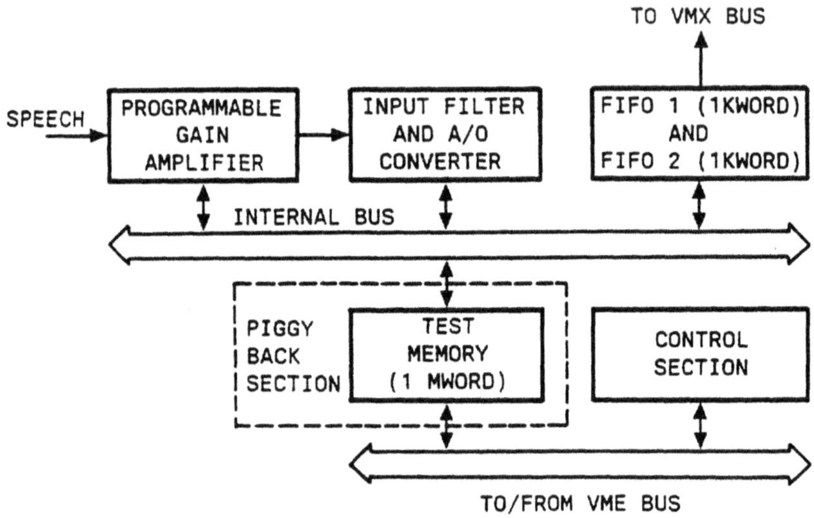

Figure 3.15: Converter block diagram

Byte addresses	Controlled function	Function type
2	VMEbus → samples memory	save-restore
4	samples memory → VMEbus	save-restore
10	gain control	standard
12	status word	standard
18	A/D → samples memory	delayed test
1A	samples memory → FIFO (many)	delayed test
1C	samples memory → FIFO (once)	delayed test
1E	FIFO control	standard

Table 3.5: Locations of the acquisition board window

frequency of 12 kHz. As a first example, for testing a DSP real time-algorithm for a predefined digital sequence, we can do the following steps: load the test data memory from the host computer via the VME bus, switch the internal bus from the converter to the test data memory and read the test data memory contents through the VMX bus for a definite number of iterations or even continuously. As a second example, when we want to test off-line a high-level implementation of the algorithms with the data generated by the real hardware, the digitized data can be stored in real time in the test data memory, whose contents can be subsequent transferred to the host computer via the VME-bus. It has also to be mentioned that this memory was also added for the case that recognition algorithms could require the further refinement of parameters extraction for some sections of speech input: however, this possibility has not been used by the algorithms actually implemented in this project.

Acquisition functions

The acquisition system functions can be distinguished into three sets:

- minimal standard functions,
- delayed test,
- save-restore.

These functions are selected by properly addressing the 32-byte window allocated for the acquisition board in the VME bus: each address controls a different function, as summarized in Table 3.5; some of these functions require only one word, while others require a sequence of words to define completely the operation to perform. Minimal standard functions allow to set first the analog input gain, then to set the board configuration through the FIFO control command word: on the basis of this command word it is possible to enable the samples acquisition on the two output FIFO and, independently, to enable or not the interrupt corresponding to the two FIFO full conditions; if interrupt is not enabled, this condition can be tested by reading the status word. All standard functions require only one word to be defined. In every case, after writing the FIFO control word, the samples acquisition in the FIFO starts automatically. Delayed test allows to fill the samples memory in a controlled way, using the address 18 of the window, then to present these samples to the FIFO several times or only once, using respectively addresses 1A or 1C of the window. These functions require a time-ordered protocol of parameters written through the VME bus on the same address, since we have to specify the first and last address of the samples buffer and in one case also the number of repetitions. With these delayed test functions it is possible to verify an algorithm repeatedly with the same samples input: in the case that we want save these samples for a new session we have to use the save-restore functions. These functions allow to save a predefined samples area in the VME bus area or to restore this samples area from the VME bus area; in this case two steps are to be followed: first a time-ordered protocol of parameters written through the VME bus allows the definition of the samples area, then VME buffer reading or writing allows the transfer of these samples from/to every VME buffer area, under master control.

3.3.3 System Configuration

The VME bus used is a well established bus, first adopted by Motorola and then standardized as IEEE P-1014 [14]: it allows a maximum transfer rate of 40 Mbytes/s on a 20 slots cabinet and can be configured for transfers of 32 bits of data maximum on a 32 bits address space maximum, although in our configuration we used 16 bits of data on

24 bits of address space only. Although a debate exists whether an asynchronous or a synchronous bus is better [16, 17], we think that an asynchronous bus like the VME is appropriate when interfacing with processors of different families, as in this case. Being a high-performance bus [15], the VME bus is provided with an arbiter, which controls the dynamic change of the bus mastership: take care to distinguish the bus master, which is a hardware dynamic concept, from the system master, which is the CPU driving the operating system, hence in our case a software static concept. In our system the arbiter is resident on the Motorola CPU; this is configured as a single level arbiter in order to simplify the implementation. Both the master and the DSP boards can become VME bus masters by suitably issuing bus requests: however the Motorola CPU board is configured for a Release on Request (ROR) behavior, that is, it takes control of the bus by default unless some other board wants it; the DSPs instead are configured for a Release When Done (RWD) behaviour, that is, they take control of the VME bus only when they really want access it: this is due to the fact that the master is the most likely board to access the bus. The system could evolve to a distributed interrupt one, since each DSP board, the converter board and the master CPU can send interrupts, and both the master CPU and the DSP boards can process incoming interrupts. However, in order to simplify the firmware implementation, all synchronizations are handled through flags, without using any interrupt at all. In each board there is some internal memory and some externally addressable memory; these last together constitute a distributed global area, addressed through the VME bus, whose map in hexadecimal bytes notation is shown in Fig. 3.16; the global area is also internally addressable from some board, hence it is biported. This architecture allows both efficiency, by quickly accessing internal memories for programs and data not shared between different boards, and ease of interaction between different boards, by using distributed global area for storing data to transmit or to share between different boards. For local data exchanges two secondary VMX buses [18] are used: the VMX bus allows a maximum of 2 masters, one primary (that is, ROR) and the other secondary (that is, RWD) and this is the configuration used by the VMX connecting the DSP2-A (primary master) and the DSP2-B (secondary master) with the VME/VMX biport memory; the VMX connecting the DSP1 with the acquisition board has only one master, the DSP1. In the implemented application the DSP DPX biport memories, although allocated in the map, are never addressed from the bus; the converter FIFO location addressed is of course only the topmost location, which can be read at all addresses reserved on the VMX map for the FIFO. After the reading of this data, the next sample becomes the new topmost FIFO location.

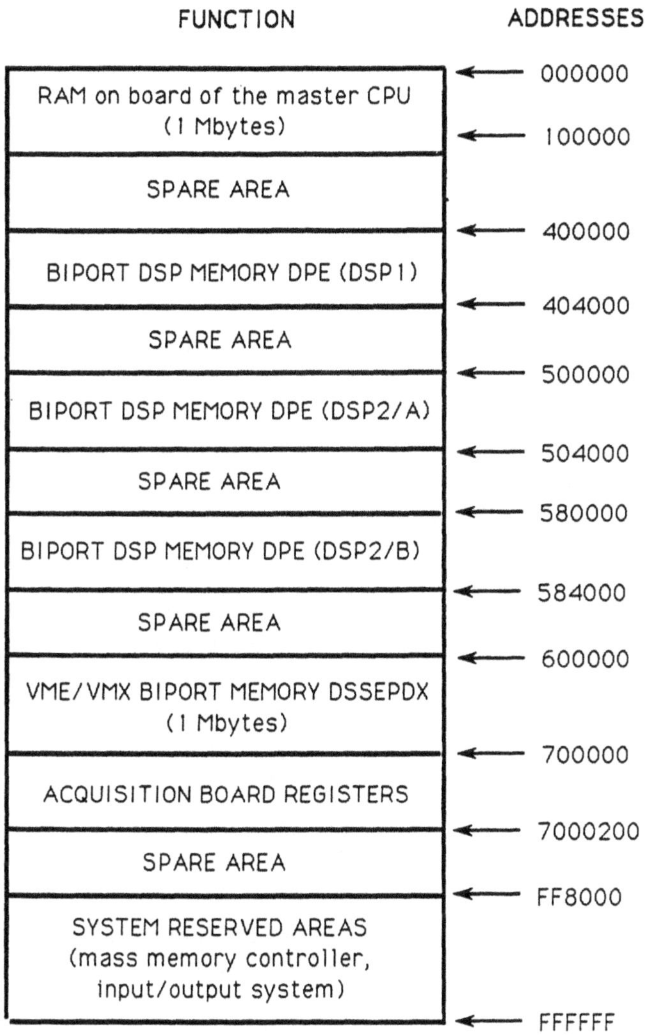

Figure 3.16: VME Addess map (hexadecimal bytes)

3.4 Firmware Blocks Details

3.4.1 Feature Extraction

Generalities

The feature extraction firmware is executed by the DSP1 and performs in each 10 msec time frame all computations from samples acquisition to spectrum vector quantization and rough classification into 6 broad phonetic classes. We were faced with the problem of implementing the firmware both accurately and quickly: this has been obtained even at the expense of some increase of memory, which is not a scarce resource in our implementation.

We added flexibility by using a suitable control area, which is configured according to the algorithm alternatives. Here the DSP1 application program is summarized in pseudo-code.

```
DSP1 board initializations;
DSP1 fixed tables initializations;
if the master wants to perform the data bootstrap for
   application dependent tables, do it;
   frame counter=0;
repeat
        repeat
             acquisition of the first samples frame;
        until the master wants to start;
        scale the samples frame;
        window the samples frame;
        evaluate FFT on the samples frame;
        group FFT into articulatory bands;
        evaluate bands logarithm;
        evaluate cepstral coefficients;
        evaluate the logarithm of the frame energy;
        if the master enables phonetic classification, do it;
        if the master enables frame vector quantization, do it;
        increment the frame counter;
        acquisition of new samples of the frame;
until the master blocks unconditionally this evolution or a
maximum number of frames has been reached and a conditional
stop was previously set by the master.
```

At the beginning the DSP1 application program initializes both the constant tables, for example the trigonometric constants used for FFT, and the application dependent tables, for example the codebook used for the spectral vector quantization and the coefficients used for the phonetic classification: by changing these tables it is possible to adapt the system to different speakers and michrophones. At system start-up these tables are all transferred to the DSP internal data area; constant tables are transferred from the DSP program area, where they have been previously compiled, while application-dependent tables are transferred from the global data area, where they have been previously loaded from the mass memory. Frame computations are performed in pipeline, so that the output of a computation is the input of the following one; these data are generally exchanged through a common pipeline area allocated in the TMS32020 internal fast block B0: this way the maximum efficiency is achieved since no data transfers are required from one routine to another; besides, the pipeline area is located in the fastest RAM area of the DSP memory hierarchy (Fig. 3.17). Scalar results (such as spectral code vector and energy code) or results used by different computations and not contiguous in the pipeline (such as frame energy) are stored as global symbols in the short fast block B2; fast block B1 is used instead for constants and intermediate results in some computations (FFT, frame classification). The DSP1 control table allocated in the VME biport area of the

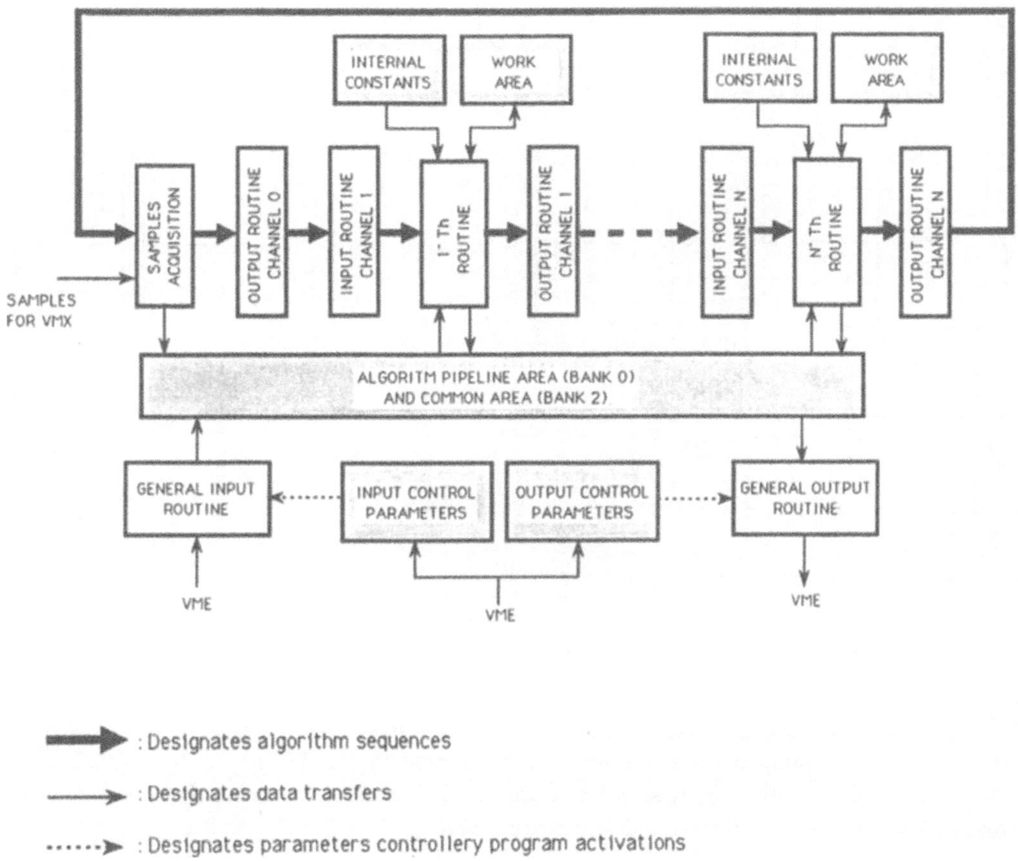

Figure 3.17: DSP1 application firmware architecture

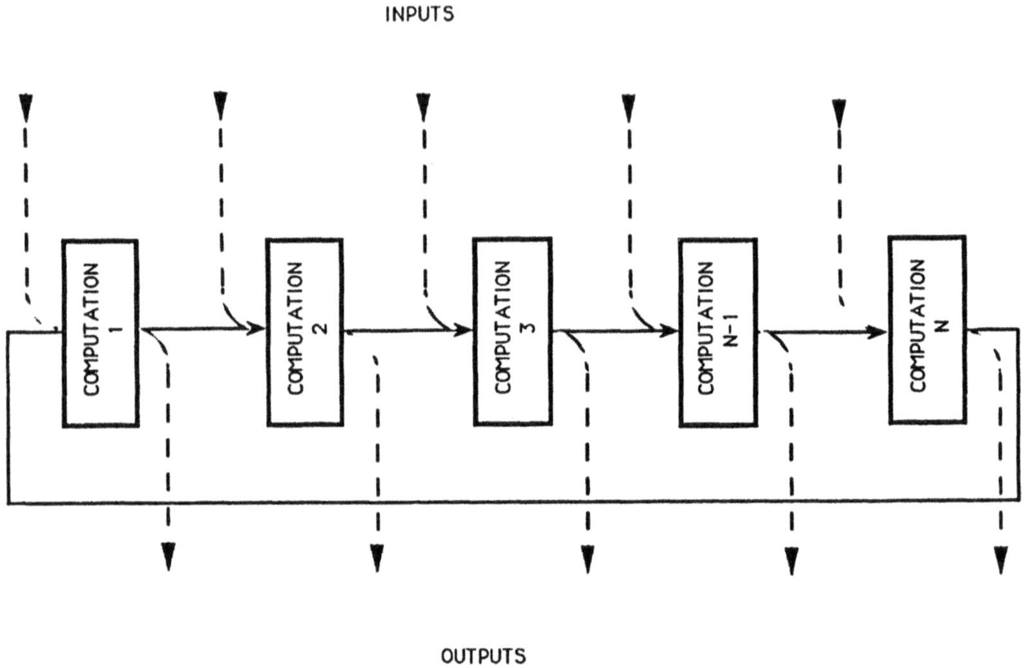

Figure 3.18: Pipelined algorithm with generalized inputs and outputs

DSP controls the algorithm evolution, the results display and the parameters input in it: in fact, by suitably setting this area, it is possible to enable or disable sections of the algorithm and to enable or disable the transfer of intermediate results from the common results area to predefined external VME areas arranged as circular buffers and vice versa. The program normally reads samples from the VMX bus, but, by suitably configuring the DSP1 control table, it is also possible to select the input from the VME bus for executing off-line experiments from files [4]. The only difference between the general input and the general output is that several kinds of output can be enabled at the same time, getting valid results on different sections of the pipeline, while only the last enabled input of the pipeline is valid, as visualized in Fig. 3.18.

DSP1 control details

The VME biport area of the DSP contains the DSP1 control block, which can be further distinguished into the general control block and the input/output control block. The general control block controls the algorithm evolution, mainly using the *COMMAND/STATUS WORD*: by properly writing it, it is possible to start and stop the

[4]This generalisation has been implemented after the end of the Esprit P26 project. In the DSP1 program released for the P26 project it was possible to input only DCTs and energies for vector quantization and coarse phonetic classification of the frame.

Channel number	Logical channel	Comments
0	input samples display	(256 words of B0)
1	normalized input samples	(256 words of B0)
2	normalized and windowed samples	(256 words of B0)
3	FFT results display	(256 w. B0, 2 w. B1)
4	band grouping results display	(36 words of B0)
5	band grouping logarithms	(36 words of B0)
6	frame energy and DCT's	(2 w. B2, 34 w. B0)
7	spectral and energetic symbols	(2 words of B2)
8	frame energy	(2 words of B2)
9	classifier results display	(2 w. B2,4 w. B0)
10	partial time duration	(8 words)

Table 3.6: Summary of DSP1 output channels

DSP1 computations. Stops can be unconditional or programmed: this last case happens when the present *FRAME COUNTER* reaches the predefined *MAXIMUM NUMBER OF FRAMES*. It is also possible to configure the *COMMAND/STATUS WORD* in such a way as to completely discard a subsection of the algorithm, for example phonetic classification. When a stop condition is met, the DSP writes suitable bits of the *COMMAND/STATUS WORD*, and hence the master can be synchronized with this stop by reading it. The output control block is partitioned into several channel areas, each of which controls the output of a specific channel. In this way it is possible to disable or enable the automatic enqueuing of intermediate internal results into VME circular buffers, whose allocation has been previously defined. In order to obtain this flexible output behaviour, we implemented a general purpose output routine, called by the DSP1 main program at points where frame results are available. This routine needs three parameters:

- the start address of the internal data to transfer to the output channel,
- the number of data to transfer,
- the physical number of the output channel.

Table 3.6 summarizes the correspondence between physical and logical channels. The channel number is associated with specific control and status words in the corresponding area of the output control block: the master enables or disables specific output channels by suitably writing the corresponding channel control word. Other than this for each channel the *START VME ADDRESS* and the *END VME ADDRESS* define the allocation of the output circular buffer, whilst the *VME POINTER* defines the address of the last written word; of course the master initially sets *VME POINTER=START VME ADDRESS*, then this pointer is automatically updated by the DSP1 each time the specific channel output transmits a new block of results. Other words in the output control block defines a limit address: when the *VME POINTER* overcomes it, a message is sent to the master through interrupts or flags: to be more precise there are two limit addresses, hence two overflow indicators: the green alarm (i.e. warning message) and the red alarm (i.e. unrecoverable error). Other words in the channel control block enable or disable interrupts or flags for the green and red alarms independently and specify respectively the interrupt number or the VME address of the flag. If the alarm condition is met and this condition is transmitted through a flag, the output program sets the VME address specified in the corresponding output control section: it is appropriate that the flag is allocated in the VME biport area of the receiving board for reducing the bus traffic. The output routine is summarized here in pseudo-code; the input routine is similar.

```
if the output channel is not enabled
    then exit;
    else (output channel enabled)
        if the new results buffer lenght plus the old buffer
        occupancy overflows the read alarm occupancy threshold
            then
                if the exception has to be
                notified through flags
                    then
                        sets the red alarm flag at the VME address
                        specified by the RED ALARM INFORMATION;
                    else (exception notification through interrupts)
                        programs the DSP board interrupt registers
                        with data specified by the
                        RED ALARM INFORMATION;
                exit
          else (no overflow of read alarm occupancy threshold)
                enqueues the new buffer to old results;
                updates the buffer occupancy;
                updates the buffer pointer;
                if the new buffer occupancy does not overflows
                the green alarm threshold
                    then exit;
                    else (green alarm threshold overflow)
                        if the exception has to be notified through flags
                            then
                                set the green alarm flag at the
                                VME address specified by the
                                GREEN ALARM INFORMATION;
                            else (notification through interrupts)
                                programs the DSP board interrupt
                                registers with data specified by
                                the GREEN ALARM INFORMATION;
                        exit.
```

The clear separation of input and output routines has also been useful in the program development with the simulator, since the simulator implements input and output from/to files differently than the hardware (i.e. through I/O instructions instead of through memory instructions): hence we confined in the input/output routine the differences between the real time version and the program version to be used with the simulator.

DSP1 algorithm details

The DSP1 algorithms have been implemented in such a way as to optimize both the computational speed and the accuracy. Computational speed has been achieved also at the expense of memory occupancy, given that the memory is not a scarce resource in our DSP board: in fact some routines have been straight line coded, other instead have been implemented through table look-up. As a result the programs are allocated in the first 16K words of the address space and the data are allocated in the second 16K words: hence the DSP1 board must be configured with the first 2 RAM banks of program and the last 2 RAM banks of data. Accuracy has been achieved by properly scaling intermediate results, hence performing a sort of block floating point, which is simplified by the availability of the efficient NORM instruction of the TMS32020: when comparing the quantization results of the floating point simulation with results of this TMS32020 integer implementation for a 10 000 frames speech data base, we found a difference in 0.5% of frames for spectral and in 0.03% of frames for energetic symbols [30]. Most of the time spent by DSP1 was in FFT computation and in spectral vector quantization: hence these were the computations to optimize in speed. The FFT computation has been improved both in speed and in accuracy by suitable implementation choices: speed has been obtained by considering that the 256 real point FFT is related to the 128 complex point FFT, whose real part is composed of even samples and imaginary part is composed of odd samples of the first 256 real point FFT [32], accuracy has been obtained by structuring the 128 point FFT into a first radix-2 decimation in frequency (dif) stage and 3 radix-4 dif stages [30, 33, 34, 35]. The vector quantization routine is also time consuming, since for each code vector Yk we have to calculate the Euclidean distance of it from the measured input X, that is,

$$\sum_{i=1}^{N} (x(i) - y(i))^2 \tag{3.7}$$

hence for *Ncod* code vectors, each of which represented by *Npar* parameters we have to compute $Ncod \times Npar$ differences: each difference has to be squared and orderly cumulated into different *Ncod* sums: hence, given that square and add can be implemented in the same DSP instruction, we have to perform $2 \times Ncod \times Npar$ instructions by frame: if $Ncod = 2^8$ and $Npar = 2^4$ we have to do $2^{13} = 8$ K operations per frame, hence 800K operations per second. This computation however can be rearranged as [38]:

$$\sum_{i=1}^{N} x(i)^2 + \sum_{i=1}^{N} y(i)^2 - 2 \times \sum_{i=1}^{N} x(i) \times y(i) \tag{3.8}$$

and this reduces computations, since the first term is computed once on the frame for all the code vectors, the second one is prestored, since it depends only on the symbol, and the third halves the number of computations required in comparison to the original formulation. The DSP1 computational time depends on the system initialization, that is, on the number of output channels enabled: this is quite negligible in the normal case activated for on-line recognition, when only output channels 7 and 9 are enabled, but it becomes important in the case of more extensive diagnostic displays. The DSP1

Routine name	Number of cycles	Program (words)	Constants (words)	Coding style
MAIN	300	320	*	L
INPUT	411	30	*	L
NORMA(1)	2885	40	*	L
HAMMI	2360	40	256	L
FTT's first stage	1912	120	124	L,O
NORMA(2)	2885	see NORMA (1)	*	L
FTT's other stages	2756* 2	1560	*	S,I
MIXREV	761	766	*	S
128TO256	3310	62	124	L,O
NORMA(3)	2885	see NORMA (1)	*	L
BANGROUP	1884	1518	*	S,I
LOGARI	1066	44	1024	L,D
DCT	2562	42	324	L
VECQUA	9286	94	2176+256	L/S
VECENE	83	30	I	L

Table 3.7: Summary of time and memory occupancy of DSP1 routines

computational time depends also from the spectral codebook dimensions and from the phonetic classifier complexity, which affects computational times of the corresponding routines; ordinarily we use for vector quantization 13 cepstral coefficients for 256 point code vectors and 10 coefficients in the frame classification, reaching always the real time: for larger codebook sizes, for example 512 points, the DSP1 alone can no longer keep up with real time [5]. Just to give an idea of the computational requirements, Table 3.7 gives the results obtained in the case that the phonetic classification is excluded and that the codebook size is of 128 symbols for 17 DCTs each. This table summarizes the number of 200 ns machine cycles spent in each computation, the program memory words used, the constant memory words used. A last column in the table reports eventual comments: L means looped coding style, S means straight-line coding style, I means that we used immediate constants, O means that a constant table has been allocated on the chip, D means data dependent routine: in this last case the number of cycles reported in the table represents an average value. From this table we can see also that in this case the program and data memory occupancy is not too high and the real time requirement has been meet, since the manipulations on a 10 ms time frame require less than 50K cycles (with a cycle time of 200 ns).

[5]In this case the feature extraction task must be restructured; it could be carried out by two DSP1 in parallel, for example.

3.4.2 Segmentation and Lexical Access

The segmentation and lexical access firmware is used only in the two-step approach, for quickly preselecting a subset of the whole vocabulary, named cohort, to be verified in more detail; both are presently implemented in the master Motorola CPU in Pascal language [28]. Segmentation and lexical access firmware is only cursory described in this section; for a more detailed description see Chap. 2. The segmentation program merges consecutive frames with "similar" phonetic labels to single segments; these phonetically labelled segments are then used by the lexical access for identifying the most likely subset of words. Just to summarize, the segmentation program receives from the DSP1 in a suitable VME-addressable area the first and the second phonetic frame hypotheses and the corresponding scores, and groups frames into phonetic segments. These are organized into a graph structure, in a process called "micro-segmentation": each arch of this graph is a "micro-segment", characterized by the initial and the final frame and by the first and the second phonetic hypothesis with their corresponding scores. Then the lexical access program finds the best match among all paths of the microsegmentation graph and all paths in the vocabulary word-tree; the nodes associated with a set of words of the vocabulary are called terminal nodes. To match a speech segmentation against the phonetic transcription of the vocabulary words a modified Dynamic Programming procedure is used, named 3DP [36]: it relies upon statistical models accounting for deletion, substitution and insertion errors of the segmentation step; a beam search strategy is used to reduce the number of active paths. The lexical access output is the set of words associated with the active terminal nodes when all input micro-segments are processed: this output will be used by the verification module as its input lexicon, reduced by an order of magnitude compared to the original vocabulary. The segmentation and lexical access firmware requires also the definition of some heuristic parameters, for example the number of frames used by majority voting filters (called windows), the threshold of rejection and the threshold of certainty used by the 3DP lexical access; these parameters can be adjusted by using the off-line evaluation program.

3.4.3 Markov Verifier Firmware

Generalities

As anticipated in Sect. 3.2.4, the search stage implemented is split into two levels both in the algorithm and in the data structures. The high level in fact describes the set of words under verification as a tree of diphone-like subword units, the low level instead describes these subwords as Discrete Hidden Markov Models (DHMM): Fig. 3.19 summarizes this. The high level algorithm (scanner) drives the search into its diphones tree of the word that best matches with the input utterance relying on the pattern matching performed by the low level algorithm (verifier). With this kind of partition the scanner is devoted to the more irregular, but less computation-intensive task, while the verifier is devoted to the more computation-intensive, but also regular task: in fact dynamic programming inside diphones representations can also be arranged in a vector form. Given the different characteristics of the scanner and of the verifier, we found it appropriate to implement the scanner in a general purpose CPU and the verifier in DSP technology, faster and well suited to vector computations. We also faced, of course, some earlier DSP limits in the

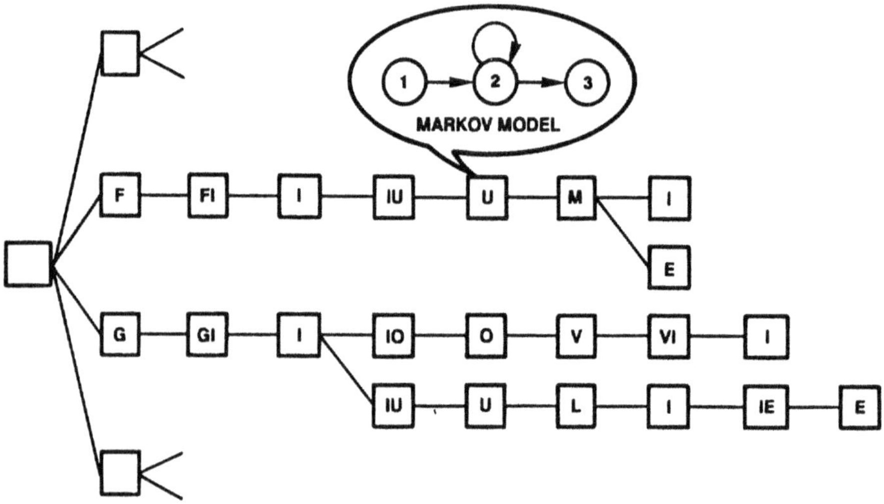

Figure 3.19: Example of the two level vocabulary description

verifier stage implementation, mainly computational accuracy and memory addressability. As far as accuracy is concerned, since no floating point was available for our DSP technology, we implemented the Viterbi algorithm only [6] by using logarithms of probabilities: this replaces multiplications with additions and, what is more important, makes the dynamic range more suitable to the integer arithmetic. As far as memory addressability is concerned, the main problem was due to the huge space required by the spectral code vector emission matrix B, whose dimensions are given by the product of the number of states with the number of code vectors. Just to give an idea of its dimensions, we can assume a number of states equal to 2^9 and a number of codevectors equal to 2^8 : in this case the B matrix requires $2^{17} = 128$ K words of data memory, which is twice the maximum addressable data memory of the DSP we used! Hence it is evident why we used the page register EXFR for extending the DSP board addressability; besides, we were forced to use an external general purpose RAM board in order to allocate the B matrix at least: in fact memory requirements for the DHMM verifier were so different that for parameters extraction we did not consider the possibility to add this extension on the same DSP board. Special care was taken to optimize the verifier speed: besides extensively using in-line coding we adopted suitably reduced DHMM. The three classes of diphones used, the silence, the stationary and the transitional ones, are described by DHMMs of 1,3 and 4 states respectively: their "complete" models will consider all transitions from the state i to the states i, i+1 and i+2, as detailed in Fig. 3.20. These complete models have been reduced by completely omitting the transitions whose probability are ordinarily very low when we use the complete models in the training: the reduced models (see Fig. 3.20a)

[6] The forward algorithm was not implemented since it is most suitable to a floating point DSP.

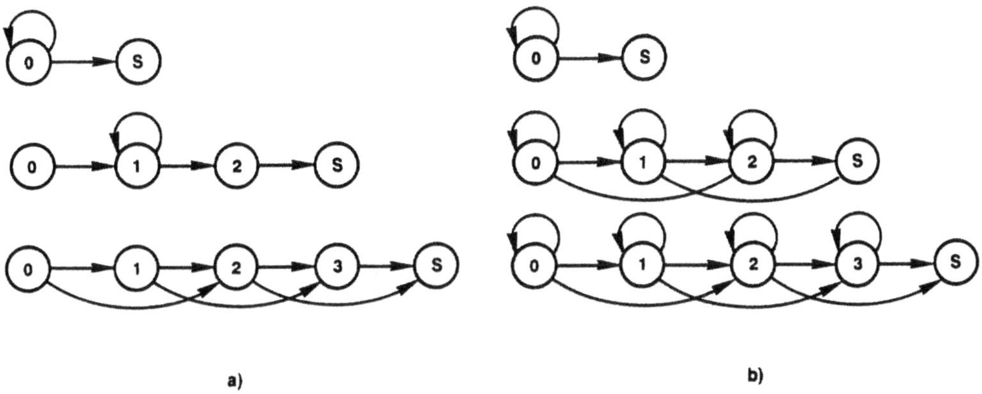

Figure 3.20: Subword unit DHMM structures used: "reduced" (a) and "complete" (b)

minimize recognition time but of course if used in the recognition they must be used also in the previous training phase by forcing the omitted transitions to 0. In the following we will detail the specific algorithm choices and data organization and placement in the verifier.

Verification stage details

The master-slave interplay is organized as anticipated in Fig. 3.7 [45]: frame-by-frame spectral and energetic codes computed by the DSP1 are broadcast from the master to all the slaves through their *BROADCAST AREA*; the master informs also the specific slave of the new nodes it must activate in this frame through the *PUSHES MAILBOX AREA*. At this point the slaves advance by one frame the dynamic programming over all the active nodes, taking into account the previous frame probabilities stored in the *WORKING AREA* and the constants in the *SUBWORDS DESCRIPTION AREA*, where for each subword the transition matrix A, the spectral emission matrix B and the energetic emission matrix C of the corresponding DHMM are stored. As a result the probabilities of the states of active nodes contained in the *SLAVE WORKING AREA* are updated; unlikely nodes become inactive. Then each slave notifies to the master through the *POPS MAILBOX AREA* information the master evaluates again the *PUSHES MAILBOX AREA* for the next frame and the computation can be iterated for the next frame. The memory on the DSP2 board uses the first board internal RAM bank for programs and the remaining 3 for data and uses an external expansion the VME/VMX biport memory. Hence six kinds of data RAM are available:

- 544 words of on-chip data RAM,
- 24K words of board internal data RAM (named PRIVATE in the following),
- 8K words of data RAM, biported with the VME bus (named DPE in the following),

- 8K words of data RAM, biported with the VMX bus (named DPX in the following: it can be used as an extension of the PRIVATE memory only),
- 1M byte of external data RAM, addressable through the VMX bus (named VME/VMX in the following);
- any other VME addressable system memory resource.

The *PRIVATE* memory is the fastest, hence its preferred use is for frequently accessed and updated data, the DPE memory is suitable for data addressed both by the master and the slave, the VME/VMX is the slower, but wider RAM area, where large data tables can be stored. The access to memories on the VME bus should be limited as much as possible to relieve the system bus. In these RAMs we have allocated the data structures detailed in the following:

- the transition matrix A, the spectral emission matrix B and the energetic emission matrix C of the DHMM and the directory TDIR of the DHMM, storing for each diphone the number of its states and the entry point into A,B,C structures: these tables constitute toghether the *SUBWORDS DESCRIPTION AREA*;
- the *BROAD* area where the master puts the frame spectral and energetic codes, the beam-search threshold and the best path distance found at the last frame: this value is used to scale down all the distances, that otherwise would monotonically increase;
- the input mailbox area *MAILIN*, that contains the number NPUSH of frame pushes and the relative *NPUSH* 4-word records (node identity, cumulated distance and backpointer of the started path, plus a spare location);
- the output mailbox area *MAILOU*, that contains the best distance found among all its paths, the number NPOP of frame pops and the relative NPOP 4 word records (with the same fields of push records);
- the *MAP* area, where at the beginning of every frame the node identity is put in correspondence with the corresponding input push, for a quick retrieval during the algorithm execution;
- the *XCOD* area, which associates with the node identity the corresponding diphone;
- the *PAGES* area, which is divided into 8-word pages for the storage of the informations on active nodes only: of these the first 6 contains the backpointers and the distances of the corresponding subword state, the last two respectively contain the node identity and the pointer to the next active page (Fig. 3.21);
- the *STACK* area, which is a stack of pointers to the free pages in the *PAGES* area: the *PAGES* and the *STACK* area toghether constitute the *SLAVE WORKING AREA*.

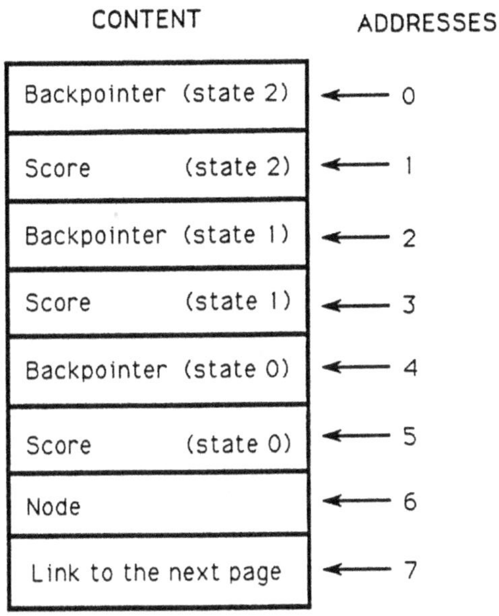

Figure 3.21: Format of a page in the PAGES area

As far as the *PAGES* area is concerned, we implemented it as a linked list of fixed length pages containing the informations on active nodes: the allocation and deallocation of pages is simplified both by the pages' linked structure and by the stack of free pages. This solution greatly reduces *PAGES* memory requirements, because we have to size this area for the maximum number of active nodes per frame only and not for the total number of nodes. Two other *PAGES* area organizations have been examined and discarded for this implementation: the static organization and the cyclic one [7]. The static organization (*PAGES* area sized on the total number of nodes) would simplify the algorithm, but would require a much larger *PAGES* area, which could find room only in the slower VME/VMX RAM extension, with a clear slowdown of the overall algorithm; the cyclic organization described in [8], in order to be effective in computational speed-up, would require a cyclic addressing mode, not available in the chosen DSP. In Table 3.8 we summarize these data structures, identifying for each the function and the size, which depends on the following parameters:

- NMOD: total of different subwords,
- NSTA: states of models in total (including the "dummy" final states),
- NCOD: symbols of vector quantization,
- NENE: quantization levels for the energy,
- NTRA: allowed transitions to a state from previous states,

Table	Function	Size (words)	Typical size and allocation	
A	DESCRIPTION	NTRA * NSTA	2372	Private
B	DESCRIPTION	NCOD * NSTA	151808	VME/VMX
C	DESCRIPTION	NENE* NSTA	1186	Private
TDIR	DESCRIPTION	2 * NMOD	256	Private
BROAD	BROADCAST	4	4	DPE
MAILIN	MAILBOX	1+4 * NPUSH	4097	DPE
MAILOU	MAILBOX	2+4 * NPOP	2050	DPE
MAP	WORKING	NIDE	4096	Private
XCOD	DESCRIPTION	NIDE	4096	VME/VMX or private
STACK	WORKING	NPAG	1024	Private
PAGES	WORKING	8 * NPAG	8124	Private

Table 3.8: Data structures used in the slave algorithm

- NIDE: different identities (i.e. nodes in the high level tree) at most used by the master,
- NPAG: pages available for the PAGES area,
- NPUSH and NPOP: maximum number of pushes and pops respectively.

To give an idea of the memory requirements, the typical memory size for the two-step application is also reported (NIDE=4096, NMOD=128, NSTA=593, NTRA=4, NCOD=256, NENE=2, NPAG=1024, NPUSH=1024, NPOP=512). The final column reports the allocation chosen for these tables in the same application: of course we have to allocate in the DPE memory the tables to be addressed both by the high and by the low level algorithm, otherwise the allocation in the private memory is preferred: this cannot be achieved only by the bulky table B, which in any case is a read-only table. The slave firmware is controlled by the master through a command register located in the DPE biport memory. A bit in this register is used for the synchronization with the master: the master sets it to start the slave and the latter resets it to notify the end of its processing. Other bits of the command register are used by the master for coding the function to be activated on the slave. The available functions are:

- the DHMM slave bootstrap;
- the configuration of allocation of slave tables;
- the software restart of the recognizer for the first frame of a new recognition;

- the single frame processing, i.e. the expansion of active paths that constitutes the main slave activity.

The bootstrap function copies TDIR, A and C areas of the DHMM from the global (VME) to the DSP private space: if the master wants to change the DHMM, it first loads it into a shared VME resource and then activates the slave bootstrap. The configuration function allows definition of the allocation of most data areas at run time instead of at link time. This way the master can choose the best memory allocation for a specific application using the same DSP firmware; for instance it can decide whether to allocate XCOD in the VME/VMX in the private area. The first case is mandatory for the two-step approach, since in this case this area must be updated by the master at run time; the second case is preferred for the single step-approach, since in this case it is fixed for the whole time and hence it does not have to be directly accessable by the master: in this case it is better to allocate it as private for maximum efficiency. The restart function clears the list of active subwords and all the work pages are made available again by suitably resetting the STACK area. The most important and heavy function is of course the single-frame dynamic programming, which is issued on a frame basis and whose evolution is detailed in the following in a pseudo-Pascal notation:

```
{ SINGLE FRAME PROCESSING FUNCTION }
allocate MAILIN in order to handle the double buffer of PUSHES;
read new values in BROAD and NPUSH in MAILIN;
for i := 1 to NPUSH do
     set MAP according to PUSH[i] identity;
point to the active page at the head of the linked list;
{ LOOP ON ACTIVE SUBWORDS }
while not end of list do
begin
     decode the subword in the page entering XCOD with the page
     identity;
     retrieve in TDIR the number of states and DHMM entry point
     for the subword;
     case number of states of
          1 : execute the D.P. routine for silence model;
          3 : execute the D.P. routine for stationary model;
          4 : execute the D.P. routine for transitional model;
          (in all these subcases take also into account pushes
           in already active pages for dynamic programming on the
           first state, suitably marking them in the PUSH list)
     end;
     if the subword is still active then
            go one step ahead in the linked list
     else
     begin
            delete the page from the linked list;
            push the page address onto the stack;
     end;
```

```
end;
{ LOOP ON PUSHES }
for i := 1 to NPUSH do
if PUSH[i] is in an inactive subword (i.e. not marked) then
begin
     retrieve in TDIR the number of states and DHMM entry point
     for PUSH[i] subword;
     update path score for current frame labels;
     if the score is under the beam-search threshold then
     begin
          pop a page address from the stack;
          initialize the page fields;
          insert the page in the linked list;
          if the number of states is 1 then
               evaluate the pop path, increment NPOP and
               write a new POP record;
     end;
end.
```

We observe that the initial allocation of MAILIN allows the maintenance of a double buffer of pushes: hence the master and slave algorithm can advance completely in parallel, without waiting for each other: in fact as pops are produced by the slaves they are processed by the master for generating the pushes for the next frame: in the meanwhile the slaves continue their computation by consuming the pushes of the present frame. We observe also that the slaves processes first all active nodes and the corresponding pushes, then process the remaining pushes which can activate new nodes. The throughput of this implementation can be roughly measured by the frame processing time of a single subword, who is about 50 us, evaluated as a weighted average of 3 possible cases (silence, stationary and transitional models): this means that a single DSP board can process in real time about 200 active nodes per centisecond frame in average: in fact the frame labels are stored in a FIFO and the DSP reads them asynchronously, averaging also higher peak values. This throughput is rather satisfactory, since the vocabulary organized as a tree of subwords and the beam search thresholding reduces the frame average of active subwords, as detailed in Sect. 3.6.

3.5 Some Details on Other System Functions

3.5.1 Program Loading and System Testing

The LOADER program drives a test and bootstrap environment in our multiDSP system. This program is implemented in Pascal language and runs in the master Motorola board; it interacts with the system console, with the system mass memory and with the "DSP ROMmed kernel". Figure 3.22 summarizes the LOADER program flow, and the corresponding console dialogue: first of all the user selects a DSP, then the user decides whether to perform or not the DSP self-test, finally a DSP program in loadable format is selected from the system mass memory and transferred first to a fixed VME addressable area, tranformed into the corresponding program RAM image, and then transferred to the internal program RAM DSP area by properly activating the DSP ROMmed loader; this procedure can be iterated for each DSP in the system, obtaining the initial system validation and the flexible loading of the intended application programs into the intended DSPs. In the self-test phase these memory tests are automatically performed in sequence:

- test of the data RAM on the TMS32020 chip,
- test of the data RAM area installed on the DSP board,
- test of the DPX memory on the DSP board,
- test of the DPE memory on the DSP board,
- test of a section of the data RAM VME addressable through the DSP board,
- test of the DSP RAM program area.

Although these tests are not exhaustive, they have enough coverage, since also the DSP chip must work properly to drive the memory test algorithm.

3.5.2 Acquisition Firmware Details

The acquisition program ACQUIP displays the words to be uttered in the training phase and stores on the system mass memory the frame energy and 17 DCTs for every word; this program is implemented in PASCAL language and runs on the master of the minimal system configuration reported in Fig. 3.23. In this configuration the DSP1 board reads through the VMX the samples of every word uttered, then extracts features through the ESPTPARA program; these parameters are then stored in mass memory files. To initialize the system the operator must first load the DSP1 program through the program LOADER; then he must load and run the ACQUIP program, who executes these steps:

- a) it chooses the vocabulary to utter and then chooses the starting word of the acquisition session,
- b) it initializes the A/D and DSP1 boards for acquisition, ENERGY + 17 DCT computation and results memorization on a 1Mbit RAM board,
- c) it displays the word to be presently uttered,
- d) it allows the operator to define the start and end points of the utterance, through a suitable keystroke activation, then automatically discards initial and final low energy records,
- e) it can display frame energies of the uttered word before storing the results on the mass memory,
- f) it stores on the mass memory a file of frame results for each word uttered, assigning it a name automatically,
- g) it repeats steps (b) to (f) until all words of the vocabulary are uttered or the operator terminates the acquisition session.

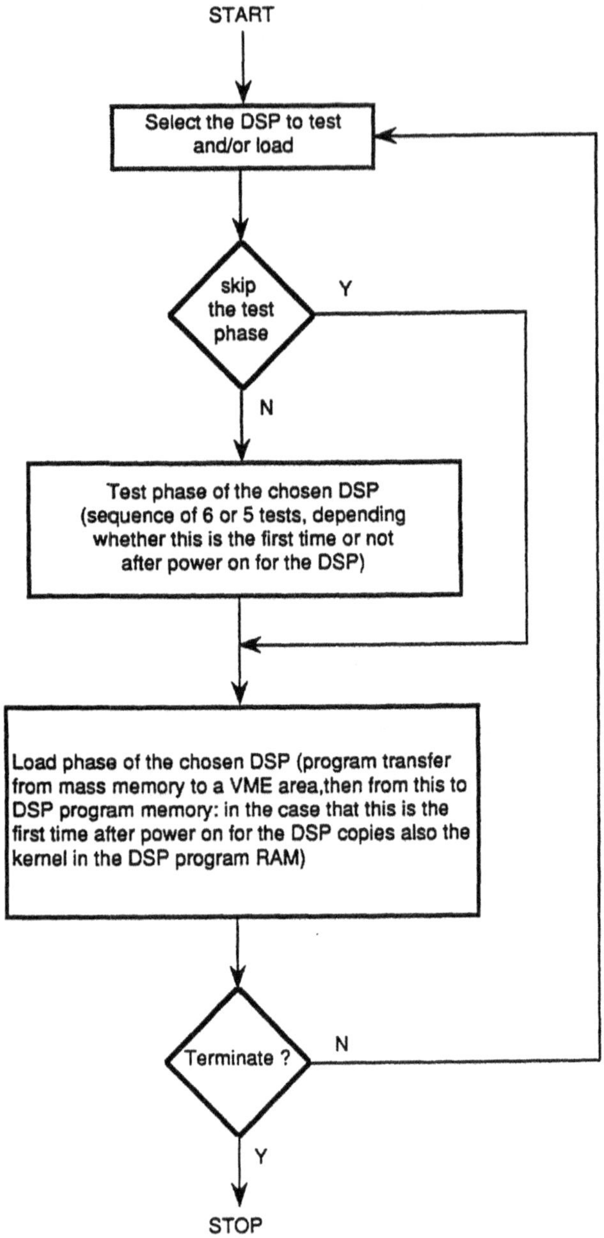

Figure 3.22: Loader program flow

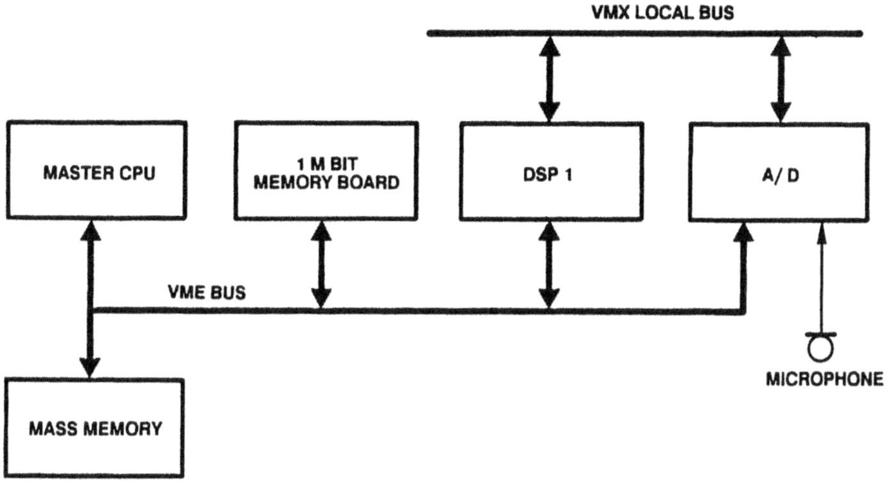

Figure 3.23: Minimal system configuration used for the acquisition

3.5.3 Parameters Training Environment

Figure 3.24 summarizes the VAX/VMS training environment. We did not duplicate the parameter training functions, already developed in the VAX/VMS environment, but we integrated this environment in our VERSADOS system: in fact the VERSADOS system is first used for real time speech acquisition until parameter files extraction; then these files are transferred in order through an RS232 serial connection to the VAX/VMS host, where system parameters are trained; then the computed parameter files can be transferred again through the same connection from the VAX/VMS to the VERSADOS system. The following parameters are evaluated from the DCT plus energy frame by frame files:

- the codebook, for the real time vector quantizer,
- the phonetic classifier parameters,
- the matching costs, for the real time lexical access module,
- the hidden Markov models (HMM) for the real time verifier.

At the end of the training, these parameters are suitably transformed from the format used in the VAX/VMS system to the format used in the final VERSADOS system, and then transferred to this one.

3.6 System Evaluations

3.6.1 General Considerations

The recognition system has been evaluated using a large test vocabulary of 1008 Italian words oriented to a geographical data base access; this test vocabulary is described by a tree connecting 5295 subwords. The different subwords selected for the Italian language

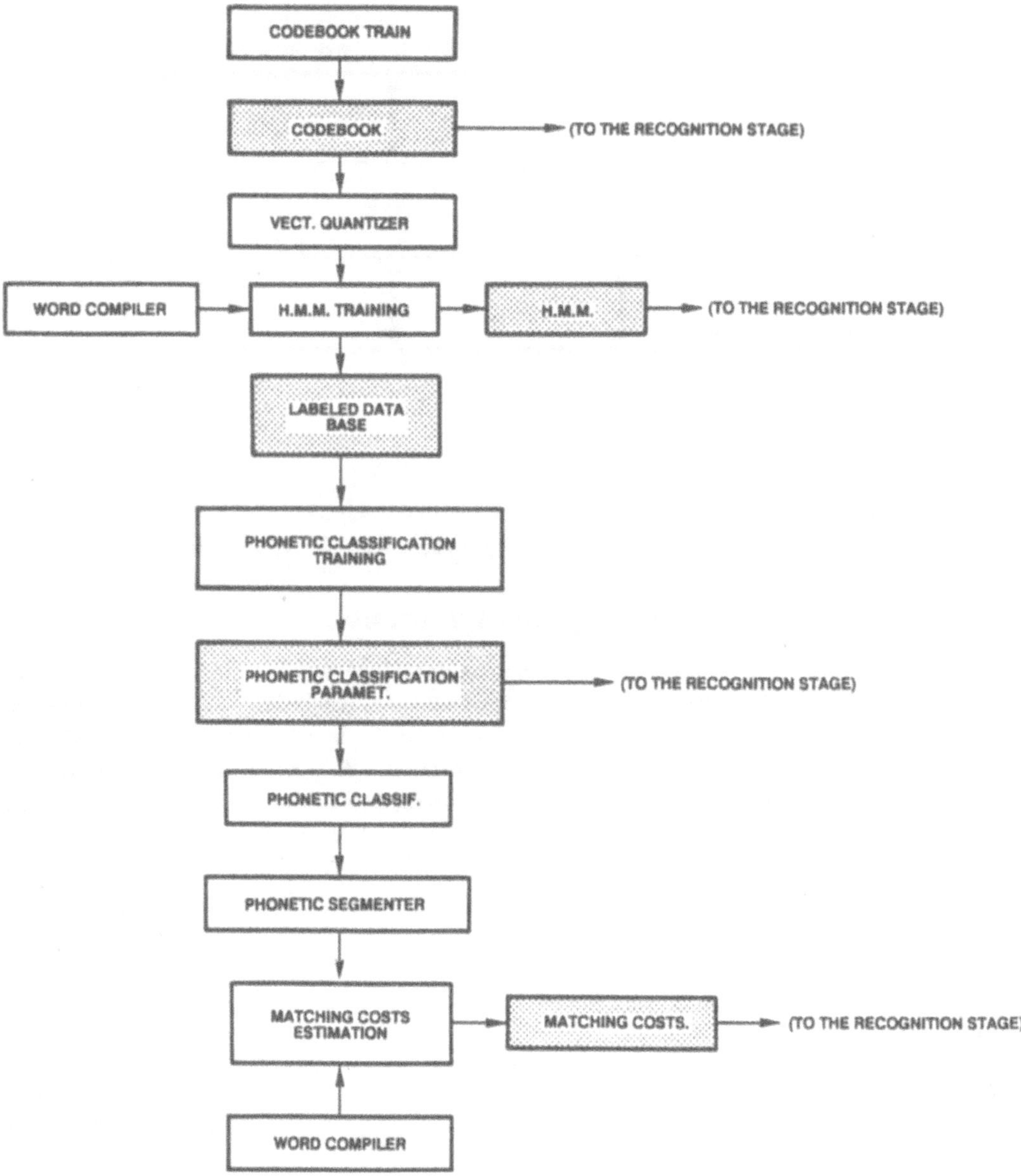

Figure 3.24: Training environment

are 125, of which 101 represent transitional sounds, 22 stationary 3-state sounds and 2 stationary 1-state sounds [5]. For continuous speech testing this vocabulary has been arranged into 100 test phrases, with a total of 578 words. The system has been trained with a phonetically balanced vocabulary of 1105 words: hence in this vocabulary each subword appears at least 12 times and in different contexts in order to have a statistically significant training. In the following we will distinguish experiments carried out on the 2 DSP, 12.5 MHz CPU system, called the basic system, experiments carried on the 2 DSP, 20 MHz system, called the intermediate system, and experiments carried on the 3 DSP 20 MHz CPU system, called the final system.

3.6.2 Single-Step Isolated Words Recognition

Single-step isolated words recognition has been characterized in the basic system version only and for two subcases (Tab. 3.9):

- single-speaker training,
- multi-speaker training.

Single-speaker training has been verified against the same speaker; multi-speaker training has been performed on a set of 6 male speakers and has been verified against a different male speaker; the codebook size is of 144 code vectors for the single speaker and of 272 code vectors for the multiple speakers. For both cases we report results obtained with a standard beam search threshold value (20) and with a more tolerant one (25): in the former case we found a better system throughput, in the latter instead a little increase of the recognition accuracy and a sensible decrease of the throughput. This second case in fact recovers some correct words in the hypothesized cohort by augmenting the number of possible paths and hence the computational load of the system. Given that the basic system configuration can follow in real time a recognition task with 200 average active subwords per frame at most, we can say that for the standard beam search threshold the basic system can follow in real time both the single speaker and the multispeaker case: hence in this case the basic system configuration (2 DSP 32020 + 1 CPU 68020 at 12.5 MHz) is already satisfactory.

3.6.3 Two-Step Isolated Words Recognition

Table 3.10 summarizes accuracy and throughput results obtained for the basic system configuration in the two-step recognition algorithm for isolated-word recognition: we consider 6 subcases, combining 2 possible values of the beam search threshold (55 and 67) and 3 possible certainty factors (1, 2 and unused) for the three-dimensional dynamic programming (3DP) lexical access module [6]; by contrast the beam search threshold of the detailed DHMM verification is 20 for all subcases. For each of these subcases we measured 8 parameters, 3 of them pertinent to the hypothesization stage and the other 5 pertinent to the verification one. The best accuracy of the hypothesization stage is obtained with the larger 3DP beam search threshold and without the certainty factor: in this case the correct word is included in the 97.5% of cases in the cohort generated by the lexical access stage; the average size of this cohort is larger than in the others subcases, however it is an

	Single - speaker		Multi - speaker	
	Beam search threshold			
	20	25	20	25
Average cohort size	3.7	5.7	6.7	13.4
Inclusion rate into the cohort	98.10%	99.00%	94.10%	97.60%
Inclusion rate up first score	91.80%	92.00%	82.20%	83.40%
Inclusion rate at second score	4.10%	4.30%	7.30%	8.40%
Inclusion rate up second score	95.90%	96.30%	89.50%	91.70%
Inclusion rate at third score	1%	1.30%	1.40%	1.40%
Inclusion rate up third score	96.90%	97.60%	90.90%	93.10%
avg. active subwords per frame	139.3	250	182	308.5
Max. active subwords per frame	755	1255	1800	2412
Average word recognition time	0.9 s	1.3 s	1.1 s	1.7 s
Average utterance duration	0.95	0.95 s	0.85 s	0.85 s

Table 3.9: Experimental results of the single-step isolated word recognition

order of magnitude less than the whole vocabulary size. The following verification stage then reduces the average final cohort size from 82.6 words to 2.7 with a corresponding small decrement of the correct word inclusion rate from 97.5% to 96.3%. The time spent in the hypothesization stage is bigger than the time spent in the verification one, while the verification stage is rather underutilized: in fact the first stage has been programmed in Pascal language in a general purpose microprocessor, whereas the second one has been programmed in assembly language in a DSP processor. We could reduce this "hypothesization stage bottleneck" of the present implementation by transporting if not all then at least some computations of the hypothesization stage (e.g. frame segmentation) into a DSP processor. This has not be pursued since the throughput was already satisfactory for the 1 k-word intended application; for larger vocabularies however this improvement could best exploit the potential throughput of this algorithm.

3.6.4 Single-Step Continuous Speech Recognition

Table 3.11 summarizes results obtained for the single-step continuous speech recognition system on 100 test phrases oriented to the vocal access of a geographical data base; four evaluations, related to four different speakers, are reported. We performed two experiments for each speaker, the first one with the standard beam search threshold (20) and the second one with a more relaxed beam search threshold (22.5): in the second one we measure some accuracy improvement with a sensible increase of the computational load. As far as accuracy is concerned, we have to point out that the measure of total missing words is rather conservative because most of these are short nonfunctional ones; their absence does not affect the understanding level accuracy [3]. Since in this case the

	Lexical-Access 3DP Beam-search threshold 55.0	67.0	Lexic. Access Certainty factor
Correct word incl. in the cohort after hypoth	92.60%	96.20%	1.0
Average cohort size after hypothesization	37.9	58.8	
Average elapsed time due to hypotesization	0.4 s	0.7 s	
Correct word incl. in the final cohort	91.50%	95.20%	
Average cohort size after verification	2.5	2.6	
Average elapsed time due to verification	0.2 s	0.3 s	
Average number of active subwords	22.2	27.5	
Maximum number of active subwords	233	298	
Correct word incl. in the cohort after bypoth	93.60%	97.40%	2.0
Average cohort size after hypothesization	48.1	80.8	
Average elapsed time due to hypothesization	0.6 s	0.9 s	
Correct word incl. in the final cohort	92.30%	96.10%	
Average cohort size after verification	2.6	2.7	
Average elapsed time due to verification	0.3 s	0.4 s	
Average number of active subwords	25.8	31.6	
Maximum number of active subwords	290	319	
Correct word incl. in the cohort after hypoth	93.70%	97.50%	NON USED
Average cohort size after hypothesization	49.1	82.6	
Average elapsed time due to hypotesization	0.6 s	0.9 s	
Correct word incl. in the final cohort	92.4	96.30%	
Average cohort size after verification	2.6	2.7	
Average elapsed time due to verification	0.3 s	0.4 s	
Average number of active subwords	27	34	
Maximum number of active subwords	290	319	

Table 3.10: Experimental results of the two steps isolated word recognition (verification stage beam search threshold = 20)

	Phrases with all uttered words in the lattice	Percentage of phases with all uttered in the words lattice	Total missing words in all lattices	Percentage of phrases with some words missing in the lattice	Average lattice hypothesized words	Average active subwords	maximum active subwords
beam search thresh. =20	86	86%	14	2.42%	371.5	429	1949
1^ speaker							
beam search thresh. =22.5	90	90%	10	1.73%	698.9	560.6	2238
beam search thresh. =20	95	95%	6	1.04%	270.9	301.3	1522
2^ speaker							
beam search thresh =22.5	95	95%	6	1.04%	472.7	397.4	1800
beam search thresh. =20	83	83%	17	2.94%	489.6	551.3	1879
3^ speaker							
beam search thresh. =22.5	87	87%	13	2.25%	884.2	710.6	2172
beam search thresh. =20	86	86%	15	2.59%	206.7	410.6	1816
4^ speaker							
beam search thresh =22.5	93	·93%	7	1.21%	405.8	539.5	2177

Table 3.11: Experimental results of continuous speech recognition on 100 phrases with 1008 words vocabulary pronounced by four different speakers

average number of active subwords per frame is substantially larger than 200, we cannot presume to have a real time system with the basic configuration, hence we improved our system first to an intermediate configuration (with a faster CPU), then to a final configuration, with two DSPs in parallel for DHMM verification; Table 3.12 compares these three configurations from the point of view of the throughput. The final system configuration already seems adequate, since with the standard beam search threshold the ratio of the total recognition time to the net utterance time after phrase endpointing is 1.67; in these conditions the user perceives the system as nearly real time since he adds to the net utterance time the starting and ending silences. We can verify however that by doubling the number of DSPs performing the DHMM verification the system speed is not doubled: this is also due to the present program implementation. We have singled out some small program modifications which can improve this point [46].

3.7 Conclusions

We have developed a multiDSP open architecture for signal processing, presently equipped with a master CPU and three DSPs, one for feature extraction and the other two for DHMM verification: in this configuration we have implemented three different kind of recognition algorithms:

- the single-step isolated-word recognition (with full search on the whole vocabulary),

		Basic system CPU 12.5 Mhz DSP2-A only	Intermed system CPU 20 Mhz DSP2-A only	Final system CPU 20 Mhz DSP2-A and DSP2-B
Avg. speech utterance duration		3.7 sec.	3.7 sec.	3.7 sec.
Avg. recognition time	beam search =20	10.5 sec.	8.52 sec.	6.2 sec.
	beam search =22.5	13.6 sec.	11 sec.	8.1 sec.
Recognition time/Utterance time	beam search =20	2.8	2.3	1.67
	beam search =22.5	3.7	3	2.2

Table 3.12: Throughput results of continuous speech recognition system on 100 phrases of words into the 1008 words vocabulary

- the two-step isolated-word recognition (with a first hypothesization stage and a second refinement stage),
- the single-step continuous-words recognition.

For all of all these implementations a rather satisfactory recognition accuracy has been measured in the speaker-dependent case with a thousand-word Italian vocabulary, with a high quality head-mounted microphone input. The input utterance is presently limited by an initial keystroke for the single word application and by an initial and a final keystroke for the continuous speech application. The isolated words case is handled in real time both in the single and in the double step approach; for wider vocabularies than we used the double step approach would be more appropriate: in the latter case, however, to have a better load balancing between the hypothesization and the verification steps it would be advisable to transport from the Motorola CPU to the TMS32020 at least the most computation-intensive sections of the hypothesization stage. In the connected word case the present configuration performs the recognition in 1.5 to 2 times the net input utterance time, and hence it is perceived by the user as operating in real time, taking into account the console interaction [7]. The system has also been adapted to a German and to a French vocabulary and, although a formal evaluation has not been performed in these cases, similar accuracy and performances could be expercted. Our real time implementation shows that present technology is already adequate for speech recognition tasks of some complexity in a not-too-expensive system. We think also that the overall

[7] In fact between the two keystrokes we have an initial silence, the utterance and then a final silence.

architecture presented (i.e. the task partition and allocation, the intertasks dialogue, the hardware organized as common bus, local plus distributed biport memory) can be retained for further more ambitious goals, while by using new-generation faster and architecturally richer DSPs, supporting also floating point computations, we could insert in this framework algorithmic improvements required for telephone input and speaker independence, excluding any keystroke interaction [24].

Bibliography

1. Y. Kawakami, H. Ishizuka, M. Watari, H. Sakoe, T. Hoshi, T. Iwata: "A microprocessor for speech recognition", *IEEE Journal on Selected Areas in Communications*, vol. 3, pp. 369-376, March 1985

2. R.E. Owen: "A VLSI dynamic time warp processor for connected and isolated word speech recognition", *Proc. of the ICASSP '85*, pp. 985-988, Tampa, Fla., March 1985

3. G. Quenot, J.L. Gauvain, J.J. Gangolf, J. Mariani: "A dynamic time warp VLSI processor for continuous speech recognition", *Proc. of the ICASSP '86*, pp. 1549-1542, Tokyo, Japan, Apr. 1986

4. J.R. Mann, F.M. Rhodes: "A wafer scale DTW multiprocessor", *Proc. of the ICASSP '86*, pp. 1557-1560, Tokyo, Japan, Apr. 1986

5. R. A. Kavaler, M. Lowy, H. Murveit, R. R. Brodersen: "A Dynamic Time Warp Integrated Circuit for a 1000 word speech recognition system", *IEEE Journal of Solid State Circuits*, vol. 22, pp. 3-14, February 1987

6. S.G. Glinski, T.M. Lalumia, D.R. Cassiday, Taiho Koh, C. Gerveshi, G. A. Wilson, J. Kumar: "The Graph Search Machine: A VLSI architecture for connected speech recognition and other applications", *IEEE Proceedings*, vol. 75, pp. 1172-1184, Sept. 1987

7. R. Cecinati, A. Ciaramella, G. Venuti, C. Vicenzi: "A dynamic time warping custom integrated circuit for speech recognition", *Proc. of the EUSIPCO '86*, The Hague, The Netherlands, pp. 1215-1218, Sept. 1986

8. R. Cecinati, A. Ciaramella, L. Licciardi, G. Venuti: "Implementation of a dynamic time warp integrated circuit for large vocabulary isolated and connected speech recognition", *Proc. of EUROSPEECH '89*, pp. 565-568, Paris, France, Sept. 1989

9. A. Albarello, R. Breitschaedel, A. Ciaramella, E. Lenormand, R. Pacifici, J. Potage, J.P. Riviere, N. Scheibel, G. Venuti: "Implementation of an acoustical front-end using the TMS32020", *Proc. of the Digital Signal Processing Conference*, Florence, Italy, September 1987

10. C. Erskine, S. Magar: "Architecture and applications of a second generation digital signal processor", *Proc. of the ICASSP '85*, pp. 228-231, Tampa, Fla., March 1985

11. K.S. Lin, G.A. Frantz, R. Simar jr.: "The TMS32020 family of digital signal processors", *IEEE Proceedings*, vol. 75, pp. 1143-1159, Sept. 1987

12. D.B. Roe, A.L. Gorin, P. Ramesh: "Incorporating syntax into the level-building algorithm on a tree-structured parallel computer", *Proc. of the ICASSP '89*, pp. 778-781, Glasgow, UK, May 1989

13. R. Bisiani, T. Anantharaman, L. Butcher: "BEAM: an accelerator for speech recognition", *Proc. of the ICASSP '89*, pp. 782-784, Glasgow, UK, May 1989

14. S. Chatterjee, P. Agrawal: "Connected speech recognition on multiple processor pipeline", *Proc. of the ICASSP '89*, pp. 774-777, Glasgow, May 1989

15. W. Fisher: "IEEE P1014 - A standard for high performance VME bus", *IEEE Micro*, vol. 5, pp. 31-41, Febr. 1985

16. D. Gustavson: "Computer buses - A tutorial", *IEEE Micro*, vol. 4, pp. 7-22, Aug. 1984

17. VME Bus Manufacturers Group: *VME Bus Specification Manual.* [with VME Revision B, August 1982, and VMX Revision A, October 1983]

18. P. Harold: "Powerful local buses join the VME bus", *EDN*, pp. 199-208, Apr. 18, 1985

19. M. L. Fuccio, R. N. Gadenz, C. J. Garen, J. M. Huser, B. Ng, S. P. Pekarich: "The DSP32C: AT&T's second generation Floating Point Digital Signal Processor", *IEEE Micro*, vol. 8, pp. 30-48, Dec. 1988

20. P. Papamichalis R. Simar, Jr.: "The TMS320C30 Floating Point Digital Signal Processor", *IEEE Micro*, vol. 8, pp. 13-29, Dec. 1988

21. E. A. Lee: "Programmable DSP architectures: Part I", *IEEE ASSP Magazine*, vol. 5, pp. 4-14, Oct. 1988

22. E. A Lee: "Programmable DSP architectures: Part II", *IEEE ASSP Magazine*, vol.6, pp. 4-14, Jan. 1989

23. A. Dinning: "A survey of synchronisation methods for parallel computers", *IEEE Computer*, vol. 22, pp. 66-77, July 1989

24. ESPRIT II Project N.2218 (SUNDIAL). Technical Annex

25. D. MacGregor, D. Mothersole, B. Moyer: "The Motorola MC68020", *IEEE Micro*, vol. 4, pp. 101-118, Aug. 1984

26. VERSADOS Operating System - Technical Documentation

27. C. Huntsman D. Cawthron: "The MC68881 floating point coprocessor", *IEEE Micro*, vol. 3, pp. 44-54, Dec. 1983

28. G.W. Cherry: *Pascal Programming Structures for Motorola Microprocessors.* Reston Publishing, Prentice Hall, 1982

29. M. Ajmone Marsan, G. Balbo, G. Conte: "Performance models of multiprocessor systems", *MIT Press Series in Computer Systems*, Chapters 9 and 10, 1986

30. A. Ciaramella, G. Venuti: "Vector quantization firmware for an acoustical front end using the TMS32020", *Proc. of the ICASSP '87*, pp. 1895-1898, Dallas, Tex., Apr. 1987

31. F.J. Harris: "On the use of windows for harmonic analysis with the Discrete Fourier Transform", *IEEE Proceedings*, vol. 66, pp. 51-83, , Jan. 1978

32. E. O. Brigham: *The Fast Fourier Transform*, Sect. 10-10, pp. 163-171. Prentice Hall, 1974

33. L.R. Morris: "Structural considerations for large FFT programs on the TI TMS32010 DSP microchip", *Proc. of the ICASSP '85*, pp.42.13.1-4, Tampa, Fla., March 1985

34. P. Kabal, B.Sayar: "Performance of fixed-point FFT's: rounding and scaling considerations", *Proc. of the ICASSP '86*, pp.6.3.1-4, Tokyo, Japan, Apr. 1986

35. S. Prakash, V.V. Rao: "Fixed point error analysis of Radix-4 FFT", *Signal Processing*, vol.3, pp.123-133, Apr. 1981

36. K. H. Davis, P. Mermelstein: "Comparison of parametric representations for monosyllabic word recognition in continuously spoken sentences", *IEEE trans. ASSP*, vol.28, pp. 357-366, Aug. 1980

37. A. Kaltenmeier: "Acoustic/phonetic transcription using a polynomial classifier and Hidden Markov Models" *Proc. of the Montreal Symposium on Speech Technology*, pp. 95-96, Montreal, Canada, July 1986

38. P. Capello, G. Davidson, A. Gersho, C. Koc, V. Somayazulu: "A systolic vector quantization processor for real time speech coding", *Proc. of the ICASSP '86*, pp. 41.1.1-4, Tokyo, Japan, Apr. 1986

39. P. Laface, G. Micca, R. Pieraccini: "Experimentals results on a large lexicon access task", *Proc. of the ICASSP '87*, pp. 809-812, Dallas, Tex., Apr. 1987

40. M. Cravero, R. Pieraccini, F. Raineri: "Definition and evaluation of phonetic units for speech recognition by Hidden Markov Models", *Proc. of the ICASSP '86*, pp. 42.3.1-4, Tokyo, Japan, Apr. 1986

41. L. Fissore, E. Giachin, P. Laface, G. Micca, R. Pieraccini, C. Rullent: "Experimental results on large vocabulary continuous speech recognition and understanding", *Proc. of the ICASSP '88*, pp. 414-417, New Jork, NY, Apr. 1988

42. L. Fissore, P. Laface, G. Micca, R. Pieraccini: "Interaction between fast lexical access and word verification in large vocabulary continuous speech recognition" *Proc. of the ICASSP '88*, pp. 279-282, New York, NY, Apr. 1988

43. L. Fissore, P. Laface, G. Micca, R. Pieraccini: "Very large vocabulary isolated utterance recognition: a comparison between one pass and two pass strategies", *Proc. of the ICASSP '88*, pp. 203-206, New York, NY, Apr. 1988

44. G. Micca, R. Pieraccini, P. Laface, L. Saitta, A. Kaltenmeier: "Word Hypothesization and Verification in a Large Vocabulary", *Proc. of 3rd Esprit Technical Week*, pp. 845-853, Brussels, Belgium, Sept. 1986

45. A. Ciaramella, G. Venuti: "Dynamic programming with hidden markov models on a TMS32020 digital signal processor", *Proc. of EUSIPCO '88*, pp. 751-754, Grenoble, France, Sept. 1988

46. A. Ciaramella, D. Clementino, R. Pacifici: "Characterization of a large vocabulary isolated words and continuous speech recognizer", *Proc. of the Eurospeech '89*, pp. 437-440, Paris, France, Sept. 1989

Chapter 4

The Understanding Algorithms

Roberto Gemello, Egidio Giachin, Claudio Rullent (CSELT)

4.1 Overview

4.1.1 Introduction

The final goal of a continuous speech understanding system is the generation of a representation of the utterance meaning, beside the recognition of the utterance words. From this representation a proper action can be taken in order to satisfy the needs of the user that interacts with the system (for instance by giving him an answer to a question). Both activities, word recognition and understanding, have to be performed and should take advantage of available knowledge about words, language and domain. Recognition must use that knowledge as a source of constraints for word disambiguation while the understanding activity is entirely based on that knowledge and requires the same effort as in the case of written natural language understanding.

The first large-scale effort to integrate recognition and understanding was accomplished within a DARPA sponsored project in the late 1970s. However, the techniques they used there for knowledge representation (mainly context-free semantic grammars) were of low complexity, as knowledge about language and domain had to be used for both recognition and understanding. That resulted in a limitation of the potentialities of the natural language understanding activity. Written natural language processing techniques, on the other hand, have considerably increased their power and flexibility in recent times, mainly in the area of representation of syntactic and semantic knowledge. New representational tools have been developed, more powerful and expressive than context-free grammars although more complex. As a general trend, keeping separate representations for syntactic and semantic knowledge is considered beneficial to better exploit specific features and regularities.

When this project was started in the mid-1980s, it was felt that taking advantage of such improved techniques could result in a more efficient way of exploiting language constraints for a system whose purpose is not limited to recognition but includes meaning comprehension. In other words, an approach has been followed in which the stress on language processing techniques is as strong as on signal processing techniques. Conceptually, the basic philosophy was to start from a state-of-the-art system for written language understanding and extend its capabilities to deal with word lattices rather than definite word sequences.

A word lattice is a collection of *Word Hypotheses* pertaining to a single utterance. Each

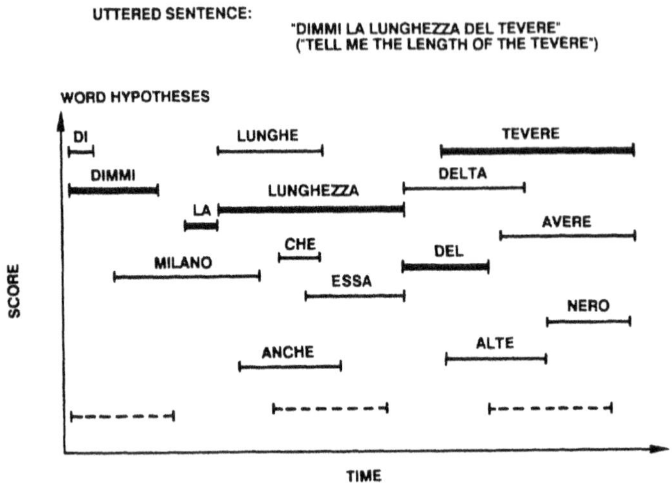

Figure 4.1: Structure of the lattice of word hypotheses

word hypothesis is characterized by a score reflecting the quality of the match between the observed signal and the word model (Fig. 4.1). The number of word hypotheses has to be high enough to contain the correct hypotheses (i.e. those corresponding to the actually uttered words). Improvements in signal processing and pattern recognition techniques have produced (and will furthermore produce in the future) a reduction in the number of wrong word hypotheses and an improvement in the reliability of their scores; thus, the importance of having well-grounded natural language processing tools will become even greater in perspective, and we expect that the approach that has been followed in this project will allow us to converge towards a global system using the best characteristics of the two different classes of techniques, signal processing and language processing, suitable for the two levels of elaboration.

At present the above mentioned approach is studied by different reasearch groups. Among them are the groups at University of Erlangen-Nuernberg [5], Philips [28], Siemens [29], CNET [2], Carnegie Mellon University [19] and SRI [40]. A common element of most of these approaches is the use of a lattice of word hypotheses as the input of the understanding stage (this feature has also been used in some systems aiming essentially at recognition only [27], [25]). These approaches to speech understanding are characterized by a more declarative way of representing syntactic and sèmantic knowledge compared to the projects developed in the previous decade, but in our opinion there are still some critical aspects that require new solutions. The final part of the introduction discusses two important aspects that have been analyzed in our research and whose solutions represents the most innovative aspect of SUSY (Speech Understanding SYstem), the system developed at CSELT.

A convincing answer to the problem of an effective integration between syntactic knowledge and semantic knowledge is still to come. The problem is from one side to maintain independent and highly declarative representations for both semantic and syntactic knowledge and from the other to use them in an integrated way in order to exploit constraints as soon as possible. While this aspect is important for written natural language understanding [6], it is vital for speech, where the search space is very large, being the non-determinism of parsing added to the uncertainty of input data.

Another crucial problem is that of control, that is, of selecting the appropriate analysis actions in order to achieve high efficiency while keeping sufficiently small the probability of incorrect understanding. While some speech understanding systems based on the use of semantic grammars were really concerned about the problem of control (e.g. [42]), now this aspect seems to be underestimated. That is not surprising, as an increased complexity of the representation formalisms for syntax and semantics makes a formal control policy hard to reach.

This section aims at giving a general overview of the problems outlined above and the solutions proposed by our system [14, 13, 30]. The section will cover the essentials about language modeling, parsing and control.

4.1.2 Some Basic Requirements of a Parser for Speech

The understanding stage needs to detect, in the lattice, the best scoring sequence of word hypotheses covering the whole utterance and coherent according to the models of the language and of the domain. The presence of word hypotheses spread all over the utterance instead of a sequence of words requires a new kind of parser whose main features are related to a very high flexibility in the control strategy. Some features of the parser are the following:

- It is important to have powerful control strategies based on the combination of word scores. An efficient parser must take this aspect into account.

- Due to the limitations inherent in the recognition stage, a "tolerant" parser is required:

  - Contiguous word hypotheses may slightly overlap, and gaps may exist between them.
  - Very short words (e.g. articles) are normally difficult to detect by the recognition level and may be missing from the lattice; if they do not convey essential semantic information the parser should not rely on them to understand the sentence.

  These requirements argue against a left-to-right parser.

- The parsing strategies must be suitable for parallelism. Only a highly parallel machine can perform speech understanding in real time. See [3], [16] for a discussion of a possible way of exploiting parallelism from the parsing strategies adopted by SUSY.

- Syntactic and semantic knowledge must be separately defined and used in a joint way. Such separation allows a reduction of the time required for an expert to define an application for a new domain, i.e. to declare all the knowledge required to adapt the speech understanding system to the new domain.

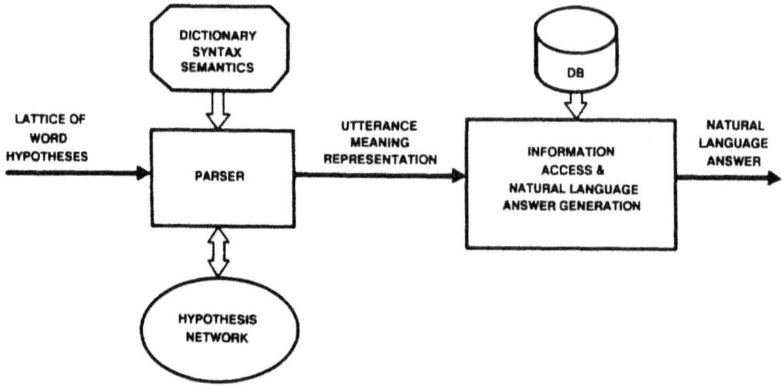

Figure 4.2: Simplified overall architecture of the understanding module

A simplified scheme of the understanding system is shown in Fig. 4.2. A recognition level, that makes no use of syntactic and semantic knowledge, generates a lattice of scored word hypotheses that constitutes the input data to the understanding level. The uttered sentences are questions aimed at extracting information from a data base pertaining to a given domain (Italian geography).

The lattice is processed by a parser that recognizes the most probable word sequence and generates an internal formal representation for the meaning of such a word sequence. The formal representation, a conceptual graph of domain concepts connected by relations, is used to extract the required information from the database. Starting from such information a natural language answer is generated and given to the user.

The parser is based on the use of a dictionary and on a set of syntactic and semantic pieces of knowledge that are briefly outlined in the following section. The internal structure generated and used by the parser is a network of phrase hypotheses. So the parser activity consists in continuously generating new phrase hypotheses that represent alternative or cooperative paths of the parsing activity. For a number of reasons that should become clear later on, we are not dealing with chart parsing [41], even if some commonalities do exist, mainly the fact that both generate and use a network structure.

4.1.3 Knowledge Sources from Dependency Rules and Conceptual Graphs

The parser, during its activity, makes use of the following different kinds of knowledge:

- A dictionary, where each domain word is characterized by its morphological and semantic features.
- A set of dependency rules (Dependency Grammar formalism, [20]) augmented with rules for controlling morphological agreement conditions; dependency rules must constitute a subset of the language sufficient to cover the application.
- A set of caseframes [11] expressed using the conceptual graphs formalism [37]. They describe domain knowledge and are represented by domain concepts connected by conceptual relations.

Starting from the last two knowledge bases and from additional syntax/semantics mapping knowledge, an integration of syntax and semantics is performed, generating items called *Knowledge Sources (KS)* in the following.

Each KS integrates different types of knowledge. The main body of knowledge is a problem-solving structure where the different subproblems are not independent but constrained by different kinds of knowledge: temporal constraints, morphological and grammatical constraints, etc.

This structure results from a "compilation process" that integrates the semantics of one or more caseframes (expressed by conceptual graphs) with the structure of one or more dependency rules. This compilation process is performed off-line through the use of "mapping" information, that relates implicit grammatical relations of the dependency rules with the semantic relations of conceptual graphs.

A KS is also characterized, in addition to the problem-solving structure, by the remaining morphological, syntactic and semantic information. This knowledge is transformed off-line into special structures suitable to perform efficiently the activity of constraints propagation. All the other kinds of knowledge used by a KS are expressed through procedures. For instance a procedure (that makes use of thresholds) represents knowledge about the recognition system word-spotting characteristics; it is used to impose constraints on the allowed gaps and overlaps between word hypotheses.

The problem-solving structure of the KS permits them to be seen as constituting a Deduction System, i.e., roughly speaking, a system able to run forward and backward. The actions that a KS can perform when it is triggered can be described in terms of five operators; some of them behave in a forward fashion, some in a backward fashion, while others allow the generation of new deductive processes and the integration of different ones.

The description of dependency rules, conceptual graphs and their integration into KSs is contained in Sects 4.2, 4.3 and 4.4.

4.1.4 The Importance of Control Strategies

The word lattice is usually characterized by a lot of spurious word hypotheses intermixed with the correct ones (i.e. those that correspond to words really uttered and covering the given time interval). Some incorrect word hypotheses may also happen to have a better score than the correct ones.

Two reasons for an effective control strategy

In a problem-solving approach to speech understanding two main problems arise:

- There is a risk of erroneous understanding, that is, spurious word hypotheses of the lattice can lead to incorrect solutions before the right one is found. So the utterance can be incorrectly understood.
- The search space is very large, adding the non-determinism typical of the parsing to the uncertainty of input data. The whole search space cannot be explicated.

The elimination of incorrect solutions requires a method for comparing solutions so that the best one can be selected as the correct one. A number, called the *Quality Factor*, is assigned to them; this number depends only on the word hypotheses involved in the solution. So a formal probabilistic method assigns a number, the quality factor, to combinations of word hypotheses, starting from their scores and from their time intervals. Among the tested methods of assigning quality factors, the following one has been selected:

- Score density with or without shortfall [42].

As regards the second problem, quality factors must be used also to direct the search, in order to find the solution long before having to perform an impossible exhaustive search. From the probabilistic point of view the most natural way of dealing with scored input data is to start with the best word hypotheses and trying to combine them together until a solution is obtained. The problem is the necessity of exploiting constraints from domain knowledge as soon as possible to drastically reduce the combinatorial activity. On the other hand a good way of exploiting domain constraints is a bottom-up parsing strategy guided by the quality factors of the word hypotheses in the lattice. But this approach is inadequate when the search space is very large due to a great amount of noise. In fact dangerous bottlenecks cannot be avoided if expectations are not considered. As an example consider a situation where a low-level constituent, part of the solution, has to be formed using a very bad word hypothesis WH1 (this could easily happen). The problem is that the solution can have a good quality factor (the bad score of WH1 is balanced by the good scores of the other word hypotheses) but it can be delayed by WH1 because a lot of word hypotheses having a score better than WH1 but really worse than the quality factor of the solution have to be considered.

The role of expectations: Integrating top-down and bottom-up parsing strategies

An important feature of SUSY to cope with these difficulties is the possibility of creating expectations at the highest levels. This is always possible in our approach because each KS has been obtained from a caseframe and can be triggered by a word that represents the caseframe header. So a good word hypothesis always has a KS that can be activated by it. In addition the use of dependency rules perfectly suits this point: each node of a dependency tree is taken by a word, and such a word hierarchy perfectly allows the generation of high-level expectations.

The informal conclusion is that in some cases good word hypotheses cause the parsing to proceed bottom-up while in other cases they first create expectations (goals) and then cause the parsing to proceed top-down, looking for word hypotheses when necessary in order to perform single backward steps. The acquisition of a new word hypothesis during a top-down step usually worsens the quality factor of a subgoal (incomplete phrase hypothesis), delaying its processing, while in the meantime other phrase hypotheses will be processed. Integration among bottom-up and top-down steps is vital: having to solve an incomplete phrase hypothesis, a check is made to see if a suitable complete phrase hypothesis has already been generated and vice versa.

Deduction instances and search

The understanding of an utterance is completed when a solution S involving a certain set w1,...,wn of word hypotheses is obtained. The quality factor resulting from w1,...,wn is supposed to be the best one among the possible solutions. Such solution can be represented by a *Deduction Tree*: the AND tree whose nodes are facts (complete phrase hypotheses) and (sub)goals (incomplete phrase hypotheses). Following the informal guidelines of the previous section, a solution is obtained starting from one or more initial word hypotheses (the best ones) and then performing predictions, bottom-up and top-down steps and joining constituents.

Let us consider the simple case of a single initial word hypothesis that performs a prediction generating a goal that is solved through a sequence of top-down steps. From the probabilistic point of view, new word hypotheses are connected one by one (assuming they satisfy all the required constraints) to the initial one until the solution is reached, connecting all the word hypotheses w1,...,wn. We call this activity a *Deductive Process* and each intermediate step is called *Deduction Instance* (DI). Some steps consist in adding a new word hypothesis to the existing ones, others represent only activities performed by a KS that do not involve the acceptance of a new word hypothesis. The OR alternatives of the overall search process are taken into account by different DIs. Each deduction instance can be represented by its deduction tree and it is characterized by a quality factor obtained applying a selected probabilistic method to the word hypotheses of the deduction tree.

A similar situation happens when bottom-up steps are considered. In this case DIs have deduction trees whose nodes are all facts; they are called fact DIs while the others are called goal DIs. A single deductive process leading or not (if it fails) to a solution is a sequence of DIs.

Joining deduction instances

The optimal result would be obtained if the quality factors corresponding to the sequence of DIs worsen gradually in quality, converging to the quality factor of the solution. The required integration among bottom-up and top-down steps can be obtained by merging together two deductive processes that have previously evolved independently: from two DIs a new DI is generated.

With some simplifications the whole deductive activity can be seen as a search in a state space. A state is a deductive process at a certain point of its evolution, i.e. a deduction instance. Operators can be applied on these states performing single prediction, bottom-up, top-down or merge steps. To each state a quality factor is also associated, then a best-first search can be performed; the state priority is given by the DI quality factor. On each state all the possible operators are applied.

The description of the conceptual foundation of the parsing strategy, the definition of deduction instances and the global strategy is contained in Sect. 4.5. In addition Sect. 4.7 illustrates the solutions to the problem of short words missing from the lattice.

4.1.5 Control Strategy and Operators

The control of the deductive activity is carried out by a *Deduction Scheduler* that at every cycle selects the best item among the remaining word hypotheses and the DIs (phrase hypotheses generated so far and inserted into a network called the *Hypothesis Network*). All the items have a priority given by their quality factors (in the case of a DI) or by their scores (in the case of a word hypothesis). Each goal DI is also characterized by a Current Subgoal, selected among its unsolved subgoals. If the deduction scheduler selects a DI, the *Deduction Cycle* is entered, otherwise the *Activation Cycle* is performed.

The activation cycle is executed when the best DI has a quality factor worse than the score of the best word hypothesis. In that case such a word hypothesis is extracted from the lattice, and the *activation* operator is applied, making predictions. Given a KS the operator decides if it can be triggered by the given word hypothesis; if so a DI is generated and inserted into the hypotheses network. Quality factors are assigned to the new DIs. Conceptually this operator creates expectations.

In the deduction cycle the selected best DI is given to the KS. The activities performed by the KS when it is triggered can be described in an abstract way through five operators that represent the process of generating new hypotheses starting from others. The characteristics of the triggering DI define which operator is applicable. Each operator application represents an alternative continuation of the deductive process leading to the selected DI. The operators are described in Sect. 4.5.13.

4.1.6 Representing Deduction Instances with Memory Structures

An aspect that has to be considered when representing DIs with memory structures is to reduce the amount of memory required and to properly organize the memory structures in order to simplify operators application.

The most trivial way of implementing deduction instances would be using an explicit deduction tree for each of them, but to keep memory occupation within reasonable limits it is necessary to make DIs share common parts, if any. A natural choice is to use AND-OR trees; unfortunately, a problem arises when constraint propagation is required, as in our case: the AND-OR trees representation assumes the OR alternatives to be independent from one another, but that is not true if constraints propagation has to be taken into account.

In order to continue to take advantage of the use of AND-OR trees even when constraint propagation has to be performed, a new memory representation has been devised and a limitation is imposed on the possible topologies a deduction tree may assume. In this way the sharing of similar but differently constrained parts is possible, at the cost of some limitations on the ways deductive processes can go on. Such limitations do not compromise integration among top-down and bottom-up activities. The allowed tree topologies are called Canonical and the resulting DIs are called Canonical DIs.

A Canonical DI can be put into a one-to-one relation with one of its particular substructures. This substructure, which is a one-level AND subtree, has been called the Representative of the DI because the information provided by it is sufficient to carry out the application of an operator on the CDI. In other words, as far as an operator application is involved, we can use the Representatives instead of the whole CDI. Representatives are implemented by a memory structure called *Phrase Hypothesis.*

Section 4.6 describes these problems and, in addition, the application of the operators on the memory structures is illustrated in detail.

4.1.7 Implementation, Development System and Results

SUSY was first implemented on a Symbolics 3600 Lisp Machine using the Common Lisp language; later on it was implemented into the C language on a SUN. A suitable Development System has also been implemented, to provide a flexible and comfortable environments to work with. The Development System permits a comfortable development of the required algorithms and, thanks to its special purpose debugger, their relatively easy debug and modification. The Development System has also proved very useful for the insertion of the knowledge bases into the system. Section 4.8 provides a brief description of the experimental results.

4.2 Representation of Syntax

4.2.1 Introduction

Syntactic and semantic knowledge has often been viewed in a speech understanding system as a set of constraints for improving the recognition activity. Thus this knowledge has been completely integrated within the recognition algorithms, which are mainly involved in statistical/phonetic processing of input signal. For this reason the syntactic and semantic knowledge involved has often been quite simple and not sufficiently flexible and linguistically powerful (e.g. semantic grammars, often finite state automata). As already discussed in Sect. 4.1, we wanted to move in a different way, taking more into account the natural language understanding aspects in the case of speech too.

In SUSY a domain independent speech recognition activity is followed by a syntactic and semantic based understanding activity. The latter activity is performed by a natural language processing system that is able to deal with a lattice of lexical hypotheses, that is, the output that a recognition stage can give using only phonetical-lexical knowledge.

The syntactic and semantic knowledge bases of this natural language understanding system must be flexible, i.e. easy to maintain, modify and expand; and linguistically powerful enough. Since they are used in a system that deals with a highly ambiguous input, they must be sufficiently tight and exploit those constraints (syntactic, semantic and agreement constraints) which can make easier the recognition and the understanding of the uttered sentence. The claim for flexibility has been satisfied by employing separated syntactic and semantic knowledge bases. This way, it is possible to better exploit specific representations that address respectively syntax and semantics, and ease of maintenance is insured because, if for example the semantic domain has to be changed or modified, little or no modification is required for the syntactic knowledge base.

Syntactic knowledge uses the *dependency grammar* formalism [20, 7] augmented with morphological agreement rules. Following the *semantic caseframe* paradigm [11], the semantic knowledge is represented through conceptual graphs [37] representing semantic caseframes. Although these knowledge bases are utilized in a speech understanding context, they are general and can be used by any natural language understanding system, e.g. for written language processing.

The two knowledge bases are seen as independent sources of knowledge; on the other hand, for the sake of efficiency, it is necessary to use both syntactic and semantic constraints during the parsing process. To obtain this result, the dependency grammar rules and the conceptual graphs are compiled off line into a set of *Knowledge Sources*. To make the off-line compiler work properly, a further source of knowledge had to be introduced. This type of knowledge describes the relations between syntax and semantics and it is sometimes called *mapping* knowledge.

4.2.2 Interaction Between Syntactic and Semantic Knowledge

A basic decision that has been taken from the beginning and that we think to be very crucial in developing a speech understanding system is the possibility of defining syntactic and semantic knowledge as two independent activities. Of course this is a goal that can not be achieved completely: there is no clearly defined division between syntax and semantics. In addition, there is the problem of efficiency for the parsing algorithms: in the context of speech (i.e. when the parser must operate on a lattice of word hypotheses) it becomes highly inefficient to perform a syntactic analysis first and a semantic validation afterwards. That means that syntax and semantics must interact tightly at processing time in order to exploit the constraints as soon as possible.

The possibility of defining syntactic knowledge and semantic knowledge as two independent activities is important because it reduces the effort required to adapt a given understanding system to a new semantic domain. In fact the use of a semantic grammar, where syntax and semantics are melted together, requires a complete redefinition of all the rules (usually context-free rules) that constitute the knowledge base of the system; in addition the number of rules could be very large if a broad coverage is required.

When syntactic and semantic knowledge are declared separately, there is the above mentioned problem of making syntax and semantics interact in a way suitable to reduce the search activity by best exploiting constraints. A possible solution could be to make this interaction to happen at processing time [6] while another possibility is based on a certain amount of compilation (or merge) of the two different kinds of knowledge into a unique structure [39].

An important problem that has to be faced in both cases is the need of knowledge about the relationships between the syntactic structures and the semantic ones. Either implicitly or explicitly such knowledge must be used to properly constrain the search using both kinds of information together. This knowledge is called *mapping* knowledge in the sense that it represents a kind of mapping between syntax and semantics [8]. Mapping knowledge will be described in Sect. 4.4 devoted to illustrate the compilation of conceptual graphs and dependency rules into knowledge sources.

Let us now summarize the basic points concerning knowledge representation and parsing:

- Different representation formalisms are used for syntax and for semantics.

- The syntactic formalism is a dependency grammar augmented with morphological agreement rules.
- The structure that is used at processing time (called the knowledge source, KS) must contain both syntactic and semantic knowledge.
- At processing time syntactic and semantic constraints are exploited simultaneously: i.e. the generated hypotheses must be syntactically and semantically correct.
- Knowledge sources must permit the probabilistic control strategy outlined in the summary and described in detail in the following sections.

The remaining part of Sect. 4.2 will be devoted to describing syntactic knowledge while following sections will illustrate semantic knowledge and the off-line compilation process.

4.2.3 Dependency Grammar

Definitions

Let us consider a dictionary V; each word of the dictionary is characterized by a lexical category. Let C be the set of these categories. We define a Dependency Grammar (DG) to be the couple

$$DG = \{C, R\}$$

where C is the set of lexical categories and R is a set of rules of the kind:

$$X_0 = X_1 X_2 ... X_k * X_{k+1} ... X_n \qquad \text{with} \qquad X_i \in C; n \geq 0$$

where X_0 is called the *governor* and $X_1, ..., X_n$ are called *dependents* of the governor X_0. The symbol * shows the governor position with respect to its dependents. The ordered sequence $X_1, ..., X_k$ is made up of left dependents while $X_{k+1}, ..., X_n$ is made up of right dependents.
The subset R_t of R includes the rules of the kind

$$X_i = *$$

These rules are called *terminal* rules because their governors have no dependents.

The result of parsing a sentence using this grammar model is a set of dependency trees. They have the following features:

- Each node of the tree (not only the leaves) is associated with a word of the input sentence.
- The sons of any subtree root are divided into two ordered sets called left and right dependents.
- The input sentence can be obtained by projecting the nodes of the dependency tree, that is, visiting it in symmetric order (left subtrees, root, right subtrees)
- The root of the dependency tree is the governor of the sentence.

An example

Let us consider
a) the lexical categories:

	(English Translation)
VERB = {vide}	saw
NOUN = {battello, binocolo}	boat, binoculars
ART = {un, il}	a, the
PREP = {con}	with
PROPER-NOUN = {Giovanni}	John
ADJ = {grosso}	big

b) the rules:

1) VERB = PROPER-NOUN * NOUN NOUN
2) NOUN = ART ADJ *
3) NOUN = PREP ART *
4) ART = *
5) ADJ = *
6) PREP = *

Then the sentence:
Giovanni vide un grosso battello con il binocolo
(John saw a big boat with the binoculars)
can be generated (and therefore be parsed) by applying the rules shown in Fig. 4.3.

Relations between dependency grammar and context-free grammar

A dependency grammar (DG) is equivalent to a context-free grammar (CFG) with word dependency information attached to each production. That is, the right-hand side of each rule of the grammar must have a "distinguished symbol" that plays the role of main symbol. The governor of the phrase associated with that rule is the governor of the phrase that is associated with the distinguished symbol. All other words that are part of the phrase associated with the production are dependents, at some level, of this governor. The example in Fig. 4.4 intuitively shows the correspondence between the two formalisms.

Remarks on dependency grammars

The DG follows the Head-Modifier linguistic paradigm. According with this paradigm each phrase is made up of a main word (Head) and some other words relating to the head which modify its role (Modifiers). Recursively the main head modifiers of a sentence will be, in their turn, heads of component phrases, and so on. The linear structure of the input sentence can be delinearized into a head-modifiers tree (dependency tree).

This general paradigm can be interpreted in many ways: for example from a syntactic point of view heads and modifiers can be defined on the grounds of syntax; from a semantic point of view concepts can be thought as modified by other concepts.

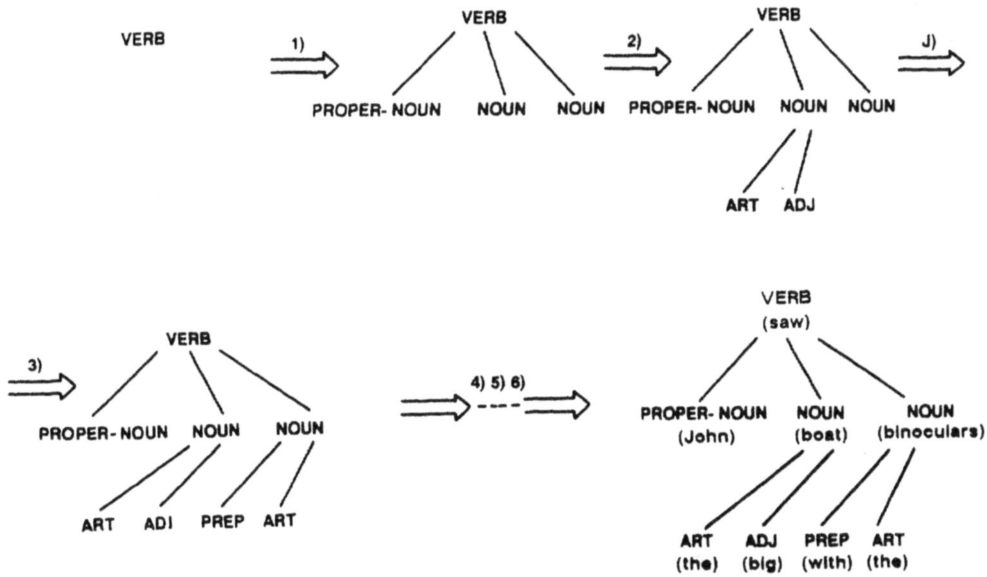

Figure 4.3: Example of dependency parse tree

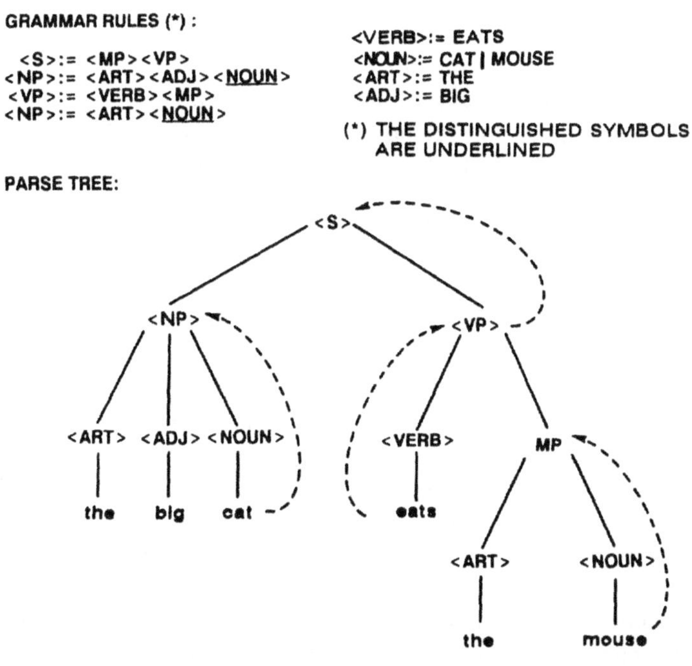

Figure 4.4: Correspondence between dependency and context-free parse trees

The most interesting feature of DG is the concept of governor (head): every phrase parsed using a DG must include a word with the role of head; if this word is not found, the phrase is not recognizable by the DG. This feature can be exploited when a parsing strategy operating on complex inputs is required, as in the case of speech, where a lattice of word hypotheses all over the utterance constitutes the system input. This feature of the dependency grammar formalism is the main reason why this formalism has been selected among others for continuous speech parsing. In fact it allows the generation of expectations even at the highest levels, starting from word hypotheses with good score. This point will be illustrated in detail during the description of the parsing control strategy.

The problem of finding a dependency tree on a sentence segment can be reduced to three subproblems:

1. finding the governor;
2. finding the left subtrees;
3. finding the right subtrees.

The problem of finding the governor is primitive and can be solved simply by searching in the input words. Of course, if that problem is not solved (that is, there is no proper governor), it is worthless to go on with the others. Thus the need to find a governor can give a useful heuristic for cutting down at every step the search activity required for sentence parsing.

4.2.4 Morphological Agreement Rules

The standard dependency grammar model does not allow us to constrain the generation of dependency trees on the grounds of morphological agreement. In fact, dependency rules consider only lexical categories and do not take into account the morphological variables (e.g. person, number, gender, etc.). For example, the rule

VERB = ART ADJ NOUN * ART ADJ NOUN

allows the recognition of the sentence: "Il grossi gatta cacciano i piccola topi" ("The big cat chase the small mice"), that is incorrect from the point of view of morphological agreement (notice that in the Italian version of the example the mistake is much more evident).

In the case of natural language parsing this fact does not necessary constitute a problem: the purpose is to understand, not to check the syntactic correctness of the sentence. But morphological agreement represents additional constraints in the case of speech understanding to reduce the amount of parsing activity. In fact, by exploiting morphological agreement constraints, the possible combinations of words can be reduced. Dependency rules should then be augmented with some mechanisms for morphological agreement. We define these mechanisms as *agreement rules*. These rules are associated 1:1 with the dependency rules. Then, if a dependency grammar DG generates the language L, the same grammar augmented with a set of agreement rules C, DG_C, will generate a language L_C, with L_C included in L.

Structure of agreement rules

Let us describe the structure of the agreement rules that augment the dependency rules. Each agreement rule relates to a dependency rule. To each element (governor or dependent) of the dependency rule (lexical category) is associated a set of agreement constraint conditions pertaining to morphologic features meaningful for that element. For every feature the constraints can be:

1. Constant constraints, e.g. GENDER = MASCULINE

2. Variable constraints, e.g. GENDER = ?X

As natural language is characterized by phenomena of agreement between words that are not close in the sentence, certain feature values can be transmitted from the governor to dependents and viceversa. The set of features that can be transmitted is defined a priori. It is then necessary to decide which information of the governor can be useful for an agreement check at a superior or inferior level. This information must be transmitted backward or forward in the dependency tree.

Definition of agreement rules

Let us consider a dictionary V. Let M be the set of morphological variables for the words of V. Every morphological variable $z \in M$ has values in a set Z_i (E.g. if z = NUMBER then Z_i = SINGULAR,PLURAL).

Given a dependency rule DR

$$X_0 = X_1 ... X_k * X_{k+1} ... X_n$$

an associated agreement rule RC is an ordered n+1-tuple

$$RC = (CM_0, CM_1, ..., CM_n)$$

where CM_i is an agreement condition set referring to X_i and is defined in the following way:

$$CM_i = VM_1, ..., VM_k$$

where VM_j is an agreement condition.

Agreement conditions are couples of the kind:

$$VM_j = (z, A)$$

where:

- $z \in M$ is a morphological variable (e.g. GENDER, NUMBER, etc.)

- A can be either a variable (taking values in Z), or a subset of Z. (Z is the range of z).

For example: the following couple dependency rule / agreement rule (DR1, RC1)

DR1: VERB = ART ADJ NOUN * ART ADJ NOUN

AR1: VERB: PERSON = 3, NUMBER = ?X
ART: NUMBER = ?X, GENDER = ?Y
ADJ: NUMBER = ?X, GENDER = ?Y
NOUN: NUMBER = ?X, GENDER = ?Y
ART: NUMBER = ?Z, GENDER = ?W
ADJ: NUMBER = ?Z, GENDER = ?W
NOUN: NUMBER = ?Z, GENDER = ?W

allows the generation of the sentences of the kind of "Il grosso gatto caccia il piccolo topo" ("The big cat chases the small mouse") but not sentences of the kind of "Il grossi gatta cacciano i piccola topi" ("The big cat chase the small mice") that would be accepted without the agreement constraints. In fact the agreement rule AR1 imposes the following agreement constraints:

- number agreement among the governor (VERB) and the left dependents (ART, NOUN, ADJ);
- gender agreement among the left dependents;
- gender and number agreement among the right dependents.

Notice that Italian is richer than English in variations due to gender, number, person, etc. In the above example all the words of the sentence (nouns, adjectives, articles, verb) are involved in morphological agreement.

Morphological agreement checks

The syntactic analysis based on the dependency rules makes use of the lexical categories associated with each word. For checking morphological agreement we have to consider morphological information (features). These features are:

1. statically associated with dictionary words
2. dynamically associated with every word W (having a governor role) involved in the syntactic analysis and transmitted forward to word W from its possible governor or transmitted backward to word W from its possible dependents.

Morphological features statically associated to words

The agreement mechanism is essentially based on the morphological information statically associated to dictionary words. To each word in the dictionary is associated a set of morphological features:

$$IM = \{im_1, im_2, ..., im_n\}$$

Every feature im_j is a couple

$$im_j = [Feature, Values]$$

made up by a morphological variable and by finite set of values for that variable.

Each "value" represents a plausible value for the feature in question. For instance a word whose gender can be either MASCULINE or FEMININE can be characterized by: [GENDER = (MASC, FEM)].

Agreement check modalities

Let us consider the rule:

$$\begin{array}{cccccccl} \mathrm{A} & = & B_1, & \ldots & ,B_k & * & B_{k+1},\ldots & ,B_n \quad \text{dependency rule} \\ | & & | & & | & & | & | \\ | & & | & & | & & | & | \\ \mathrm{SA} & & SB_1 & & SB_k & & SB_{k+1} & SB_n \quad \text{agreement rule} \end{array}$$

The dependency rule can pass the agreement check if two kinds of checks succeed:

1. Constant checks on the governor and on every dependent;
2. Variable checks between every dependent Bk and the governor A.

4.3 Representation of Semantics

4.3.1 Introduction

Semantic analysis is based on the use of caseframes represented as conceptual graphs [37]. In a first phase a different organization had been used. The present approach is based on an automatic off-line compilation of dependency rules and conceptual graphs into knowledge sources that satisfies the basic requirement of permitting an efficient control strategy. The use of conceptual graphs as a starting formalism to represent semantic knowledge about both words in the domain and internal meaning of utterances is presented.

4.3.2 Word Information in the Dictionary

A word in the dictionary can present two different kinds of ambiguity:

1. Morphological ambiguity - A word can have more than one possible morphological class. For instance the Italian word *abito* can be either a *noun* or a *verb* (according to the English translations "suit" and "to live")
2. Semantic ambiguity - A word can have different possible meanings. For instance the italian word *cane* can be either an animal (dog) or one part of a rifle (cock). The different meanings can refer to words with the same morphological category (as in this example) or not.

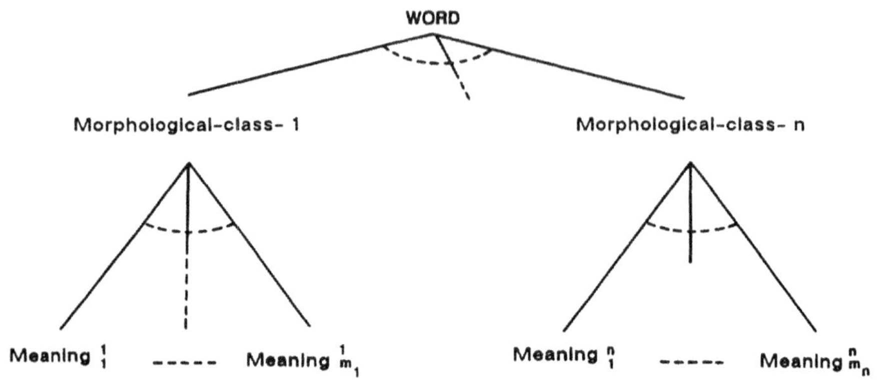

Figure 4.5: Hierarchical structure of lexical and semantic alternative word meanings

The semantic knowledge in the dictionary is organized in the hierarchical way shown in Fig. 4.5.

The semantic information associated with a word consists in a pointer to a conceptual graph representing the word "meaning" within the chosen domain. A referent is also present if the word represents some specific individual. If a word has more than one meaning in the domain, more than one conceptual graph pointer is inserted into the dictionary.

4.3.3 Caseframes and Conceptual Graphs

The central notion of caseframe is the the idea of a head concept, usually associated to a word, that is modified by a set of related concepts. Each of these modifiers plays a certain role (*case*) with respect to the head concept. Let us consider for instance the concept *bagnare* ("to wash"). Within natural language sentences such head concepts can be modified by cases. Among them: the agent of the action (AGNT), the object of the action (OBJ), the possible location (LOC), when the action takes (took) place (TIME), etc. The fillers of the cases are different concept types: the AGNT and OBJ should be generic entities having certain characteristics: the AGNT must be able to "wash" the object. Certain cases could be missing when uttering a sentence, like the TIME case.

The formalism chosen to represent caseframes is the Conceptual Graphs (CG) formalism [37], modified in such a way to fulfill our needs. Conceptual graphs are bipartite oriented graphs with two types of nodes: concepts and relations.

Concept nodes represent entities, actions or states that can be described in a natural language sentence. They correspond to intensional concepts that are connoted by words of the sentence. A concept is characterized by a Conceptual Type (Type in short) and by a referent (optional) that, if present, represents the element of the extension the node refers to. If the concept node represents a generic individual such a referent is missing. For instance an unspecified river can be represented by a node like [river] while the Tevere is represented by [river:Tevere]. A specific river that has no name but that is not generic can be represented as [river:#234].

Conceptual relations connect the concept nodes. From the caseframes point of view they corresponds to the cases.

A *type hierarchy* is defined over the conceptual types. Such a hierarchy is a partial order relation defined over the types. In the following, such relations will be represented by "<=" (less general than). Given that s, t are two types, if s <= t then s is a subtype of t while t is a supertype of s. Two operators, called minimal common supertype (mcs) and maximal common subtype (MCS), are defined over the types. Given two types t1 and t2, if w = mcs(t1,t2), then t1 <= w, t2 <= w and there is not a Type w1 different from w such that w1 <= w, t1 <= w1, t2 <= w1.

4.3.4 The use of Conceptual Graphs

Conceptual graphs are used for two purposes: internal representation of the utterance meaning and semantic representation of the relevant concepts of the domain that can be connotated by the words in the dictionary.

The internal representation of the utterance meaning, that has to be constructed in order to be able to extract the data required by the utterance, is obtained starting from the semantic representation of the words and of the domain concepts and using the syntactic knowledge that is required to correctly connect them. Let us consider the conceptual graphs that can be used to parse a sentence like "Dimmi le regioni bagnate dal Tevere" ("Tell me the regions washed by the Tevere"). The graph for *bagnare* (to wash) is

```
  ''[BAGNARE]-
(agnt) -> [FIUME]
(obj) ->  [REGIONE].''
```

This graph means that a river (*fiume*) can wash (*bagnare*) a region (*regione*). Obviously this is not the only meaning for *bagnare*, but the meanings that are not relevant inside the chosen domain are not taken into account. In this case there are other possible meanings for *bagnare* that are relevant for the domain: for instance a sea (*mare*) can wash a region or a province. That means that other conceptual graphs involving *bagnare* have to be defined to cover utterances like "Quali regioni sono bagnate dal mare Tirreno?" ("Which regions are washed by the Tirreno sea?"). To deal with the above mentioned meaning of *bagnare*, four conceptual graphs can be defined. Such graphs can be shortly represented in the following way:

```
    ''[BAGNARE]-
(agnt) -> [FIUME+MARE]
(obj) ->  [REGIONE+PROVINCIA].''
```

Here the symbol "+", when encountered by the conceptual graph compiler, whose goal is to automatically generate the KSs, causes the generation of the types *fiume+mare* and *regione+provincia* together with the related hierarchical relations:

```
''FIUME+MARE''          >  ''FIUME''
''FIUME+MARE''          >  ''MARE''
''REGIONE+PROVINCIA''   >  ''REGIONE''
''REGIONE+PROVINCIA''   >  ''PROVINCIA''
```

Such implicit hierarchical relations are added to the type hierarchy defined by the person in charge of defining the system knowledge bases. The algorithms deal directly with implicit types (lists of user defined types). An alternative way is to define a *minimal common supertype* for *fiume* (river) and *mare* (sea) as a "washing entity".and a similar procedure for *regione* (region) and *provincia* (province).

Conceptual graphs are also be used to represent the surface semantics of utterances. With the term *surface* semantics we intend to point out that only the superficial semantic structure of the utterance is represented. For certain domains such a representation is sufficient to perform the activities expressed by the utterance, but for others a mapping activity is required to map the superficial structures into deep semantic structures that do not depend on the structure of the utterance.

4.3.5 Representation of the Utterance Meaning

As anticipated above, the utterance meaning is represented by making use of conceptual graphs; more precisely, the utterance meaning results from the join of the conceptual graphs that are associated with the meaningful words in the utterance. The meaning representation for the sentence:

Quali province della Campania confinano con le regioni bagnate dal Tevere?

(Which provinces of Campania border on the regions washed by the Tevere?)

should be:

```
''[CONFINARE]-
        (agnt) -> [PROVINCIA:?x]-
                      (part-of) -> [REGIONE:Campania]
        (with) -> [REGIONE]
                      (obj) <- [BAGNARE]
                                    (agnt) -> [FIUME:Tevere].''
```

To be more precise this conceptual graph is called *abstraction*, due to the presence of a parameter (?x), called a formal parameter; in fact all conceptual graphs containing parameters are called abstractions. Abstractions are equivalent to Lambda-expressions. The denotation of an abstraction containing a parameter is the set of all the constants that when substituted for the formal parameter make the Lambda-expression (corresponding to the abstraction) true. In the example the denotation of the abstraction is exactly what we are looking for, i.e. the provinces that:

- are part of a region having name Campania,
- border on a certain region R1,
- and region R1 is washed by a river having the name Tevere.

Such a representation can be transformed into a set of conjunctive clauses that can be used to access the data in order to provide the answer.

From the practical point of view, it is sometimes necessary to postprocess the representation during the answer generation process in order to give a natual language answer tailored as far as possible to the utterance structure, and that could also require keeping track of some morphological features of the words in the utterance.

4.4 The Compiler of Conceptual Graphs and Dependency Rules

4.4.1 Introduction

This section discusses the integration of conceptual graphs with dependency rules. This discussion illustrates also the mapping information that has to be provided to allow such integration. The compilation process that generates suitable structures starting from these representations is outlined.

4.4.2 The Use of Dependency Rules

Dependency grammars have been selected as a formalism for representing syntactic knowledge for the following two main reasons:

- Dependency rules allow an easy integration of syntactic knowledge with caseframes thanks to the similar notion of governor for the dependency rules and of header for the caseframes.
- Each dependency rule requires the presence of a word with the governor role in order to be activated. Consequently all the nodes of the resulting parsing tree correspond directly to a word. That allows the creation of expectations at the highest levels in the parsing tree and that is a basic requirement of our parsing strategy.

Let us consider as an example for the whole section the sentence

Quali province della Campania confinano con le regioni bagnate dal Tevere?

(Which provinces of Campania border on the regions washed by the Tevere?)

The dependency tree that corresponds to the example is depicted in Fig. 4.6. The corresponding dependency rules used to produce this dependency tree are:

```
rs1)  verb     =  noun        *  noun
rs2)  noun     =  adj         *  pr_noun
rs3)  pr_noun  =  prep        *
rs4)  noun     =  prep   art  *  verb
rs5)  verb     =              *  pr_noun
```

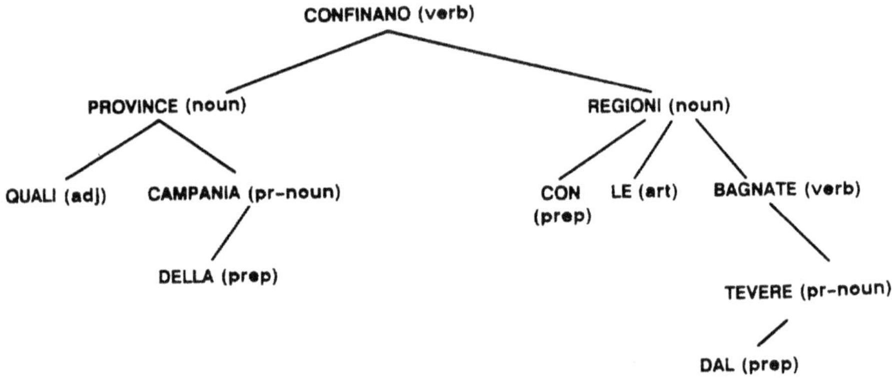

Figure 4.6: Dependency tree for the example sentence

where the symbol * represents the governor position. The associated rules for morphological agreement check are not reported for simplicity; note that rule rs4) requires a verb characterized by features TENSE = PAST, MODE = PARTICIPLE and others.

4.4.3 Integrating Conceptual Graphs and Dependency Rules - the Mapping Knowledge

The basic idea is to generate off-line, starting from dependency rules and conceptual graphs, structures (called briefly knowledge sources, KSs) suitable to allow an efficient (for speech) parsing strategy. Particularly, the probabilistic control of the search and the consequent integration of top-down and bottom-up activity is required.

Dependency rules are the starting point of the compiling activity. The question is: given a dependency rule and the whole set of conceptual graphs representing the domain knowledge, is it possible to create one or more KSs that take into account all the possible interactions of the rule with semantic knowledge?

A basic point is that the partition of knowledge between the KSs is based on locality, i.e. each partition is aimed to generate a certain class of constituents, not to contain a predefined kind of knowledge (like a partition for syntax and a partition for semantics). Then each KS must combine the time adjacency knowledge, the syntactic and morphological knowledge, and finally the semantic knowledge that is necessary to handle specific classes of sentence segments. Then, in simple words, given a dependency rule it is necessary to generate KSs able to deal (both syntactically and semantically) with those sentence segments that require that dependency rule.

A first simple example will be considered to clarify this idea. Let us consider the dependency rule:

```
rs1)    verb    =    noun    *    noun
```

(the morphologic agreement conditions are not reported here) and the conceptual graph, whose meaning is that a province can border on a region:

```
cg1)    ''[CONFINARE]-
                (agnt) -> [PROVINCIA]
                (loc)  -> [REGIONE].''
```

These two structures (rs1 and cg1) together, through the integration activity, could lead to a compositional structure of the kind:

CONFINARE = PROVINCIA <header> REGIONE

but for this purpose it is necessary to know that the agent ("agnt" of the action corresponds to the left dependent of syntactic rule rs1 and that the "loc" case corresponds to the right dependent of rs1.

Then there is the necessity of having additional knowledge, called "mapping" information, that allows the correct association of semantic types to the governors and dependents of the dependency rules.

For this purpose, each dependency rule is augmented with information about grammatical relations [8]. A grammatical relation is associated with each dependent Di, accounting for the grammatical relation existing between the governor G and the constituent having Di as a governor.

In the example the associated grammatical relations could be *subject* for the left dependent and *object* for the right dependent. In addition, grammatical relations are associated with the governor G too, accounting for the admissible grammatical relations that can involve (in the dependent condition) the constituents having G as a governor. In the example, governor G can be the governor of an utterance, i.e. it does not usually play the dependent role; this fact is expressed using the special virtual grammatical relation *sent* that could be imagined to depart from the header of a complete sentence. After this augmentation, dependency rule (rs1) looks like:

```
rs1)    verb       =        noun          *        noun
       (sent)           (1 subject)            (1 object)
```

The meaning of the two ones in front of the grammatical relations will be explained in a moment. Now the *mapping knowledge* associates one or more conceptual relations to each grammatical relation, together with their directions. In the case of the example the mapping knowledge is:

```
subject -->  agnt+
object -->  loc-
```

The augmentation of the dependency rule (rs1) together with this association is sometimes called *mapping rule*.

In the mapping the "+" or "-" sign at the end of a name represents the plausible direction of the conceptual relations. A "+" sign means that the conceptual relation is leaving the concept associated to the governor to enter the concept associated to the dependent while a "-" sign means that it has the opposite direction. So the governor of the dependency rule has to be associated with the header of the conceptual graph (cg1) (*confinare*) and not to one of the dependents. In the example all the relations must leave the node that corresponds to the governor.

Syntactic rule (rs1) is also characterized by the fact that the two dependents must both be part of the same conceptual graph. This fact results from two numerical labels "1" at the beginning of each list; they state that both dependents must be part of the same graph. That is not always the case, as we will see in a second example.

4.4.4 Combining Different Conceptual Graphs

Given a dependency rule it is necessary to examine all the conceptual graphs of the domain in order to consider all those that can satisfy the requirements expressed by the mapping rule associated with that dependency rule.

For efficiency reasons it is not desirable to have a KS for each (set of) conceptual graphs that can correspond to the selected dependency rule. So many (set of) conceptual graphs can be grouped together to generate a single KS, and consequently a single compositional structure. This result is obtained by making use of explicit supertypes or of implicit supertypes (through the use of the "+" operator). Consider for example the case of two conceptual graphs:

```
''[PROVINCIA+FIUME+LAGO]-
        (part-of) -> [REGIONE].''

''[ISOLA]-
        (part-of) -> [MARE].''
```

They state that a province is part of a region and that an island is a part of a sea. Consider a syntactic rule and a mapping rule like:

```
rs2)               noun        =   art       *        noun
   (subject, object)                       (attr)

                      attr   -->  part-of+
```

(This rule deals with sentence segments like " ... the lakes of the regions [that] ...".. The resulting compositional structure could be:

ISOLA+PROVINCIA+FIUME+LAGO = J <header> MARE+REGIONE

This compositional structure does not provide sufficient constraints to semantic analysis: in fact it is not possible for a province to be part of a sea (at least not directly in questions). To overcome this kind of problem each KS is augmented with the whole list of conceptual graphs that took place in the generation of the compositional structure. This list of graphs is internally represented through suitable structures that are used at run time during the propagation of semantic contraints.

4.4.5 A More Complete Example

It is necessary to explain that the governor does not need to correspond to the header of conceptual graphs; in the case of the previous example the header does correspond to the governor but this is not always the case; this new example describes this situation. Now let us consider the syntactic rule:

```
rs2)    noun   =  prep art   *   verb
```

characterized by the following restrictions on morphological features (now they are essential to explain the example):

```
NOUN:     gen = ?x;  num = ?y;
PREP:     type = simple;
ART:      gen = ?x;  num = ?y
VERB:     mode = partic.;  tense = past;  gen = ?x;  num = ?y;
```

and the conceptual graph saying that rivers wash regions:

```
''[BAGNARE]-
     (agnt) -> [FIUME]
     (obj)  -> [REGIONE].''
```

The above mentioned dependency rule takes into account sentence segments like: "... con le regioni bagnate dal Tevere" ("... with the regions washed by Tevere".. Now these two structures together should lead, through integration, to a compositional structure like:

REGIONE = <header> BAGNARE

and the mapping rule could be:

```
    noun      =     prep    art    *      verb
  (object)                             (agnt-compl)

         agnt-compl   -->  obj-
         object       -->  loc
```

This rule states that in the case of a syntactic structure of a noun modified by a relative clause, the relative clause could be the object ("obj") as in the example. Now the relation is exiting from the concept corresponding to the dependent and coming into the concept corresponding to the governor.
The restrictions imposed on the verb features by the dependency rule are essential. In fact, in the case of a relative clause having finite mode (like indicative mode and present tense), the conceptual relation involved could be "agnt", as in sentence segments like: "... con i fiumi che bagnano ..." ("... with the rivers that wash ...").

The relation "loc" that results on the left side of the mapping rule represents one of the possible semantic relations that the sentence segment treated by the rule can have with respect to a higher level constituent. The relations on the left side are simply used during the parsing process to check that only admissible connections are performed. When the sentence segment "con le regioni bagnate dal Tevere" is generated only the semantic relations on the left of the mapping rule ("loc" in our case) are permitted to be used to connect this segment to another possible supersegment, like our sentence.

4.5 Parsing - Conceptual Level

4.5.1 Introduction

The input structure of the understanding level is a lattice of word hypotheses, i.e. a set of hypotheses about words characterized by the following information:

1. Hypothesized word,
2. Time interval of the word instance,
3. Score.

Two are the objectives of the understanding process: to complete the recognition activity and to understand the meaning of the utterance.

From one side the understanding level must complete the recognition process. It has to detect a sequence of word hypotheses in the lattice such that the following requirements are satisfied:

1. The word sequence must be both syntactically and semantically correct, i.e., more precisely, it must be compatible with the system knowledge about the language and about the application domain.
2. The word hypotheses of the sequence should cover exactly the time interval of the uttered sentence. Ideally, they should not overlap and no gaps among them should exist. From a practical point of view, due to uncertainties in word spotting, a certain number of overlaps and gaps between adjacent word hypotheses normally exist and must be considered. Thresholds can be used to define the accepted level of imprecision.
3. Among all the sequences of word hypotheses that satisfy points 1 and 2, the sequence with the best word hypotheses (the word hypotheses with the best scores) should be preferred. For this purpose a score can be assigned to a solution starting from the scores of the word hypotheses involved.

From the other side the system has to understand the meaning (in the fixed domain) of such word sequence, that is, to generate a formal representation of the meaning itself. From that representation the system will be able to perform the activity requested by the utterance (in our case the extraction of the desired information from a data base).

Both objectives imply the use of syntactic and semantic knowledge. For the first objective (to complete recognition), that knowledge is used as a source of constraints to solve the uncertainties that still remain after the recognition level has used phonetic and acoustic knowledge to perform the first part of the recognition activity. For the second objective, syntactic and semantic knowledge is used to generate the meaning of a word sequence in the same way used by natural language interfaces. Within natural language interfaces syntactic and semantic knowledge has to be expressed in a declarative way; in addition the process that uses that knowledge is usually non-deterministic (i.e. a search activity is required).

Our aim is to perform these two activities in an integrated way: only one declarative representation of syntactic and semantic knowledge is given to the system and it is used for both objectives; in addition the two activities are completely merged.

4.5.2 Lexical Component and Model Component

The "solution", that is, the sequence of word hypotheses that has to be selected as the correct one, must satisfy constraints of many different kinds. These constraints express two different types of knowledge. First, there is the knowledge that refers to the word hypotheses by themselves: time interval and score. This kind of knowledge is used to verify the conditions outlined by previous points 2 and 3 (i.e. to have a temporally acceptable solution with the best overall score). Second, there is the knowledge about the language and the application domain. This knowledge is used to satisfy the conditions requested by point 1 (i.e. to guarantee the syntactic and semantic correctness of the solution).

We refer to these two kinds of knowledge as the lexical component and the model component respectively. The two components must cooperate, and indeed the system uses them in a joint way. Notwithstanding this, we believe that it is reasonable to keep a clear distinction between them, at least for descriptive purposes. In fact, not only different techniques are employed in order to exploit the different constraints pertaining to the two components, but they have different goals. The task of the lexical component would be to put together word hypotheses into "chunks" so that time constraints are satisfied, and its ultimate goal is to reach a solution having a good score without having to combine too many word hypotheses. It is not per se concerned with the problem of checking the syntactic and semantic correctness of the chunks it is getting through. Conversely, the goal of the model component is to put together word hypotheses according to a model of the language and of the application domain, until a syntactically and semantically acceptable solution is found.

The correct sequence of word hypotheses has to satisfy the constraints pertaining to both components, and they can play different roles in different approaches to the problem. In the next section the importance of using scores to guide the search is outlined.

4.5.3 Importance of a Score Guided Search

When the lattice generated by the recognition level contains many word hypotheses (of the order of a hundred times the number of actually uttered words) and when the model of the language and application domain are not limited to toy examples, then the analysis techniques have to use the scores associated with the word hypotheses to guide the search. In other words it is not possible to have an analysis mechanism that tries first to recognize all of the possible sequences of word hypotheses (i.e. that satisfy all the constraints) and that afterwards considers their scores for the selection of the best sequence. In fact the amount of search that would be necessary to detect all of the possible sequences of word hypotheses is extremely high, since the non-determinism of syntactic/semantic knowledge is added to input uncertainty: the presence of many word hypotheses instead of a few words.

The search has to be directed towards the best sequences from the very beginning, and only a small part of the implicit search space should be examined. The search should then start from the best word hypotheses, should be directed by the scores of the word hypothesis sequences and should stop when an acceptable solution has been detected (i.e the probability of finding better solutions by continuing the analysis is sufficiently low). Starting from this assumption, many different approaches to the problem are still possible.

4.5.4 Search from the Point of View of the Lexical Component

From the point of view of the lexical component, the main objective of the analysis is to start from the best word hypotheses and to expand them with other word hypotheses until an agglomerate covering exactly the whole utterance time interval is generated. New agglomerates can be obtained either by adding new word hypotheses to old agglomerates or by joining together old agglomerates. As a special case an agglomerate can be a sequence of adjacent word hypotheses (the term adjacent is used taking the gaps and overlapping thresholds into account) but in general word hypotheses do not need to be adjacent.

Each agglomerate is characterized by a quality factor. A quality factor is assigned starting from the scores and time intervals of the word hypotheses making up the agglomerate. The quality factor constitutes an evaluation of the probability that the set of word hypotheses is actually present in the uttered sentence. Details of the ways this quality factor can be obtained are not included in this section.

Control strategy of the lexical component

The basic control cycle consists in selecting the best agglomerate (the term "agglomerate" includes a single word hypothesis as a special case) and processing it to generate new agglomerates. Only the agglomerate with the best quality factor is selected at every control cycle. The selected agglomerate can generate new agglomerates either by having a new word hypothesis added to it or by being joined with other agglomerates. A quality factor is then assigned to them and the control cycle is repeated.

What is really done starting from the selected agglomerate depends on the search performed at the model component. In fact the selected agglomerate can only be expanded in a way consistent with the syntactic and semantic knowledge of the model level. Note that at the beginning there are only word hypotheses: a special case of agglomerates. Section 4.5.10 will illustrate the control strategy of SUSY.

4.5.5 Relations with the Model Component

If the lexical component takes control of the parsing process in this way, there would be the problem of validating the agglomerates from the point of view of the model component: if an agglomerate does not satisfy syntactic and semantic constraints it has to be eliminated (or still better, it should not be constructed).

A selected Island is expanded in all the possible ways to find adjacent (on the left or on the right) word hypotheses and the model component is used to validate them. In other words, the phrase hypotheses generation follows left or right expansion and there is no attempt to follow the model: time adjacency is guiding the search, not grammar rules.

The verification of an agglomerate of word hypotheses requires inferences at the model level, and efficiency considerations suggest the use of model knowledge to decide which agglomerates are worthwhile trying to generate (use of syntactic/semantic predictions). In SUSY the model component does not have a slave role, i.e. it is not used just to verify the correctness of the agglomerates that are going to be generated on grounds of time adjacency. Instead it has a key role in deciding the direction of expansion of an agglomerate and can override the time adjacency criterion, leading to the presence of "holes" within an agglomerate (i.e. an agglomerate is not necessarily a sequence of words, so agglomerates are not like Islands in [42]).

To be consistent as far as possible with the lexical component point of view it is necessary for the model component to be able to generate internal structures (complete and incomplete constituents) that can be associated with agglomerates (i.e. set of word hypotheses). Such internal structures, completely validated by both the components, have a quality factor (the quality factor of the agglomerate) and a precise syntactic/semantic characterization. They are called phrase hypotheses in the following, until a more precise definition is given in Sect. 4.5.8. The quality factor of a phrase hypothesis (as well as the score of a word hypothesis) will be taken into account by a control strategy derived from a refinement of the "naive" control strategy of the lexical component.

4.5.6 Relations with some Former Systems

Our main effort has been to adapt the basic ideas of chart parsing to the peculiarities of continuous speech: mainly the large search space involved. The system that has given us some good ideas about an effective and efficient way of controlling the search was HWIM. The final system is very different from it, expecially when the representation and use of syntactic knowledge is concerned, but nevertheless it does share a few commonalities.

An idea shared with HWIM and other systems is that constraints pertaining to the two components have to be used in a joint way during the analysis. Another common point is the conceptual idea that a solution can be obtained by formulating hypotheses, each of them validated by the presence of a set of word hypotheses satisfying all the possible constraints (to be more precise, that there are no evidence of constraints not satisfied). Hypotheses can be expanded, two hypotheses can be joined together, etc., until a hypothesis is based on word hypotheses covering the time interval of the whole utterance.

The HWIM system considers central the role of the lexical component to formulate hypotheses, while the model component is mainly used in a passive way to check the correctness of these hypotheses. In SUSY, phrase hypotheses are formulated on the grounds of linguistic knowledge; the time intervals of the word hypotheses are still checked every time a new word hypothesis becomes part of a phrase hypothesis but time intervals do not guide the search. In this respect, our approach shows some similarities with Hearsay's [10], though, unlike Hearsay, we stress the importance of a formal control as a means to cut down the search and to avoid the generation of incorrect solutions.

Another significant difference between our system and HWIM concerns the structure of the phrase hypotheses. HWIM always expands a phrase hypothesis either on the left or on the right, then the result is that hypotheses are "islands", i.e. sequences of adjacent words. In SUSY the word hypotheses supporting the phrase hypotheses are not necessarily adjacent.

4.5.7 The Model Component

We have seen that syntactic and semantic knowledge is represented, after the compilation phase, by knowledge sources (KSs). Now let us see the knowledge sources at the lowest level of detail, at what is necessary to describe the control strategy in a precise way. Each KS is characterized by a number of *slots*: one of them is called the *header* and has to be filled by a suitable word while the others are called *filler slots* and have to be filled either by words or by phrase hypotheses provided by other KSs. A slot of a KS is called *terminal* if it has to be filled by a word, while it is called *non-terminal* if it has to be filled by the results provided by other KSs. Then the header slot is terminal while the filler slots can be either terminal or non-terminal. The parsing process consists in the generation of phrase hypotheses that result from the activity of filling the slots of the KSs.

A simplified view: the problem solving paradigm

Just for generality and for the sake of simplifying the control strategy description, we are going to describe the parsing process as a problem solving task: the problem P of filling completely a KS (i.e. of generating a phrase hypothesis by such KS) can be decomposed into the subproblem P* of filling the *header slot* and into the subproblems P1, ..., Pn of filling the *filler slots*, if any.

This fact can be written as: P := Ph, P1, ... , Pn.

Of course, for what we have seen, the subproblems are not at all independent one from the others: constraints of morphological, syntactic and semantic nature need to be propagated and controlled each time a subproblem has to be solved. In addition temporal constraints have also to be propagated and verified because header and filler slots must follow the pattern of adjacency of the KS.

Thanks to this generalization, the abstract description of the control strategy we are going to describe can be applied not only to spech but also to other signal understanding fields (like sonar, vision, etc.). As the constraint propagation and control techniques are specific to the application field (speech, vision, ...) we will describe the control strategy without taking them into account here: in other words the description is simplified as far as possible to make it clearer.

The knowledge source partition

From now on the problem solving structure of a KS is abstractly represented in the form:

1) C := C1, C2, ... , Cn.

where the meaning is: in order to classify a certain word sequence as being of class C (the class of constituents detected by the KS) it is necessary to classify n word sequences as being of class C1, C2, ... , Cn. Some of such sequences will be constituted by just one word (header or terminal fillers) while others will have to be classified by other KSs (which have their own problem-solving structure).

Now it is necessary to spend a few words about what the Ci classes are. When a KS wants to fill one of its non-terminal filler slots, it is necessary to activate some other KSs (or even itself, recursively). Which other KSs have to be involved in this task? The answer to this question requires a static (off-line) partition of the KSs into classes. Such a partition takes into account both syntactic and semantic knowledge and is performed off-line by putting into the same class those KSs that contribute to detecting similar constituents. Coming back to the symbols in 1), they refer to two types of classes: one type, called a *terminal symbol* is used to indicate terminal problems, i.e. sets of words that can solve the subproblem, like the subproblem of filling the header slot. The other type, called *non-terminal symbol* corresponds to the classes of the previously mentioned KS partition. While terminal symbols can be matched directly against the word hypotheses, the non-terminal symbols require the KS activity.

As far as the second type is concerned, what it is really relevant here is that:

1. Given a class Ci it is possible to know which KSs are of that class, i.e. can classify word sequences as members of the class.

2. Given a class Ci it is possible to know which KSs have *non-terminal fillers slots* of that class, i.e. can use a word sequence of class Ci to fill one of their non-terminal filler slots.

Knowledge sources, facts and goals

This aspect corresponds more or less to the forward and backward control activity in a problem solving system.

We use the term *fact* to indicate an instance of a KS whose slots have been completely filled. Every fact has an associated symbol (class): in the simplified view the only way to obtain a fact of class C is to apply a KS of class C.

A KS is called *terminal* if it has only terminal slots: it relies only on terminal symbols. A terminal KS can be applied considering only the word hypotheses. In our case a KS contains at least one terminal symbol: the symbol C* that represents the header slot. *Non-terminal* KSs rely at least on a non terminal symbol; C* is the most important concept while the others act as modifiers.

If KSs are only applied forward, then only facts are generated during the search activity. But if they also run backwards, then goals have to be dealt with too. In our approach the presence of a "best first" search that considers the scores of the word hypotheses requires the presence of both search strategies and a complete step-by-step integration among them. By now for *goal* of class C we mean simply the objective of finding out one (or more) facts of class C. A goal can then be represented by a complete description of the constraints that those facts have to satisfy. The next section will define the deduction instances (DIs) and their use as basic items for the control of the search activity. The definition of DIs is necessary in order to better understand the relations among goals and facts and their roles in a real best first search.

4.5.8 Deduction Instances

We have seen that the activity of finding a solution can be viewed as a search process. The part of the implicit search space that is being incrementally explicited could then be represented as an OR tree of nodes. These nodes are called *Deduction Instances* (DIs). Each node represents an intermediate step of a deductive process possibly leading to a solution. Each DI can be represented by an AND tree, called a *deduction tree*, whose nodes can be facts or subgoals. A deduction tree corresponds roughly to a parse tree in the case of speech; more precisely a parse tree corresponds to a DI which is a fact. In the case of a goal DI the parse tree has some parts missing.

In a classical goal-driven search the steps involved are decomposition of a goal into subgoals and verification of a terminal subgoal against the input data. In the case of a data-driven search, input data can be grouped together to form deduced facts, deduced facts are also be grouped together to form new deduced facts and so on until a solution is possibly reached.

Let us consider the case of a single deductive process that led to a solution. If a single strategy (data-driven or goal-driven) is used, then, starting from an initial state, the final state (solution) has been reached through a single path of states (DIs). In our approach the need of a best first search requires goal-driven steps to be mixed with data-driven steps. To allow such integration among data-driven and goal-driven activities, special operators (called merging operators) are able to join two different paths together. In other words two DIs can be merged to form a new DI.

Deduction instances are the basic items managed by the control strategy and conceptually constitute the Deduction Instances Data Base (DIDB). DIs can be grouped into two classes: *fact* DIs and *goal* DIs. Fact DIs represent facts of a certain class (according to the definition of Sect. 4.5.7) while goal DIs are characterized by at least one still unsolved subgoal. One of the subgoals of a goal DI has to be defined as a Current Subgoal (CS) when such a goal DI is inserted into the DIDB. A goal DI does not represent simply a subgoal (the current one) but a subgoal together with its context from which linguistic constraints are obtained through constraints propagation activity. In other words a DI represents the whole deductive process leading to the current situation, not only the description of a certain subgoal to be solved.

It is important to operate on DIs and not simply on facts and subgoals for the following reasons:

1. The best first search requires a single state to be characterized by a priority in order to decide, at each control cycle, what state has to be treated first (we are

not interested here in what operator has to be applied first). Such priority should not be based on heuristics but has to be the result of a domain-dependent function applied to the word hypotheses that have supported the deductive process until now. In other words each DI must correspond directly to an agglomerate (as introduced in Sect. 4.5.4) and its quality factor can be used to perform the best first search previously described.

2. The use of DIs makes simpler and more formalized the integrated search strategy: two DIs can be merged to generate a new DI.

3. Constraint propagation activities during the deduction makes every DI extremely specific: subgoals of the same class get different constraints when inserted in different contexts.

We have seen that constraints coming from both the lexical and model components are exploited during search. Then the only agglomerates that are generated are those containing the word hypotheses on which a fact DI relies on or those that currently support a goal DI.

The presence of a 1:1 relation between DIs and agglomerates allows the kind of control strategy previously outlined in Sect. 4.5.4 to be applied directly on the DIs: at each control cycle the best DI (the DI with the best quality factor, i.e. the DI that corresponds to the agglomerate with the best quality factor) is selected and the operators are applied on it to generate new DIs.

4.5.9 Activation: Scores and Quality Factors

Before illustrating the control strategy in a formal way it is necessary to better discuss some conceptual points about the creation of expectations. A problem that the control strategy has to deal with is the decision whether to continue to deal with a given DI or to decide to try to generate new DIs starting from a certain new word hypothesis.

As each DI is supported by a given agglomerate and has a given quality factor, a reasonable solution is to compare the quality factor of the best DI with the score of the best (remaining) word hypothesis. To do that, the quality factor of a DI and the score of a word hypothesis need to be comparable. In SUSY they are, as we use the density method of combining together the scores of the word hypotheses of the agglomerate supporting the DI.

Let us call S1, S2, ... , Sn the scores of the word hypotheses and t1, t2, ... , tn their time intervals. Then the quality factor of a DI supported by an agglomerate of such word hypotheses will be:

QF = (S1*t1 + ... + Sn*tn) / (t1 + ... + tn)

This formula shows that, if all the WHs had the same time interval, the quality factor would have been just the mean value of their scores: QF = (S1 + ... + Sn) / n.

Given the possibility of comparing scores and quality factors, our solution is to start a new deductive process when the score of the best remaining hypothesis in the lattice is better than the quality factor of the best DI produced so far. Some systems select just a certain number of word hypotheses as "seeds" to be used to create Islands; the critical aspect is that the optimal number cannot be calculated in advance as it depends on the single utterance. We experimented with this strategy initially, but experiments performed by changing the number of word hypotheses activated during the initial phase showed that this approach was not really efficient.

This intermixed procedure, as far as we know, is original and has the great advantage of solving a common problem: while the word hypotheses involved in the solution have, together, an acceptable quality factor (the quality factor of the solution), there could be just one or two uttered words that have a bad score, due, for instance, to mispronunciation, reduced length, enviromental noise, or to a more critical acoustic model for word(s). The reasons why this procedure, together with the other specific features of the control strategy, can solve the above mentioned problem will be clear later on: we try now just to give an informal justification. The word hypotheses in the lattice can be classified into two sets: those that are used to create expectations, i.e. to generate new deductive processes, and those that are not. By now it is sufficient to anticipate that, under certain conditions that will be only partially fulfilled, the first set is composed of those word hypotheses that have a score better than the solution quality factor, while those of the second set have a score worse than the solution quality factor.

The informal conclusion is that what is really good (word hypotheses in the first set) is used incrementally as "seeds" to generate new deductive processes, while the "garbage" is only used during the deductive processes to expand the DIs. The process of selecting a certain word hypothesis to start a new deductive process is called *activation*. The next subsection describes such phase.

The ACTIVATION operator

The first question is: what does it mean to begin a new deductive process starting from a word hypothesis and how is it possible to do that? We can think of an *activation* operator that, starting from the word hypothesis, can generate a set of DIs. Given our KSs, the activation of a word hypothesis means to trigger all the KSs that have a *header* slot that can be filled by such word hypothesis and then to generate for each such KS a new DI, where the header slot is filled with the word hypothesis (i.e. the DI is supported by the word hypothesis). What it is really important, in our opionion, for a continuous speech understanding system, is to allow such possibility: it is for this reason that a header based knowledge representation formalism has been selected.

Let us see now the activation phase using the problem solving paradigm. Every word hypothesis is characterized by a set of features. The *activation* operator, using the values of these features, can know what terminal symbols correspond to this word hypothesis. KSs with terminal slots corresponding to these terminal symbols are then activated and fact or goal DIs are generated. Each DI is characterized by having that terminal subgoal satisfied, i.e. the terminal slot filled. If the KS has only one terminal slot, then a fact DI is generated, otherwise a goal DI is generated.

Note that in our current implementation no pure goal DIs are allowed; that means that goal DIs have to be supported by at least one word hypothesis. This decision is due to two main reasons:

1. A pure goal DI is not supported by any word hypothesis, so it is not possible to assign quality factors to this kind of DIs, and no best first search can really be performed.

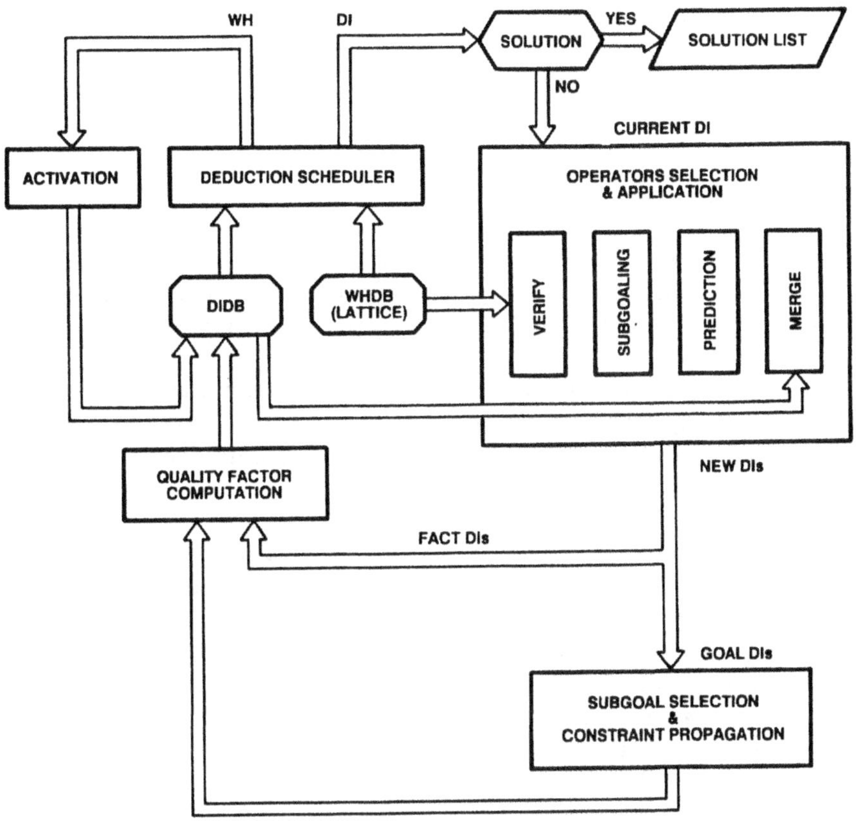

Figure 4.7: Architecture of the score-guided control strategy (DIDB = Deduction Instances Data Base; WHDB = Word Hypotheses Data Base).

2. KSs are characterized by the header presence and the problem of finding the header is a terminal problem. In our implementation, when a goal is decomposed into subgoals by applying a rule, then the terminal subgoal corresponding to the header has to be solved before other subgols of that DI can be treated.

4.5.10 Control Strategy

The control of the deductive activity is carried out by a *deduction scheduler* (DS, see Fig. 4.7). At each control cycle the deduction scheduler looks into the *deduction instances data base* (DIDB) and into the *word hypotheses data base* (WHDB).

At the beginning no DIs exist in the DIDB, so they are generated starting from the best word hypothesis through the application of the *activation* operator.

At each control cycle the quality factor of the best DI contained in the DIDB is compared with the score of the best word hypothesis contained in the WHDB (lattice in the case of speech). The best item between these two is selected.

If the deduction scheduler selects a word hypothesis, the activation phase takes place, generating new DIs that will be inserted into the DIDB. Otherwise, if a DI is selected, the deduction phase takes place. Five different activities are performed during the deduction phase. The activities are:

1. *Solution test*: If the DI constitutes an acceptable solution it is stored in a solution list. If the strategy is optimal the analysis can terminate. Otherwise the analysis can go on until the resources are consumed and then the best solution is selected. Section 4.5.11 better describes this point.

2. *Operators selection and application*: According to the type of the selected DI one or more operators are applied on it, creating a set of new DIs. A description of the possible operators and their application conditions is contained in Sect. 4.5.13.

3. *Subgoal selection*: A current subgoal (CS) is selected for each new goal DI. This is obtained through the application of a *subgoal selection function* that can use time adjacency considerations and heuristics derived from linguistic knowledge to perform its task.

4. *Constraint propagation*: when a new goal DI is generated adding a word hypothesis or a deduced fact to a previous goal DI, constraints are propagated inside the new DI starting from the word hypothesis or fact. Constraints are only propagated towards the subgoals that are likely to become the current subgoal. Constraint propagation is quite important as it makes the subproblems description more specific. The constraints involved are time constraints, reducing the range where word hypotheses can be located, syntactic and morphological constraints limiting the lexical features of the candidate word hypotheses and finally semantic constraints.

5. *Quality factors computation*: The function described in Sect. 4.5.9 combines the scores of the involved word hypotheses to obtain a *quality factor* (QF) for the DI. For the goal DIs only the word hypotheses that support it until now are considered. Remember that at least one word hypothesis is always present to support a goal DI.

4.5.11 Optimality and Efficiency

The analysis strategy is characterized by optimality if the first solution S (an agglomerate covering the whole utterance and being a plausible sentence in the given domain) selected by the scheduler has a quality factor such that no other solutions obtainable later on by continuing the analysis process can have a quality factor better than that of S.

If the search strategy is optimal the analysis process can stop as soon as a solution is selected by the scheduler. Otherwise the analysis process should continue until the resources are exhausted; at that point the solution with the better quality factor (if any) is selected. Optimality is of course a desirable characteristic, but optimality does not always mean efficiency: a non-optimal search strategy, making use of heuristics, can lead to a solution with a smaller amount of search activity than that required by an optimal search strategy. Nevertheless an at least near-optimal search strategy has to be pursued, otherwise there are excessive risks of expanding almost all the implicit search space or of accepting an incorrect solution.

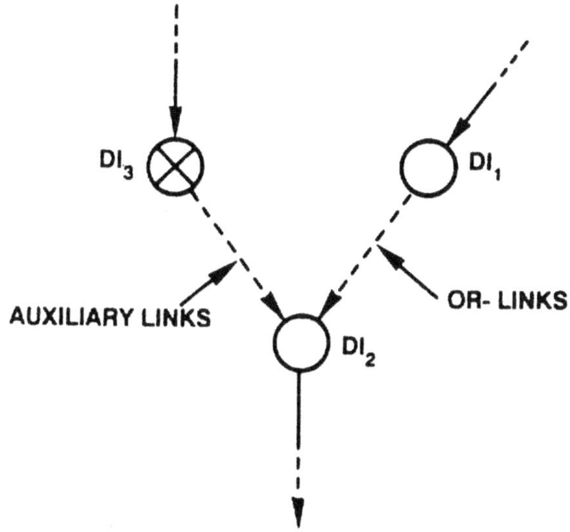

Figure 4.8: Deduction instances and search space. Deduction instances are states of a search spaces: in addition to the traditional OR links, auxiliary links (represented by dotted lines) have been introduced for the application of the *merge* operator to a goal DI1 that involves a fact DI3

4.5.12 The search space and the specialization relation

The entire search process can be represented by a forest of OR-trees whose nodes are the DIs and whose links (called OR-links) relate to the operators application. The *activation* operator generates new OR-trees; in fact new independent deductive processes can begin as a consequence of the activation of a word hypothesis. An OR-link connecting two DIs, DI1 and DI2, means that an operator applied on DI1 gave DI2 as a result.

A relevant aspect is that the various trees of the forest are not completely independent from one another. When the *merge* operator is applied to a certain DI1 to generate a new DI2, a DI3 of another OR-tree has to be considered, given the intuitive meaning of the word merge. A new kind of link (indicated with dotted lines in Fig. 4.8) is added to the traditional OR-links: the *merge* operator generates a new DI that is connected both by a traditional OR-link and by an *auxiliary link*. An auxiliary link going from DI3 to DI2 means that the generation of DI2 has been the result of the presence, in the DIDB, of DI3 but that DI3 has not been the immediate cause for the generation of DI2: it is only a precondition.

With the exclusion of the *activation* and *prediction* operators, when a new DI is generated by an operator application, one of the involved links (either OR-link or auxiliary link) can be seen as a *specialization relation* between the two DIs. We say that a specialization relation holds between DI1 and DI2 if DI2 is more specific than DI1 and they are connected by an OR-link or by an auxiliary link. This situation happens when DI2 has acquired new supporting word hypotheses or because one of its non terminal subgoals has been decomposed into subgoals in a certain way.

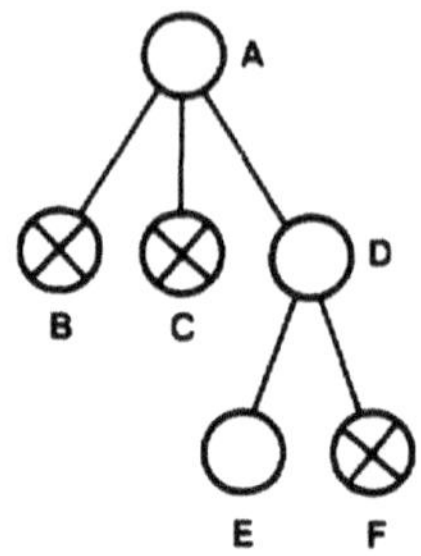

Figure 4.9: A DI that corresponds to the application of two KSs

The *prediction* operator is an exception: in fact, though it generates new DIs from a fact DI and hence can be represented by arcs in the OR search graph, it does not actually specialize the fact DI it is applied to - indeed, a fact cannot be specialized at all, but only inserted into a new context.

The specialization relation is treated with greater details in the following sections, where the various operators are explained.

4.5.13 Description of the Operators

The operators applied during the deduction phase are: *subgoaling*, *verify*, *prediction* and *merge*. When a certain DI is selected by the scheduler, one or more operators are applied to it. Which operators are applied depends on the selected DI. In the following we describe each operator, indicating the characteristics that the DI must have in order to apply the operator itself.

We recall here what was said in Sect. 4.5.7: a terminal subgoal can be directly matched against the input data (word hypotheses in our case) and a terminal KS is characterized only by terminal slots. By the way, we recall that in SUSY each KS is characterized by at least one terminal slot (the one corresponding to the header).

In the figures that will be used to indicate the operators applications, the following conventions will be applied:

1. DIs are represented by trees (AND trees) according to the problem-solving structure of the applied KS.

2. The daughter nodes of a given root node represent the symbols of the KS filler slots while the root itself represents the class of the KS. Figure 4.9 represents a DI example.

3. The crossed nodes of a DI represent *fact* nodes while non crossed ones represent *goal* nodes; if a goal node has no daughters than it has still not been expanded into subgoals (a KS has still not applied to it).

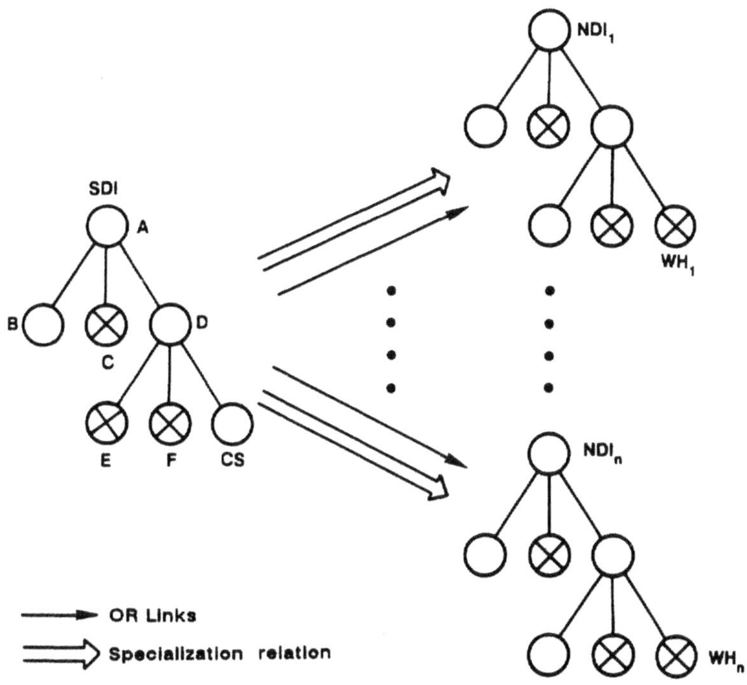

Figure 4.10: Application of the *verify* operator to a goal deduction instance SDI. n new goal deduction instances are generated.

The VERIFY operator

Type of operator: MONADIC.
Starting DI: GOAL DIs with current TERMINAL subgoal.
Generated DIs: GOAL or FACT DIs.

The *verify* operator is applied to a goal DI characterized by a terminal current subgoal CS (see Fig. 4.10). The *verify* operator checks if the current terminal subgoal CS of the selected deduction instance SDI can be solved by some of the word hypotheses in the lattice. In the case of our knowledge representation formalism (the KSs), the terminal subgoal is usually the problem of filling the *header* slot of the KS. To do so, it matches the subgoal description of CS (resulting from the propagation of constraints from the rest of the DI) against the word hypotheses in the lattice. Let us suppose WH1, ... , WHn to be word hypotheses able to satisfy subgoal CS. For each of them the *verify* operator generates a new DI to be inserted into the DIDB. Each of these new DIs represents a new step of the deductive process that led to the SDI. The new DIs can be either goals or facts. The *verify* operator is then working mainly at the lexical component: the new DIs differ from SDI for having a new word hypothesis to support them in addition to those that support SDI.

Thinking of the search space, these new DIs are directly connected by OR links to the starting SDI. A *specialization relation* exists between the SDI and each of the new DIs.

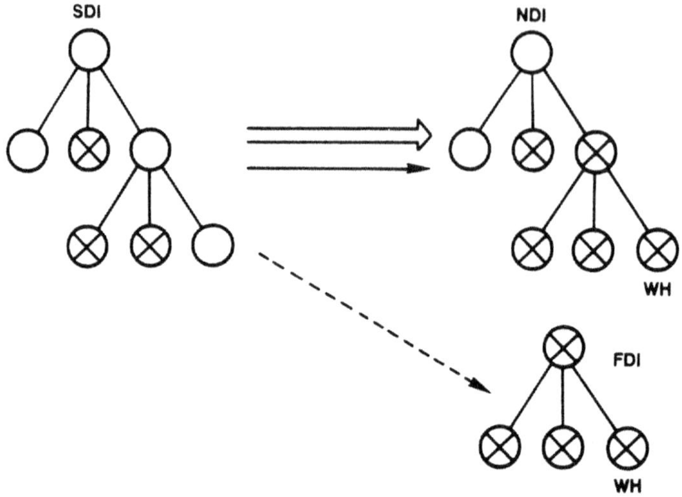

Figure 4.11: Application of the *verify* operator. A "by-product" fact deduction instance, FDI, is generated

In fact the new DIs have acquired new pieces of evidence at the word level; then a choice has been made, making them more specific than SDI.

Another activity is also performed by the *verify* operator: in addition to the above mentioned new DIs, other fact DIs can be generated (see Fig. 4.11). That happens when the solution of the current subgoal CS leads to the generation of a new fact by the triggered KS (i.e. when the KS can generate a new complete constituent). The new fact from one side has been generated in the *context* of SDI, so it must be part of NDI, but from the other side, it now becomes a "free" fact FDI, that could be used by other DIs or on which it can be applied the *prediction* operator. Note that these additional new fact DIs are not connected through OR-links to the starting SDI; they are connected to SDI by an auxiliary link, as the SDI was a precondition for the generation of these new DIs.

A fact of this kind, like FDI, is extracted from the context in which its generation took place: its score is computed taking into account only the word hypotheses making up the fact itself while the other word hypotheses supporting the starting SDI are not considered.

The SUBGOALING operator

Type of operator: MONADIC.
Starting DI: GOAL DIs with current NON-TERMINAL subgoal.
Generated DIs: GOAL DIs.

This operator is directly applied on the DI selected by the scheduler. This SDI must be a goal DI characterized by a non terminal current subgoal CS (see Fig. 4.12). The *subgoaling* operator triggers all the KSs that can decompose (sub)goal CS into subgoals (i.e. all the KSs that can hope to fill the filler slot CS); refer back to Sect. 4.5.7 if

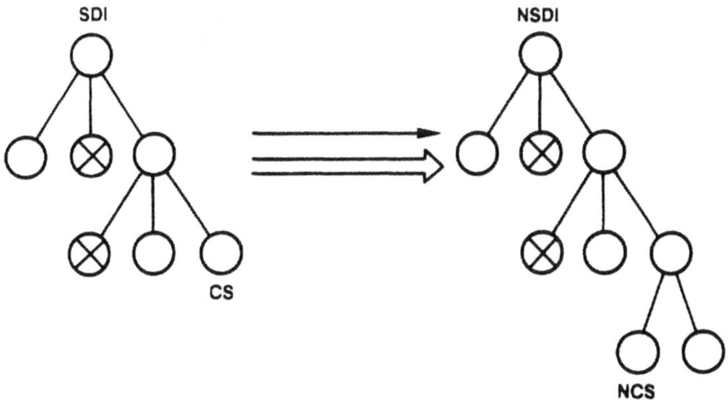

Figure 4.12: Application of the *subgoaling* operator

necessary. For each successful application of a KS a new NSDI is generated, characterized by subgoal CS decomposed into a set of subgoals that have (all) to be solved. In other words, node CS in the deduction tree of NSDI is not a leaf any more but has become an AND subtree corresponding to the problem-solving structure of the applied KS.

In regard to the search space, these new DIs are connected by OR links to the starting SDI. There is also a specialization relation between the starting DI and each of the new DIs. In fact the new DIs are more specific than the starting one: the current subgoal has been decomposed in a certain fixed way and then a decision at the model level has been taken.

The *subgoaling* operator works only at the model component: no word hypotheses are taken into account during its application. An interesting consequence is that the new DIs, generated when the *subgoaling* operator is applied on a certain SDI, all have the same quality factor as the SDI. Then they will be selected at once by the scheduler (they are as good as the starting DI).

On these new DIs the *subgoaling* operator could be applied again, but this does not happen in our implementation: it is not possibile to continue the search at the model component without any further support from the lexical component.

What really happens in practice is now described: we have seen that when a new DI is generated by applying the *subgoaling* operator on a certain SDI, the current (sub)goal CS of SDI is decomposed into a set of subgoals. Among these subgoals there is always at least one terminal subgoal NCS (the one related to the header slot) that is chosen as the current one by the subgoal selection function and is immediately solved by applying the *verify* operator on NSDI. Then the sequence *subgoaling + verify* is applied directly to SDI without having to insert into the DIDB the new DIs obtained by the application of the *subgoaling* operator alone. That is also important because often there are no word hypotheses that can satisfy that terminal subgoal (NCS). In such case no DIs are generated at all.

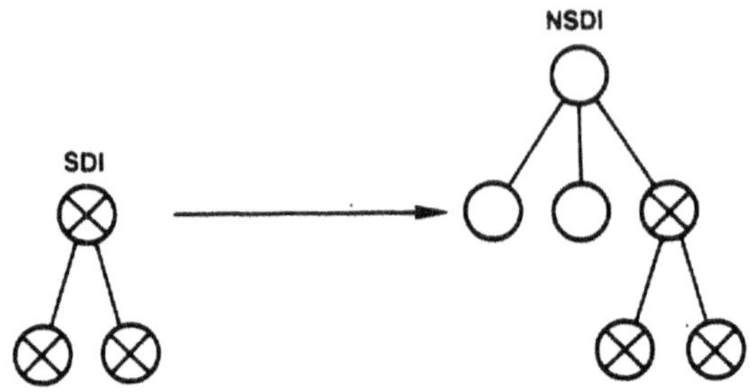

Figure 4.13: Application of the *prediction* operator. No specialization relation holds between SDI and the newly generated NSDI

The PREDICTION operator

Type of operator: MONADIC.
Starting DI: FACT DIs.
Generated DIs: GOAL DIs.

The *prediction* operator is applied on a selected fact deduction instance SDI. New goal DIs are predicted starting from SDI (see Fig. 4.13). The *prediction* operator triggers the KSs characterized by having SDI able to satisfy one of their filler slots. For each applicable KS a new DI is then generated. If the triggered KS had only one non-terminal filler slot, then a fact DI instead of a goal DI would be generated, but in our case only goal DIs are generated, as all the KS have at least the header slot that is a non-terminal one.

The *prediction* operator, as well as the *subgoaling* operator, works mainly on the model component. In fact the new DIs have the same supported word hypotheses as the starting SDI.

In regard to the search space, the new DIs are connected to the SDI by OR links. Note that there is no specialization relation between SDI and the new DIs. In fact we see each of the new DIs as a generalization step of the deductive process that led to SDI: a new root goal is generated and will be treated.

From another point of view we could see the *prediction* operator as a way of generating new deductive processes (i.e. new OR-trees in the search space) in a way similar to the *activation* operator. But we prefer the first view as it is more consistent with the whole theory and with the *merge* operator in particular.

The MERGE operator

Type of operator: DYADIC.
Starting DI: FACT or GOAL DIs (see below).
Generated DIs: GOAL or FACT DIs (see below).

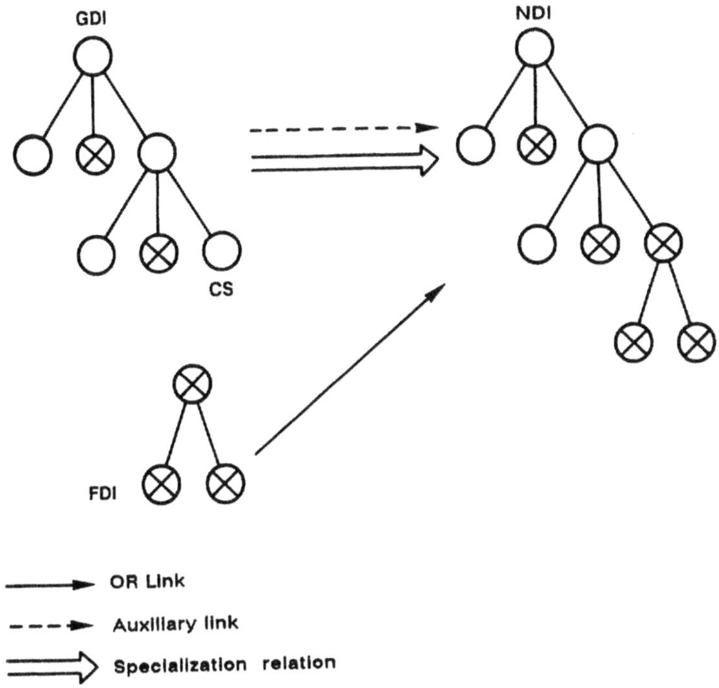

Figure 4.14: Application of the *merge* operator to a fact FDI (selected by the scheduler) and a goal GDI

The *merge* operator is a dyadic operator. Thus it is applied on two DIs; the first one is selected by the scheduler (the best DI) while the second one is extracted from the DIDB.

The merging process represents a way to join together two different paths of the search space. In fact, starting from two DIs that have evolved independently one from the other, a new DI is generated. This new DI is supported by the union of the word hypotheses that support the two DIs on which the operator has been applied; in addition the deduction trees of this new DI results from the union of the derivation trees of the two starting DIs.

From the point of view of the lexical component, the application of the *merge* operator can be seen as a way to group together two different agglomerates that have been previously generated. It is similar to the proposed islands collision mechanism in the case of HWIM.

A first point is which DIs are chosen to be merged with the DI selected by the scheduler. The set of candidate DIs depends on the characteristics of the starting DI and on some system parameters that control the amount of merging to be performed. A function that provides the set of DIs to (try to) be merged with the selected DI is defined in the system.

There are two possibilities: in the first one (see Fig. 4.14), the selected DI is a fact (FDI) that it is going to be merged with a goal (GDI); in the second case a selected goal (GDI) has to be merged with a fact (FDI). The two cases differ only for which item (FDI or GDI) has the best quality factor and then has been selected by the scheduler.

Let us suppose that the subgoal CS of GDI can be solved by FDI. The resulting NDI can be either a fact DI or a goal DI. A fact DI is generated if the deduction tree of the goal DI contains only one subgoal.

In regard to the search space, things are not as simple as before: two deductive paths are joined together and both are necessary for the solution. As we said before, NDI is connected to the selected DI (FDI or GDI) by an OR-link. In this way the search activity is still represented by OR-trees; in addition an auxiliary link connects NDI to the other DI. This one has not been selected by the scheduler but it has nevertheless been chosen as a candidate for being merged with the selected one.

There is still a specialization relation either between GDI and NDI or between FDI and NDI. In other words the specialization relation could be either the OR-link or the auxiliary link: the *merge* operator application can cause the specialization of either the selected DI or of another DI. In Fig. 4.14 the specialization relation corresponds to the OR-link.

Merging fact DIs with goal DIs allows a better integration among goal-driven and data-driven search activity and allows also the use of previous results of the search activity

4.6 Parsing - Memory Structures

4.6.1 Introduction

In the previous section we have described from a conceptual point of view the inferential activity of the lattice parser of SUSY. Deduction instances (DIs) have been introduced as the basic conceptual items managed by the parser. They are deduction process instances on which the operators are applied. At each control cycle the best DI is selected by the scheduler and the proper operators (*prediction, subgoaling, verify, merge*) are applied to it. An activation phase is performed when the scheduler selects a word hypothesis instead of a DI (i.e. when the best word hypothesis is better than the best DI).

One problem that has to be solved when using DIs and integrating forward and backward search activities is to reduce the amount of memory necessary to represent DIs and to properly structure the DIDB in order to simplify operators application (the *merge* operator, mainly). This section deals with these aspects and proposes a suitable structure for the DIDB: a *hypothesis network* making use of two main classes of links.

4.6.2 Representing DIs with Memory Structures: Some Problems

The deduction instances in the DIDB are represented by their *deduction tree* (DT). The most trivial way of implementing a DI would be of course to use an explicit DT for each of them. This solution, however, is not acceptable because of the large number of DTs that should be stored in the memory.

To reduce memory occupation it is necessary to make DIs share common parts, if any. For instance, when two or more deduction instances are generated starting from a certain DI, their memory representations have some common parts. The most natural type of representation that meets such requirements are AND-OR trees. Unfortunately, a problem

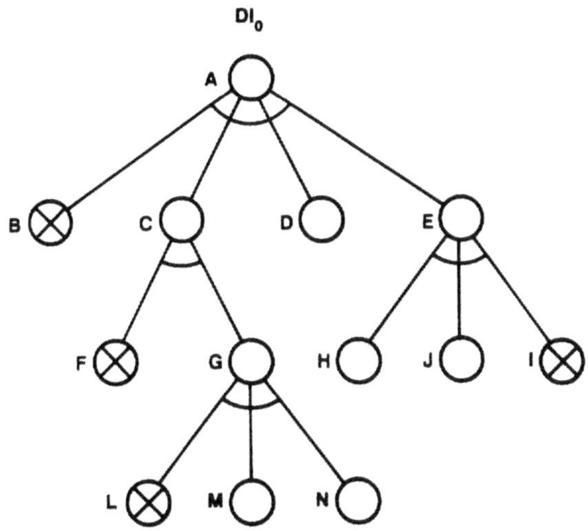

Figure 4.15: A deduction instance (DI0) represented by an AND tree

arises when constraint propagation is required: the use of AND-OR trees should assume the OR alternatives to be independent, but this is not true if constraint propagation has to be performed. An example will clarify this statement.

Let us consider the goal deduction instance DI0 depicted in Fig. 4.15. Let M be the current subgoal of DI0, and let us suppose that it is a terminal one and that it can be solved by two different word hypotheses WH1 and WH2. Two new DIs , DI1 and DI2, can thus be generated. The new situation, which makes use of AND-OR trees, is depicted in Fig. 4.16.

Now, suppose that N is selected as the current subgoal of DI1. Unfortunately, N belongs to both DIs: DI1 and DI2. Since they are distinct and endowed with different word hypotheses, the constraints that have to be transmitted to N are different in the two cases and it could happen that a fact DI of class N can satisfy, say, DI1 but not DI2, for it is compatible with the constraints derived from WH1 but not with those derived from WH2. This means that different constraints have to be propagated to subgoal N. This can be done by splitting subgoal N in two subgoals N1 and N2 and associating them with WH1 and WH2 respectively. The new situation is depicted in Fig. 4.17: two alternative subtrees with roots G1 and G2 have to be generated. This second kind of use of AND-OR trees to represent DIs is the one really used by SUSY.

The situation becomes even worse if we consider the other subgoals that could be chosen as current ones. Let us consider, for instance, subgoal J. If J were chosen as the current subgoal of the deduction instance DI1 of Fig. 4.16, different constraints would

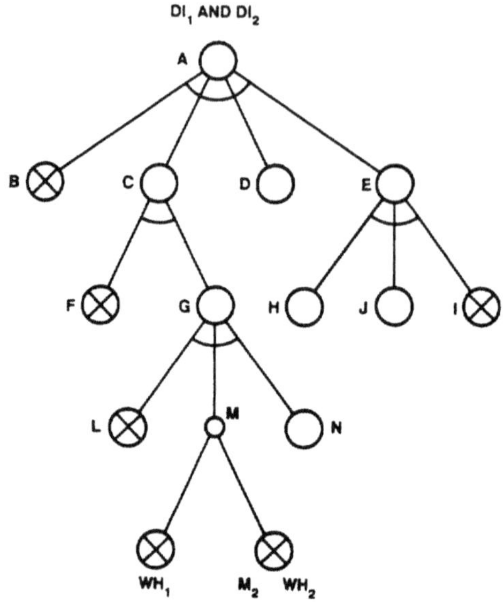

Figure 4.16: Two deduction instances DI1 and DI2 represented by using in a first way AND-OR trees: problems of constraint propagation

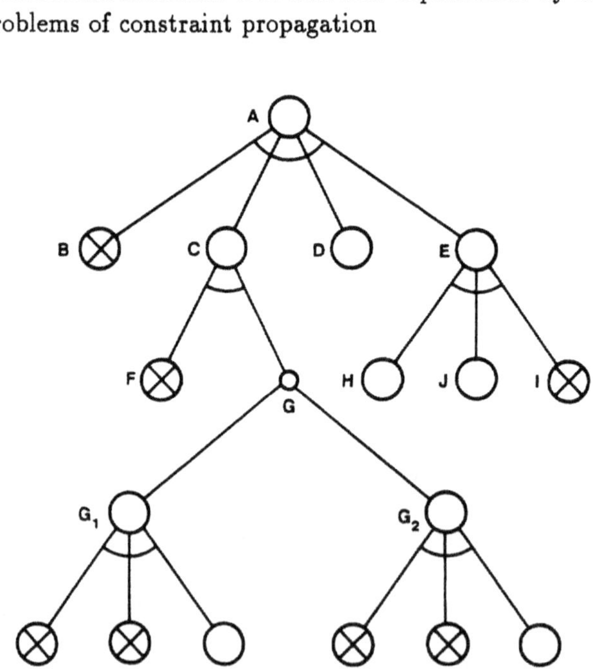

Figure 4.17: Two deduction instances DI1 and DI2 represented by an AND-OR tree duplicating subgoal G into G1 and G2

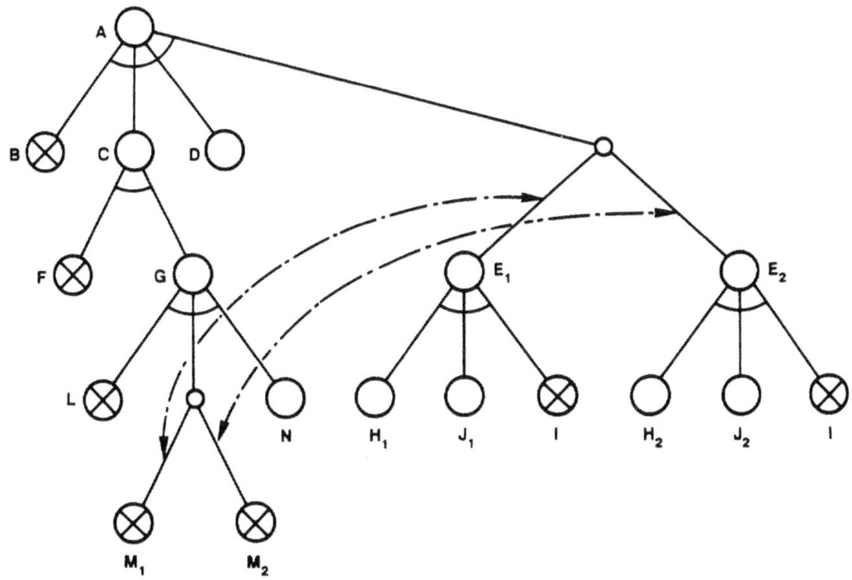

Figure 4.18: The AND-OR tree should represent four DIs but only two must be represented, as the OR alternatives are not independent. For this purpose two complete AND trees would be necessary

have to be transmitted to J, and two different subtrees having root E should be generated, corresponding to WH1 and WH2 respectively. This is shown in Fig. 4.18. Note that the OR alternatives pertaining to nodes M and E are not independent, as one would expect if one interprets this structure as a normal AND-OR tree: the alternatives only exist in couples (indicated by dotted lines in Fig. 4.18). As a matter of fact, the tree represents just two DIs instead of the four that there would be if the ORs were independent.

By applying the same reasoning to the other subgoals, it is easily seen that the whole DT has to be n-plicated into many DTs, each having its own characteristics and constraints. To keep the n DTs implicitly united in a single structure would be of no use.

In order to continue to take advantage of the use of AND-OR trees also in the case when constraint propagation has to be performed, we have studied a memory representation in which the nodes can be shared between DIs without the need to n-plicate the tree. We show that this is feasible if strong limitations are imposed on the possible ways a deductive process can go on. This, of course, results in limitations on the possible topologies of the DTs; the admissible DTs are called *canonical DTs*, and the associated DIs are called *canonical DIs*. A remarkable aspect of these limitations is that they maintain complete integration between forward and backward activities. As a matter of facts, our system never uses DIs other than canonical.

4.6.3 Canonical Deduction Instances

We define canonical DIs (CDIs) starting from the definition of canonical deduction trees (CDTs):

Definition 1:

- A DT is *homogeneous* if and only if it is a fact DT or a not yet decomposed (sub)goal. A *non-homogeneous* DT is one that is not homogeneous.

Definition 2:

- A DT is canonical if it is homogeneous
- A DT is canonical if
  - All the (sub)DTs connected to the root are canonical and
  - No more than one of them is non homogeneous.
- No other DTs are canonical.

Definition 3:

- A DI is canonical (CDI) if and only if it corresponds to a canonical DT.

For example, the DIs represented by the DTs depicted in Fig. 4.19 are Canonical; the DI of Fig. 4.15 was non canonical (its DT has two non homogeneous subtrees).

From the above definitions a consequence follows, that will be stated in the form of a theorem.

Proposition:

- *Consider a goal CDI that corresponds to a non homogeneous deduction tree: such a tree contains exactly one non homogeneous one-level AND subtree.*
  Proof - By recursion: consider the CDT associated with the CDI. If it is a one-level tree, the CDT itself is the subtree we are looking for. Otherwise, since the CDI is canonical, Definitions 1 and 2 ensure that its associated CDT has just one canonical non homogeneous subtree. Then the above discussion can be applied to this subtree, until a one-level canonical non homogeneous (sub)n-tree is found.

We call this one-level non homogeneous AND subtree the NHS. An example is shown in Fig. 4.20.

The proposition implies that there is a one-to-one correspondence between a goal CDI and its NHS. The NHS is called the representative of the CDI. In the case of a fact CDI

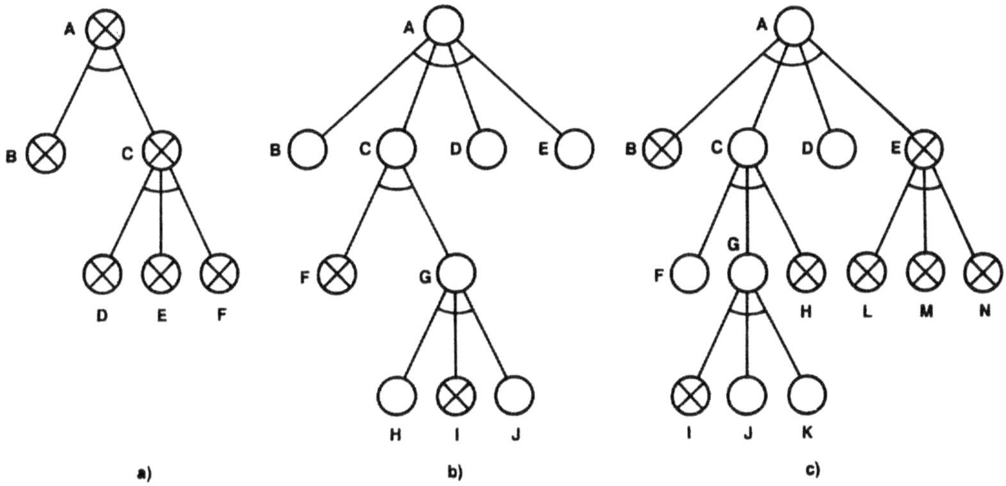

Figure 4.19: Example of canonical deduction instances, CDIs

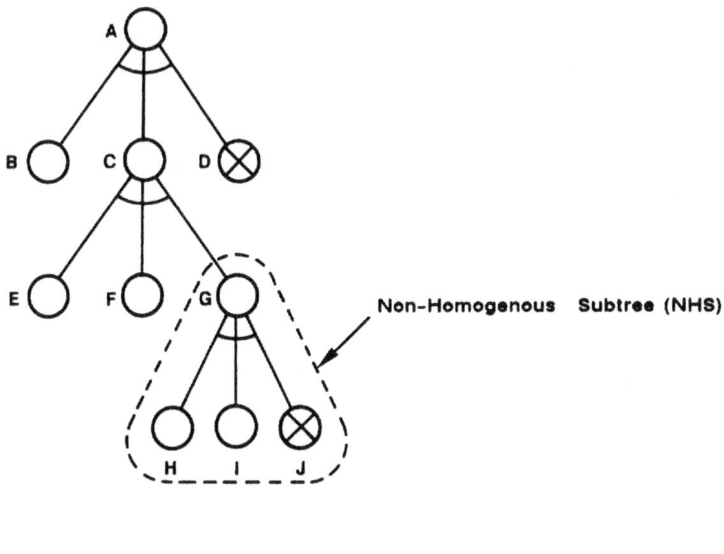

Figure 4.20: The non homogeneous Subtree (NHS) in a canonical deduction instance (CDI)

its representative is assumed to be the one-level AND subtree of its root. In this way each CDI is represented exactly by one one-level AND subtree.

We are now able to give the restriction on the way deductive processes can go on so that only canonical DIs are generated. The required restriction pertains only the subgoal selection function:

- The current subgoal of a goal CDI can be selected freely only among the leaves of its NHS.

Starting from a set of canonical DIs and applying the operators described in the previous sections, all the newly generated DIs have to be canonical.

As far as the *subgoaling* operator alone is considered, this strategy is similar to a kind of deph-first search: only when a subgoal is completely solved (a fact is generated) is it possible to treat subgoals that are ancestors or sisters of that subgoal. When the *merge* operator is involved, two canonical DIs can be merged together only if the resulting DI is still canonical.

The importance of CDIs lies in their one-to-one correspondence between DIs and their representatives one-level AND subtree. Indeed, the idea is to use somehow the above described representatives instead of the whole CDIs. More precisely, we want to characterize the representatives with all the information that is necessary to carry out an operator application to the CDI when it is selected by the scheduler. For example, temporal, semantic and syntactic constraints will be part of the necessary information, but a complete structural description of the DT will not. This objective can be practically realized by introducing a special structure called a *phrase hypothesis* (PH). How PHs are used to implement CDIs in a fashion that insures full compatibility with the use of AND-OR trees will be explained in the next section.

4.6.4 Phrase Hypotheses as Representatives of CDIs

A *phrase hypothesis* is a memory structure that implements a one-level AND subtree. A PH is, on the grounds of its definition, a non-terminal problem that has been decomposed into subproblems according to a certain problem-solving structure and where none, one or more subproblems have been solved. From the point of view of the KSs, a phrase hypothesis can be seen as an *instance* of a KS, having its slots completely or partially filled and whose aim is to fill all of them in order to complete itself. Clearly, if all the subgoals have been solved, the PH represents a fact rather than a goal. If a PH represents a fact it is said to be complete; otherwise it is said to be incomplete.

A PH is used to implement the NHS of a goal CDI, or the root one-level AND tree of a fact CDI, and stores all the information that is necessary for that CDI to be processed, when selected by the scheduler. In this way the PH acts as a "representative" of the whole CDI in the DIDB. As we have said, constraints are part of the information associated with PHs, and the canonicity of the CDI insures that the problems of constraint-conflict will not arise. Similarly, when a new CDI is generated, only one new PH is created, characterized by all the information necessary to process the new CDI and hence representing the whole new CDI in the DIDB.

We conclude this paragraph with an observation that will be resumed later on. Between PHs and KSs a n:1 relation exists, as PHs are *instances* of KSs. As there is also

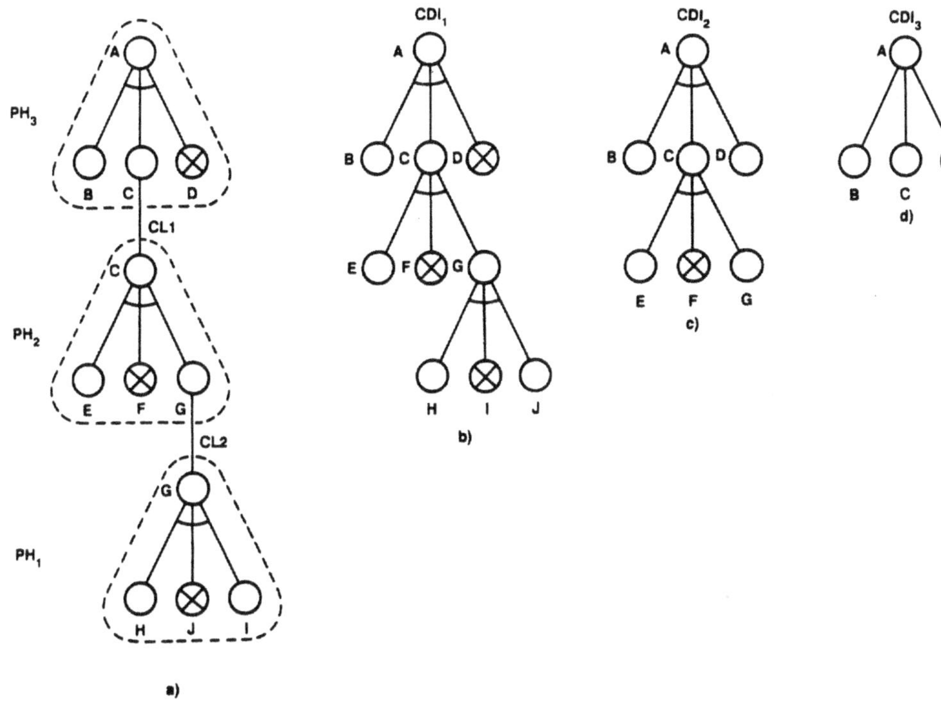

Figure 4.21: The PH-tree in a) represents the three canonical deduction instances (CDIs) represented in b), c), d)

a 1:1 correspondence between PHs and CDIs, it follows that all the fact and goal CDIs making up the DIDB can be partitioned into equivalence classes according to the KS they correspond to.

Phrase hypotheses and AND-OR trees

Phrase hypotheses can be connected by *composition links* (CLs) to form OR-trees of PHs, i.e. AND-OR trees, as PHs are AND trees. Let us consider, for example, the case of the simple tree given in Fig. 4.21a. Incidentally, this tree has no OR alternatives. Its structure strongly suggests a direct correspondence with the CDI reported in Fig. 4.21b (CDI1). However, only its node PH1 represents CDI1: the other PHs represent other CDIs. For example, PH2 represents the canonical deduction instance CDI2 given in Fig. 4.21c and possesses all the information and constraints pertaining to CDI2; similar considerations hold for PH3, which represents CDI3. Thus the PH-tree actually represents three CDIs, and not only one.

A misunderstanding must be avoided. When we said that PH1 represents CDI1 we did not mean that the other PHs are unrelated to CDI1, but only that PH1 can be used alone when an operator is applied to CDI1. The way the other PHs are interconnected by CLs gives information on the structural characteristics of the CDI. We will refer to these PHs as the component PHs of the CDI.

The interesting point is that PH-trees can have non-canonical structures, possibly with OR alternatives; since only canonical DIs are considered, no conflict will arise. An example is shown in Fig. 4.22. The PH-tree of Fig. 4.22a represents seven CDIs, three of which are reported in the figure.

Figure 4.22b shows the canonical AND tree CDT6 represented by PH6. Clearly, the subtrees deriving from the OR alternatives PH4 and PH5 are not present; thus, CL4 and CL5 were discarded. However, CL1 has been discarded too; otherwise, the resulting DT would not have been canonical. Informally, one could say that PH6 "sees" the PH-tree it is part of as lacking the PH-subtrees that would give rise to a non-canonical structure. More precisely, PH6 represents a CDI describing a deductive process in which choices have been taken at the OR alternatives and subgoal B was not yet decomposed. Note, by the way, that CL2 is not discarded in CDT6 because, PH2 being the representative of a fact, the resulting structure is still canonical.

A similar analysis may be done for the other two cases. The previous examples showed some general concepts that we now summarize:

1. PHs correspond directly to KSs.

2. A PH-tree of n PHs represents exactly n canonical DIs.

3. An incomplete PH that is part of a PH-tree represents the canonical deduction tree corresponding to the only AND tree, extracted from the OR PH-tree, that is canonical and has PH as its non homogeneous subtree (NHS): when extracting AND trees from PH-trees, the CLs are taken into consideration if and only if they do not compromise canonicity. CLs to facts do not compromise canonicity.

4. A complete PH has the same structure of an incomplete PH, the difference being that there are no subgoals but only facts. A complete PH always represents a fact CDI.

Phrase hypotheses and contexts

There is an alternative way of seeing the correspondence between phrase hypotheses and DIs. From this point of view there are two kinds of PHs: those whose root is free (i.e. it is not connected to other PHs) and those whose root is not free. The former PHs are said to be *free from context* while the latter are *within a context*.

If a PH is *within a context*, its quality factor takes into account also all the word hypotheses which are involved in such a context; in this way it is the representative of the whole context. Of course the definition of context is recursive, so in Fig. 4.23, PH1 is in the context of PH2 that is, in its turn, in the context of PH4: so PH1 is in the context of both PH2 and PH4.

The result is that the context of PH1 is the whole deduction tree and the word hypotheses that have to be considered to determine its quality factor are B,L,M,N,H,I (the solved terminal subgoals of the deduction tree). In such a way PH1 represents the *whole* DI. The constraint of canonicity means that the PHs constituting the context of PH1 (i.e. PH2 and PH4) must have the remaining subgoals (i.e those not on the context path: C and G) either already solved (like E,B and H) or still to be decomposed (like F and D).

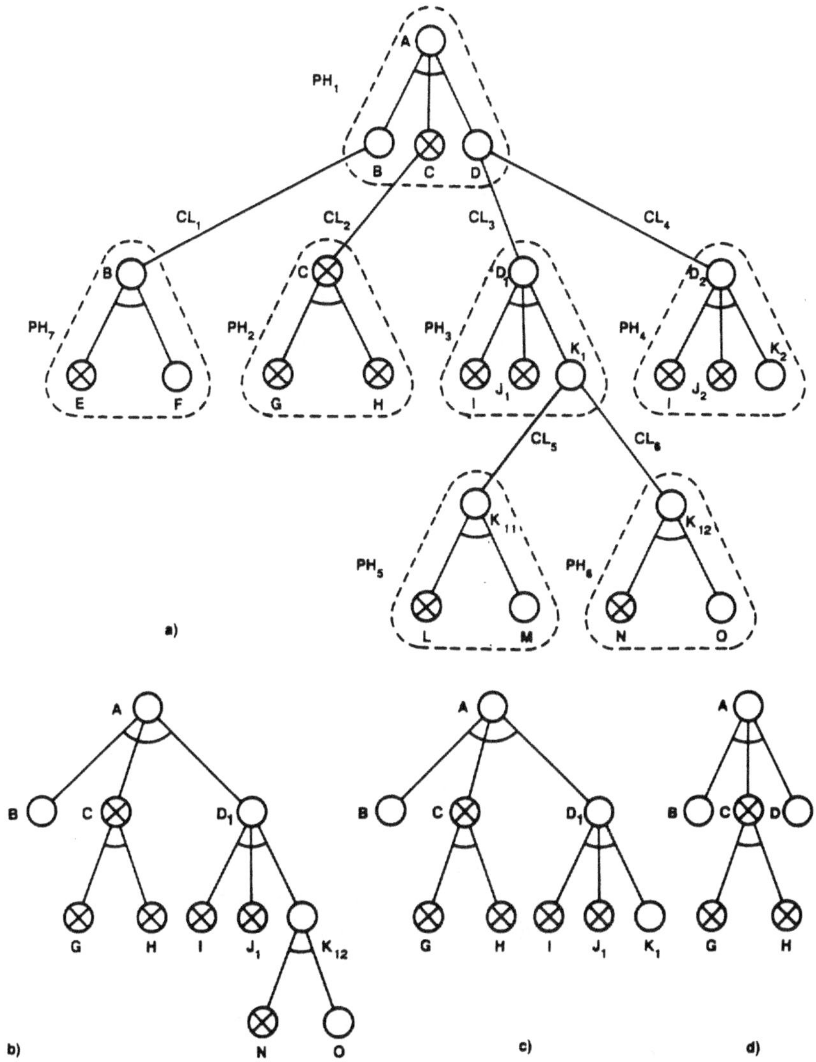

a. The AND-OR tree of phrase hypotheses represents seven canonical DIs. Seven AND trees can be extracted, each corresponding to a canonical DT.

b. The CDT corresponding to PH6. CL1 has been discarded, otherwise the DT would not have been canonical. When extracting AND trees, the CLs are taken into consideration if and only if they do not compromise canonicity.

c. The CDT corresponding to PH3. CL5, CL6 and CL1 have been discarded.

d. The CDT corresponding to PH1. CL2 has not been discarded: CLs to facts do not compromise canonicity and hence they are always considered.

Figure 4.22: Phrase hypotheses and AND-OR trees

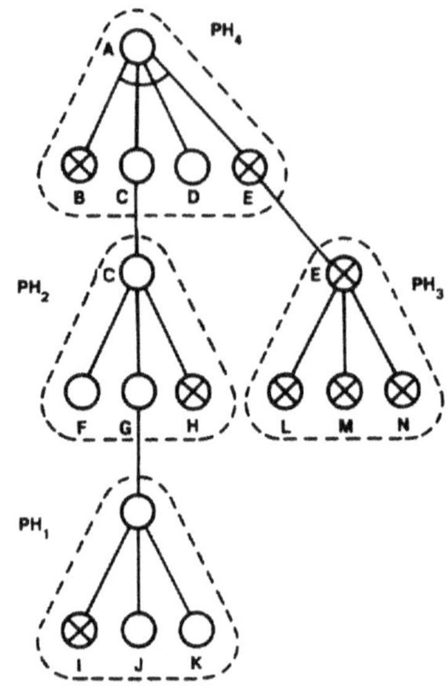

Figure 4.23: A phrase hypothesis PH1 in the context of two other phrase hypotheses PH2 and PH4

4.6.5 Search Space of CDIs and Links Between PHs

Since every goal or fact CDI corresponds to a single representative PH, it follows that the Specialization Relations between CDIs can find a mapping onto a relation between PHs. This relation is represented by links that connect PHs. Two different kinds of links exist:

1. The first type of link refers to PHs corresponding to different KSs: it is the composition link CL between PHs that we have introduced before. Until now, CLs were used to form OR trees of PHs. As we have seen, they do represent the topology of a deduction tree, that is, the structural characteristics of a DI.

2. The second type of link refers to PHs of the same KS. It is a novel link called a *specialization link* (SL). An SL is only used to represent the *specialization relation* at the memory level.

Figure 4.24 presents a simple example. The specialization relation sr between CDI1 and CDI3 is transferred to the memory level as a specialization link SL between PH1 and PH3.

To be precise, an SL between two PHs does not simply map a specialization relation, but maps an element of the transitive closure of the specialization relation. Note that specialization relations can be represented, at the memory level, also by the CLs (when the *subgoaling* and *merge* operators are involved).

The example in Fig. 4.25 shows that link SL3 between PH2 and PH5 corresponds to an element sr* of the transitive closure of the specialization relation. A specialization link between PH4 and PH5 was not inserted because PH4 and PH5 refer to different classes; anyway it is not necessary, being present link SL3. PH5' represents the by-product CDI5' obtained through the application of the *verify* operator to CDI4.

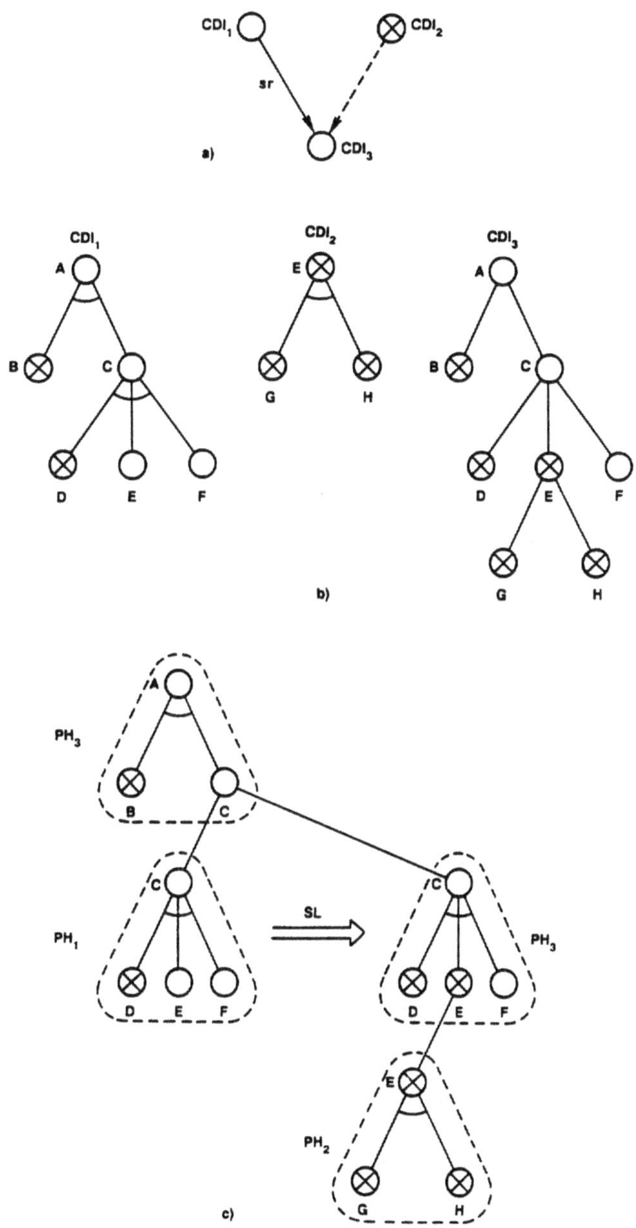

The specialization relation sr between CDI1 and CDI3 (a) has been transferred to the memory level (c) as a specialization link SL between PH1 and PH3 (representing CDI1 and CDI3).

Figure 4.24: Specialization relation and specialization links

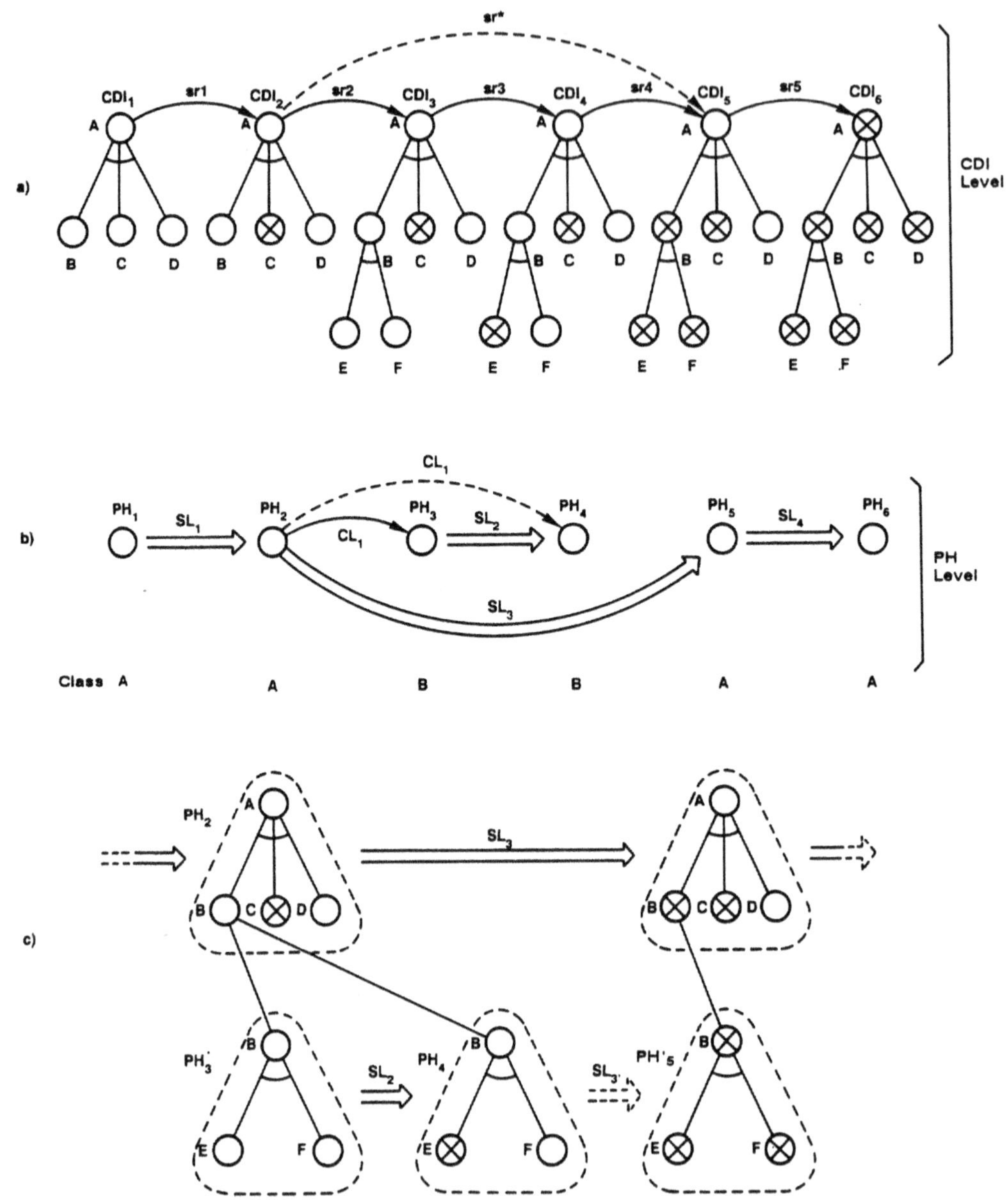

Mapping among specialization relation (a) and specialization and composition links (b). The specialization link SL3 between PH2 and PH5 (b and c) corresponds to the sr* connecting CDI2 to CDI5. For efficiency reasons link CL1' is maintained (CLs are also use to constitute the AND tree associated with CDIs).

Figure 4.25: Specialization relation and composition links

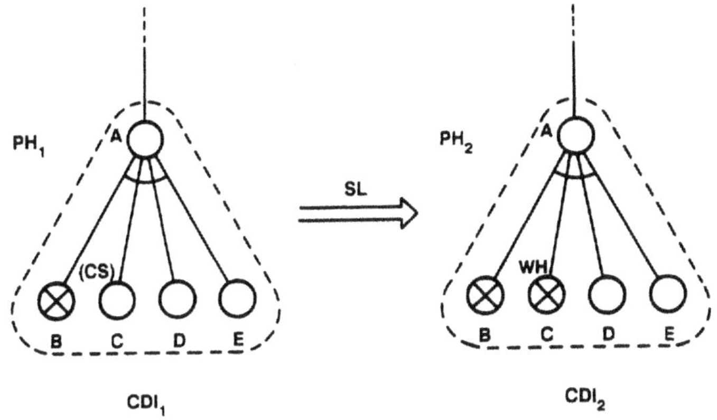

Figure 4.26: Application of the *verify* operator: no facts have been generated

Note that a link, SL3', was added between PH4 and PH5'. This link does not represent a specialization relation at all but only an auxiliary link. Links such as SL3' are added only because they are useful during the application of the MERGE operator. In Sect. 4.6.9 this subject will be discussed in further detail.

In summary, if PH1 is connected by a SL to PH2, then CDI1 (represented by PH1) is less specific than CDI2 (represented by PH2). This means that CDI1 is one of the previous states of the deductive process leading to CDI2. CDI2 does not need to be directly connected to CDI1 by a specialization relation, but they must be connected by the transitive closure of the specialization relation. In other words there must be a sequence CDI11, ... , CDI1n of deduction instances such that CDI1 is connected to CDI11, CDI11 to CDI12, ... , CDI1n to CDI2.

PHs connected by SLs form the so-called specialization trees (STs). From what was said above, each ST corresponds to a KS and the level of a PH in the ST corresponds to its level of completion: complete PHs are always leaves of STs. In Sect. 4.6.4 it was said that, as there is a 1:1 correspondence between PHs and CDIs, all the fact and goal CDIs making up the DIDB can be partitioned into equivalence classes according to the KS to which they correspond. So STs correspond directly to KSs.

The following sections deal with the application of specific operators, and they will better clarify these concepts.

4.6.6 The VERIFY Operator

The VERIFY operator solves a terminal subgoal with a suitable word hypothesis from the lattice. That subgoal could be solved in many different ways if there is more than one suitable word hypothesis. Let CDI1 be a goal deduction instance characterized by a current subgoal CS; let PH1 be the phrase hypothesis representing CDI1 (see Fig. 4.26). Let us suppose that the VERIFY operator is applied to CDI1. For the sake of simplicity, let us assume that only one suitable word hypothesis WH can be found in the WHDB able to satisfy CS.

The resolution of CS requires the generation of a new deduction instance CDI2. From the memory point of view it will be necessary to generate (at least) a new PH in order to

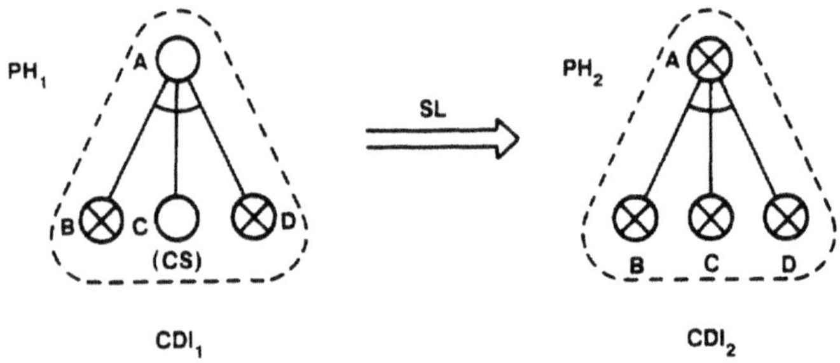

Figure 4.27: Application of the *verify* operator: only one complete phrase hypothesis (PH2) has been generated by the application of the *verify* operator

represent the canonical deduction instance CDI2. It is then reasonable to make a copy of PH1 generating a new hypothesis PH2 where its subgoal CS is solved by WH.

By now two different kinds of actions are taken, according to whether PH2 is still a goal or has become a fact:

1. In the first case (as in Fig. 4.26) PH1 and PH2 must be connected by a specialization link SL. In this case, therefore, the specialization relation between CDIs is directly mapped onto a SL between PHs.

2. In the second case (PH2 is a complete one) there are still two possibilities:

   - In the first one PH1 has no ancestors (see Fig. 4.27). In this case there would be no other goals to be solved and the deductive process associated with CDI1 can stop. The new CDI2 is then a fact DI connected by a specialization relation to CDI1. At the memory level PH2, that is a complete PH obtained in the same way we have seen before, is connected by a specialization link SL to PH1.
   - In the second one PH1 has ancestors (see Fig. 4.28). In this case PH2' is not at all a representative of a specialization of CDI1: it is only the representative of a new CDI (CDI2') that represents the by-product of the application of the *verify* operator to CDI1. Here PH2' is connected to PH1 by a specialization link SL3, though this link does not represent a specialization relation at all.

In the last case what we have to do is to run the specialization relation backward until a deduction instance CDI0 pertaining to a different KS is found. At the memory level we can do that simply following the CL of PH1. A new deduction instance CDI2 is then generated and represented by a new phrase hypothesis PH2, obtained by copying and specializing hypothesis PH0. The specialization consists in solving the current subgoal

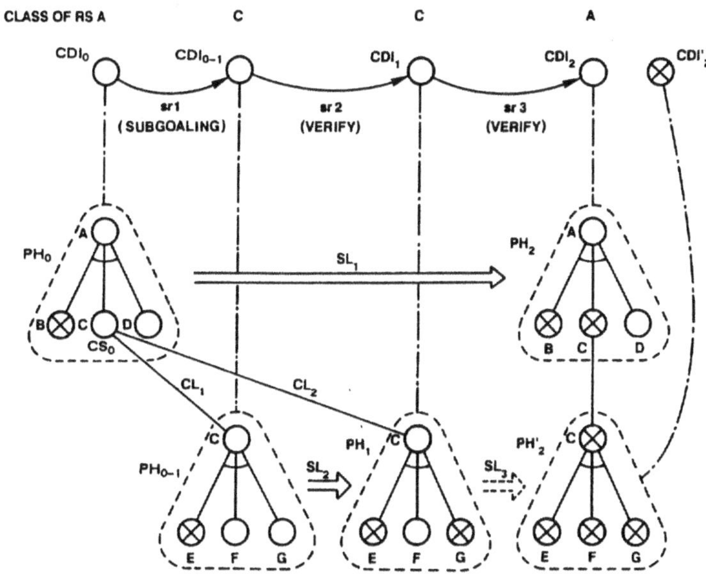

Figure 4.28: Application of the *verify* operator: more than one complete phrase hypothesis (PH2 and PH2') has been generated by the application of the *verify* operator

CS0 of PH0 with PH2' (i.e. a composition link between PH2 and PH2' is established). PH2 is then connected to PH0 by an SL.

Now, if CDI2 is still a goal, the situation is precisely the one depicted in Fig. 4.28. If instead the result is that CDI2 is a fact itself, the above mentioned procedure takes place again, until applicable.

4.6.7 The SUBGOALING Operator

We consider the application of the *subgoaling* operator to a certain CDI1 characterized by a current subgoal CS and represented by phrase hypothesis PH1.

Let us assume the result of the application to be a new CDI2 where the current subgoal CS of CDI1 has been decomposed into the subgoals (E,F,G) according to a given KS. The phrase hypothesis that represents CDI2 is PH2 (see Fig. 4.29). One of its subgoals, say F, will then be selected as the current subgoal NCS for CDI2.

CDI2 is more specific than CDI1 because it contains more information about the way of solving one of its subgoals; then it must be connected to CDI1 by a conceptual specialization relation. PH2 is connected by the composition link CL1 to PH1.

Composition link CL1 stands both as a structural way of reconstructing the deduction tree corresponding to CDI2 and as a correspondant of the specialization relation between CDI1 and CDI2 at the memory level.

4.6.8 The PREDICTION Operator

The *prediction* operator is applied to a fact CDI and generates new goal CDIs. We consider the case, shown in Fig. 4.30, in which the *prediction* operator is applied to fact CDI1,

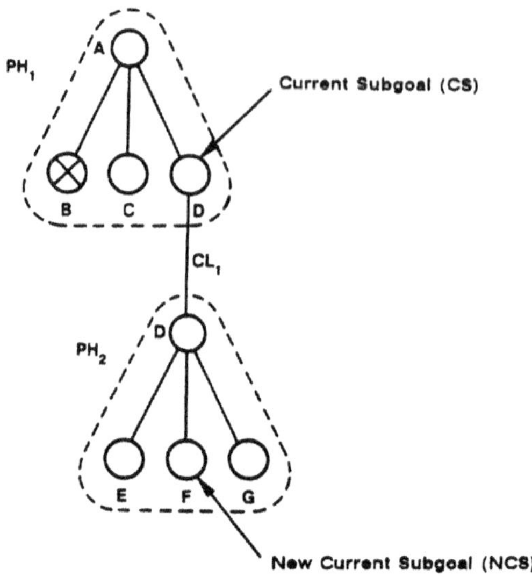

Figure 4.29: Application of the *subgoaling* operator

represented by PH1, and generates the goal CDI2.

It is assumed that the KS requires three fillers slots of class B, C and D, and that D is the subgoal solved by the fact F. Then the result of the *prediction* application is the generation of a new phrase hypothesis PH2 connected to PH1 through CL1. One of its subgoals (say B) will be selected as the current subgoal for CDI2.

Note that CDI2 is a novel goal to be pursued; thus it is not connected by a SL to any CDI. The function of the *prediction* operator is the "creation" of novel goal CDIs rather than the specialization of already existing CDIs. The CDIs generated by the application of the *prediction* operator are the roots of future conceptual specialization trees.

4.6.9 The MERGE Operator

The *merge* operator is the most complex of the four operators. Given the current deduction instance CDI1, the *merge* operator, being a dyadic operator, tries to use other CDIs contained in the DIDB to produce new CDIs.

Not all the CDIs of the DIDB are candidates for the *merge* operator application; only a subset is extracted. Let be CDI2 one of them. If the tests on the constraints conditions succeed, the *merge* operator applied on CDI1 and CDI2 generates a new deduction instance CDI3.

Merge can take place during both bottom-up and top-down search activity; the involved tasks are nevertheless the same. Fig. 4.31 represents the application of the *merge* operator between fact CDI1 and goal CDI2. CDI1 is represented by phrase hypothesis

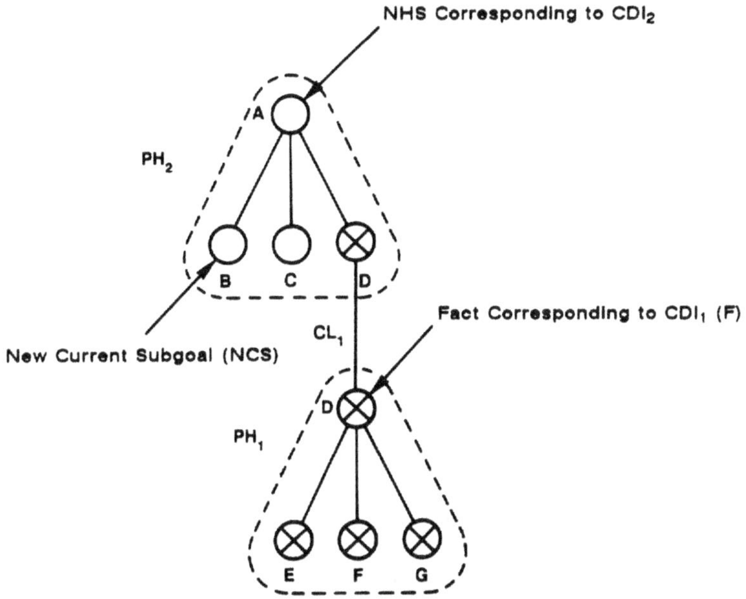

Figure 4.30: Application of the *prediction* operator

PH1; goal CDI2 is represented by phrase hypothesis PH2, current subgoal CS. The current subgoal CS of CDI2 has to be of the same class (D in figure) of CDI1. What happens is very similar to the application of the *verify* operator: a new goal CDI3 is created, substituting subgoal CS of CDI2 with the fact CDI1.

As was the case with the *verify* operator, two cases must be distinguished, according to whether PH2 completes itself thanks to PH1 or not. If not, CDI3 is a more specific goal than CDI2 (a conceptual specialization relation connects them).

Then the new phrase hypothesis PH3 of CDI3 will be connected to PH2 by a SL. This is the actual situation depicted in Fig. 4.31. Otherwise, if PH2 completes itself, beside the creation of the appropriate links, it would be necessary to follow the procedure that has been described during the illustration of the *verify* operator; we do not repeat those considerations.

It is interesting to observe that the CDI selected by the scheduler (selected CDI) could be either the fact (CDI1) or the goal (CDI2). Considering a single couple of CDIs the only difference is that the specialization relation always connects the goal CDI to the new CDI while in the search OR tree the selected CDI is always connected to the newly generated CDI (see Fig. 4.32).

How links are exploited

The compositional and specialization links are used heavily during many phases of the control cycle. We have illustrated how CLs and SLs are followed during the application

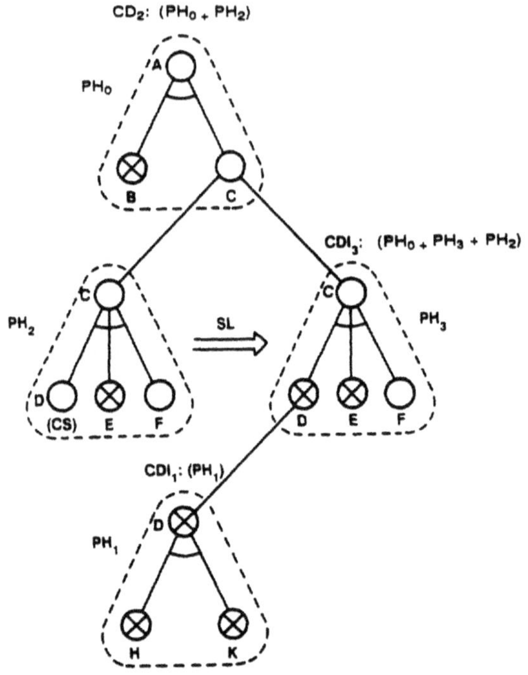

Figure 4.31: Application of the *merge* operator: merging a fact DI with a goal DI. The *merge* operator applied on CDI1 and CDI2 (represented by PH1 and PH2) generates CDI3 (represented by PH3)

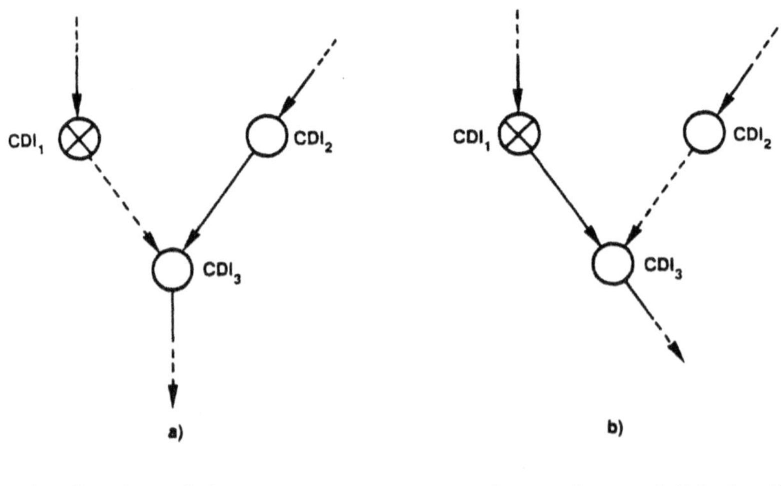

Figure 4.32: Application of the *merge* operator on a fact and a goal DI. Auxiliary links (represented by dotted lines) can start either from the goal (CDI2) or from the fact (CDI1). a) The scheduler selected goal CDI2. b) The scheduler selected fact CDI1. No differences are present at the memory level for the two cases (see Fig. 4.31)

of the *verify* operator (Sect. 4.6.6); here we shall describe the use of specialization links during the application of the *merge* operator.

1) Merge of a fact CDI with a goal CDI; the selected CDI is a fact.
All the KSs that have at least a filler slot of the class of the selected fact CDI are triggered. Then, for each KS, its associated specialization tree is scanned, searching for goal CDIs (in practice, PHs) that could be merged with the fact CDI. When a phrase hypothesis PH is found that is not compatible for merging, the specialization subtree having PH as its root is no longer searched: in fact, the PHs belonging to that subtree are more specialized than PH and thus they too will be incompatible for merging. The decision whether two CDIs (represented by their representative PHs) are compatible or not depends on the constraints propagation and check activity.

2) Merge of a goal CDI with a fact CDI; the selected CDI is a goal.
Let CDI1 be the selected goal CDI and CS its current subgoal. All the KSs are of the same class as CS. For each of them its associated specialization tree is scanned, searching for fact CDIs (PHs) that could be merged with CDI1. It is interesting to note that thanks to the "spurious" SLs (like SL3' of Fig. 4.28) candidate fact DIs can be easily found as the leaves of the specialization tree. This explains why those links were added.

4.7 Parsing - Dealing with Missing Words

4.7.1 Introduction

This section describes a method for analyzing lattices of lexical hypotheses when short, less significant words are not detected by the recognition system. The basic consideration is that some function words, like articles, some prepositions and other usually short words, are often unnecessary to understand a sentences. Thus it is possible to correctly analyze a lattice in which some of these words are missing without querying the user.

There are other cases of possibly missing words, different from the one referenced here in that the words have a significant semantic content. For example, words such as 'mount' or 'Piedmont' (that is, common or proper nouns of the domain's entities) are usually essential for understanding the uttered sentence and thus they fall into the latter category. To cope with cases in which a word of such a type is not present in the lattice of lexical hypotheses, it is necessary to correctly understand the parts of the sentence in which all of the right words have been hypothesized, to spot the zone of the utterance in which the undetected word should lie, and either to perform a more specific verification at the phonetic level, or to start an interaction with the user aimed at eliciting the information necessary to identify the word.

4.7.2 The Problem

The problem of devising a strategy able to analyze a lattice independently (at least to a certain extent) from the presence of some types of short word hypotheses cannot be eluded, because short words, consisting of one or two phonetic units, are by their nature unreliably recognized. A short word covers a very low number of states of the hidden Markov model used to represent it [17], and then its score depends heavily on random events (burst noise, defect in pronunciation, etc.) that are temporally coincident with the uttered word. The situation is different for a long word, because it covers a high number of states and thus its score depends on events spanned on a longer hystory; that is, it is more 'averaged'. In addition there is the problem of *coarticulation* between such short word and the previous and subsequent word that affect the real pronunciation of the word. So, short words happen more frequently than long words to be badly recognized or to go undetected; in addition, their scores are not always reliable. If they are undetected, a standard analysis requiring all of the uttered words to be present in the lattice simply would not work, and if they have bad scores it would encounter heavy inefficiencies.

There is also an opposite problem for continuous speech. It can happen with a certain frequency that false short words are erroneously detected, and a good score is assigned to them. That is especially true when their phonetic representation is also part of a longer word that was actually detected and if their corresponding time interval is free of significant score-degrading events. In these cases, a standard analysis would unduly delay the solution by considering such incorrect word hypotheses.

Types of frequently missing short words

In the subset of the Italian language defined by the knowledge bases of the system, short words of the type mentioned above fall into different classes.

1. Articles ("il", "lo", "la", "i", "gli", "le"). The presence of such words is, in our domain, almost always irrelevant for the correct comprehension of the utterance, provided the other words are correctly recognized. Indeed, using grammar rules it is possible to infer what the appropriate article should be, given a correct interpretation of the rest of the sentence; so this information can be used to complete recognition.

2. Prepositions ("di", "a", "da", "in", etc.). Italian prepositions come in two types, simple and articulated. Simple prepositions are usually very short and generally monosyllabic. Articulated prepositions, which group into a single word a preposition and an article (e.g. "dello" ["If the"]), are instead longer and more easily recognized. Though in some cases they are unnecessary to understand a sentence, it is not desirable to simply ignore them, because they could provide useful temporal constraints and (especially the long ones of them) carry a reliable score. So care must be taken in handling with missing prepositions. A good approach would be one that, though able to abstract from prepositions in general, takes some types of them into consideration under appropriate circumstances.

3. Auxiliary verbs ("è", "ha"). Same as articles or simple prepositions. Again, however, other (longer) verbal forms of "essere" (to be) and "avere" (to have) could provide useful constraints. The discussion is similar to that about prepositions.

4. Conjunctions ("e", "o"). For correct understanding it is of course essential to distinguish between conjunctions such as 'and' and 'or', so their loss is irrecoverable. In the present phase of the project, however, the syntactic/semantic knowledge bases do not allow the use of conjunctions.

The basic idea

The most trivial approach to the problem of short words is simply to ignore them totally, just placing a suitable temporal 'hole' in the zones where they should lie, according to the syntax rules [1]. However, such an approach presents some drawbacks because, as was said, some words could provide useful constraints or positively influence the analysis thanks to their score. Thus a more flexible strategy was devised.

The basic observations on which the strategy is grounded are the following. Short words of the kind previously considered, even when present in the lattice, are not of much aid to predict concepts, given the reduced reliability of their scores. Thus the problem of solving subgoals corresponding to this type of words should be treated differently from the 'normal' subgoals. In particular:

1. the problem should be solved 'locally' by the deduction instance (by the phrase hypothesis at the memory level) containing such subgoals;
2. the subgoal solving procedure must be able to assume, under some specified conditions, that the subgoal has been 'solved' all the same, even without finding suitable words in the WHDB. This special action, that allows it to cope with missing words, will be called *default solving* in the following; the normal subgoal-solving will be referred to as the *search solving* (i.e. that requires a search in the lattice);
3. one would still like to exploit time constraints and score contribution of words, particularly if they have a significant length;
4. the 'localization' of the problem might give rise to some non-optimalities, but this can be tolerated if quasi-optimality is preserved in practice and a significant improvement in efficiency is reached.

The approach: the JVERIFY operator

The previous considerations gave rise to a strategy based on the following principles.

1. The short words previously discussed are accounted for by a special conceptual type called *jolly* and represented as J. In the dictionary these words are linked to the conceptual type J.
2. Each time an analysis starts, the word hypotheses belonging to this category are extracted from the lattice and introduced into a separate database called the *jolly data base* (JDB).
3. A knowledge source (KS) may contain one or more slots of type J, but there are no KSs having J as the header conceptual type. That is, there can be KSs with a problem-solving structure like:

$$A = J * B$$

[1]Note that even a somewhat drastic approach like this has nothing to do with keyword-based understanding. In fact, though short words are allowed to be missing from the lexical hypothesis data base, their 'functional' role defined by syntax is preserved.

but not such as:

$$J = *$$

In other words, jollies become terminal subgoals. Consequently, the *activation, prediction* and *merge* operators are precluded from acting on words of type J. This limitation has mainly the purpose of preventing the jollies from generating by their initiative a lot of unreliable deduction instances.

4. From the previous point it follows that a jolly subgoal has to be solved during a backward step. However, the *subgoaling* operator cannot be used, because the jollies are terminal subgoals. Even the *verify* operator cannot be used as such, because it is not able to accomplish the *default solving* defined above. Thus, a new operator, *jverify*, was added. Next section contains a detailed description of the actions taken by *jverify*, when it is applied to a deduction instance. The subsequent section will discuss how to integrate the application of *jverify* in the control strategy described in the previous section.

4.7.3 How JVERIFY Works

This section is divided into three parts. The first two describe the actions taken by *jverify* during the search and the default solving, respectively. The third explains how they are integrated.

Search solving

During the search jolly solving, *jverify* operates much as *verify* works. The *jverify* operator is applied on a goal DI having a jolly (and hence terminal) subgoal J (see Fig. 4.33). The subgoal J will be characterized by its morphological, semantic and temporal constraints, in a way similar to those of a normal subgoal. The *jverify* operator checks if the subgoal J can be solved by jolly word hypotheses satisfying the constraints. Such word hypotheses are contained in the Jolly Data Base (JDB) that is extracted from the WHDB at the beginning of the analysis. Let us call JH1,....,JHn the word hypotheses of the JDB able to satisfy subgoal J For each of them the *jverify* operator generates a new DI having a new quality factor that keeps into account the score of JHi. The newly generated DIs are inserted into the DIDB and can be goals or facts. In either case, the new DIs are connected to the starting DI by OR links.

Similarly to the *verify* operator (see Sect. 4.5.13), *jverify* can generate fact DIs in addition to the above mentioned set of DIs; in such case the discussion is analogous to the one for *verify*. From the point of view of the phrase hypotheses, the situation is again very similar to what takes place during the application of the *jverify* operator.

Default solving

In the *default solving*, starting from a given DI (see Fig. 4.34) having a jolly subgoal J, a new DI is created, in which the subgoal J is labeled as solved, without associating any word hypothesis to it.

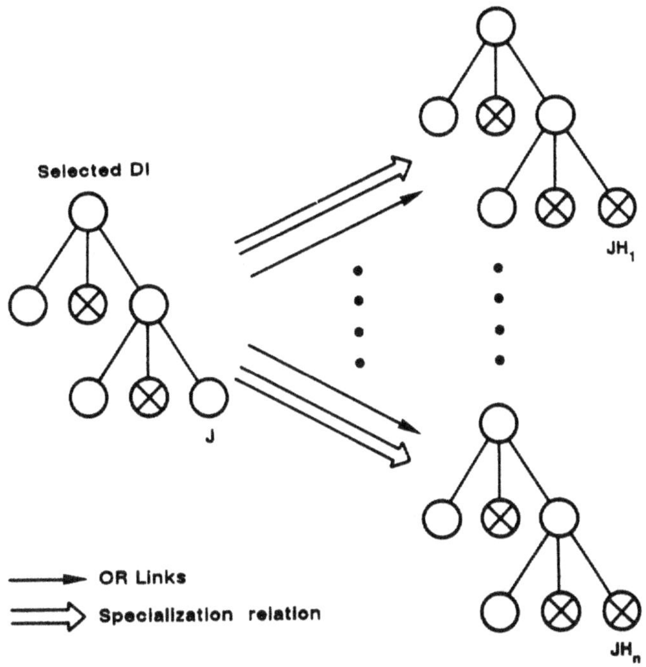

Figure 4.33: Application of the *jverify* operator - search solving

The new DI keeps the quality factor of the starting DI. It is then inserted into the DIDB and connected to the starting DI with an OR link. A new by-product fact DI can also be generated if it is the case.

Default solving, which is useful for making the analysis proceed independently on the contents of the JDB, is dangerous if used too heavily because it generates many DIs and so must be carefully controlled.

The idea is to consider totally unreliable the recognition of words shorter than a threshold Sj. Thus, if the lattice does contain some word hypotheses shorter than Sj, they are considered unreliable and discarded when the JDB is extracted from the lattice. It follows that a subgoal J is default-solved only if its time constraints are compatible with a fictitious word hypothesis of length Sj. Of course, the gap tolerances must be kept into account. More precisely, let ϵ_d be the maximum allowed gap between word hypotheses, and $((l_1\ l_2)\ (r_1\ r_2))$ the time constraint of J, where $(l_1\ l_2)$ is the range where the word hypotheses should begin and $(r_1\ r_2)$ the range where it should end.
Then, J is *default solved* if

$$max(0, r_1 - l_2) < Sj + 2\epsilon_d$$

The new time interval of node J is then restricted by assuming that a lexical hypothesis of maximum length Sj has been associated with it. Temporal constraints are then propagated to the other subgoals of the DI (like F in Fig. 4.34). Of course, acting on the threshold Sj regulates the amount of default solving. The most suitable value of Sj must be determined experimentally.

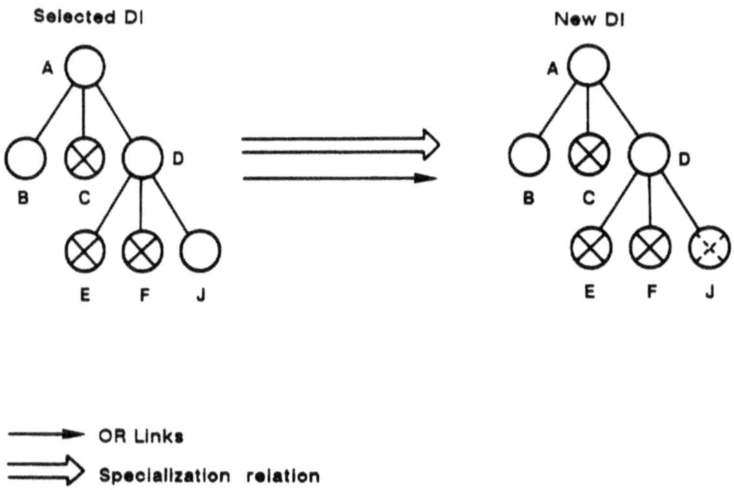

Figure 4.34: Application of the *jverify* operator - default solving

Integrating search and default solving

The most relevant aspect of a DI having a *default-solved* jolly subgoal is that it is equivalent to any DI of the same type, having the jolly subgoal solved by a word hypothesis. At the extreme, raising Sj to the value of the largest jolly of the JDB, the search solving should not be performed. It is important, then, to avoid the generation of DIs with jolly subgoals solved by search if it is possible to solve them by default.

A different problem is the following. If we know, given the syntactic-morphological constraints, that a Jolly subgoal must stand for a long word (such an articulated preposition like "dello" ["of the"], it is of no use trying to solve that subgoal by default, even if its temporal constraints would allow such an operation. Conversely, if we know that the subgoal stands for a very short word, there is no need to try the search solving.

The first problem (avoiding useless search-solved jollies) can be coped with by limiting the search for words during the search solving. This is obtained by restricting the original time interval of the unsolved jolly subgoal in order to exclude the possibility of accepting word hypotheses that generate DIs that are equivalent to those obtained by default solving.

The second problem (avoiding useless default-solving of surely long jollies) is coped with by examining the morphological constraints for the Jolly subgoal. If they make it possible to determine if the word must be definitely short or long, the search or the default solving are inhibited, respectively. Otherwise, *jverify* proceeds normally.

A flag exists for inhibiting default solving that is used, for example, in the so-called 'text mode'. In 'text mode', a fictitious lattice is built up just by typing the words from

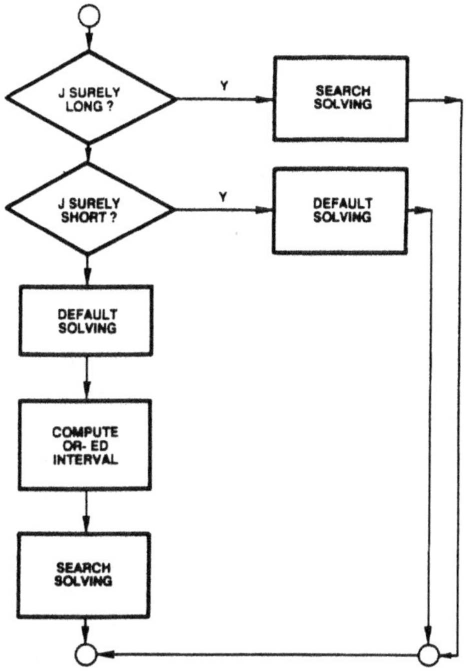

Figure 4.35: Operations performed by the *jverify* operator

the keyboard, as in a written-language understanding system. This mode is frequently used for rapidly debugging newly-expanded knowledge bases without having the need for true lattices. Since the purpose is to check semantics and syntax, the words of type jolly cannot be ignored and thus default solving has to be inhibited. The global operation of *jverify* is shown in Fig. 4.35.

4.7.4 When to Apply the JVERIFY Operator

Having defined the *jverify* operator, a way has to be devised to integrate the use of such operator within the standard control strategy described in the previous sections. Many modalities have been tested.

The simplest modality consists in submitting the *jverify* application entirely to the standard strategy, considering it just as another 'normal' backward operator. In this case, *jverify* is applied whenever the scheduler selects a DI whose current subgoal is of type jolly. Other more complex modalities have been experimented with and proved a little more effective, but we are not going into details here. Extensive data on the experiments can be found in the next section.

4.8 Experimental Results

This section presents the result of extensive experiments performed in order to assess the functionality of the whole recognition/understanding system.

As is well known, evaluating the performance of speech recognition systems is a difficult task, especially when continuous utterances are treated. This is mainly due to the difficulty of establishing standard analysis conditions and standard performance metrics.

Recently, some efforts have been devoted to defining common databases of spoken utterances, especially in the U.S.; for instance, the DARPA has supported the development of a database of continuous sentences that should be used for evaluating speaker-independent systems. However, such databases refer to a specific dictionary and hence to a specific domain; they are therefore of no use for speech understanding systems that have been developed over different domains, not to speak of different languages.

The situation is more promising on the side of performance metrics. Apart from the fact that one obvious measure exists, namely the percentage of successful sentence recognition, some other recognition measures are gradually emerging as the most diffuse in the worldwide research, if not as true standards [34, 31]. However, when the purpose is to understand a sentence rather than to recognize its words, subtler problems arise: is it still significant to rely on measures more or less related to the number of correctly recognized words, or would it be more meaningful to measure the performance on the grounds of correct understanding (i.e., in a pragmatic sense, of correct answer to the question)? In the measures presented here a compromise has been adopted: performances are given in terms of rate of correctly recognized sentences, where by "correctly recognized sentence" we mean a sentence that is composed of the words which were actually uttered, *except* for short functional ones, which - according to the treatment described in Sect. 4.7 - can be directly hypothesized by the parser without being necessarily present in the lattice. For example, suppose that the actual sentence was "Dimmi la regione cui appartiene Torino" and the recognized sentence is "Dimmi la regione ? ? appartiene Torino" (the two question marks stand for two default-completed jollies "a" and "cui"). Only one jolly ("cui") was present in the uttered sentence, but the recognized sentence is considered correct because the parser cannot discriminate between the two sentences given the information contained in the lattice. However, if the actual sentence was "Qual e' la regione bagnata dal Po" ("Which is the region washed by the Po") and the recognized sentence is "Qual regione bagna il Po" ("Which region does the Po wash"), the recognized sentence is considered incorrect even if from the semantic viewpoint both ask for the same information and would give rise to the same answer.

A final difficulty is the following: in order for any type of measure to make sense, the degree of freedom with which words can follow one another must be given. In other words, it is needed a measure of the coverage of the language model used for the recognition/understanding activity; if the model is strict and the corresponding allowed language is of small size, the error rate will be lower than in the case of a wider model. Traditionally, such a measure has been based on information-theoretic principles, and has been represented by quantities like *entropy* or *perplexity* [36, 21]. Such quantities are satisfactory and relatively simple to use when the language model is of statistical nature (such as n-grams or word n-tuples models) or is based on a regular or context-free grammar; however, they cannot be easily extended to language models of considerable complexity like the one employed here, and it is necessary to use equivalent measures that allow sufficient ease of computation at the expense of precision.

This section is organized as follows. We first describe the way that has been followed to determine the language coverage, then we discuss the experimental results; a careful examination of error types is also presented. Next the result of the jolly treatment is discussed. Experiments on the optimality degree of the system are successively presented, with reference to the discussion contained in Sect. 4.5. The strict correlation between the computational load and the average score of the sentence (or of the score of the worst word hypothesis) is also shown in this section. Finally, some specific causes of errors, crucially depending on some system parameters, are described in detail and the trade-off between their elimination and the increase of average computational load is discussed.

4.8.1 General Performance Results

The coverage of the language model

There are several difficulties that make the concept of perplexity inapplicable for the determination of the coverage of the language model employed in this system. First, the model does not take into account the probabilities of the different sentence constituents, but only discriminates admissible phrases from inadmissible ones. Then assumptions should be made on how probabilities would have to be distributed in order to obtain a meaningful measure. Alternatively, perplexity could be computed starting from quantities like maximum entropy; however, the derivation of maximum entropy requires knowledge of the number of sentences of any length [36]; now, apart from the fact that the model allows in principle sentences of infinite length and therefore some adjustments to the derivation method should be made in order to take account of this fact, the computation of such numbers is extremely heavy because of the constraints, of both syntactic and semantic nature, that relate even distant words in consequence of the particular formalisms used in the system.

A pragmatic approach has therefore been followed in order to evaluate the language coverage. The goal of the computation was the average branching factor of the hypothetical tree resulting from the union of the derivation trees of all the sentences allowed by the language model. (Under some assumptions, its meaning is comparable to that of perplexity.) Since it would be too computationally expensive to obtain such a tree explicitly, because of the long-distance constraints reminded above, the computation proceeded in two consecutive phases. In the first one, a relaxed version of the language was used, which does not include long-distance constraints. The average branching factor of this relaxed language was computed. Then, an experimental phase allowed a measurement of the extent to which the actual language prunes the relaxed one; this was done by randomly selecting sentences allowed by the relaxed language and testing if they were also allowed by the actual one. Such a measure, suitably averaged on the sentence lengths, permitted us to obtain a correction factor on the branching factor of the relaxed language and eventually to determine the branching factor for the actual language. The resulting value was about 35.

Recognition algorithm	Correct identification rate %	Failures %							Average number of DIs
		Resources	Missing words	Gap	Overlap	Better-scored sentence	Non optimality	Total	
1-step VIT	78	0	4	5.5	5.5	5.5	1.5	22	357
1-step FORW	78	1.5	7	2.5	3.5	5	2.5	22	250
2-step VIT	77	1	8	4	1	7	2	23	331
2-step FORW	79	1	8	3	1	5.5	2.5	21	238

Table 4.1: Experimental results - 1st phase

Performance results

Performance results are given in terms of correct sentence recognition/understanding rate, as specified above. We recall that by "correctly recognized sentence" we mean a sentence that is composed of the words which were actually uttered, except from short functional ones.

A first program of experimentation was carried out in order to evaluate the understanding algorithm on a large set of lattices as well as to assess the relative advantages of different methods used at the recognition level. Four recognition procedures were tested in the course of the experiments; about 80 lattices per procedure (speaker dependent) were parsed and the result of each parsing were analyzed in detail. The four recognition methods were 1) VIT, 1-step (use of Viterbi decoding algorithm, and no word preselection - i.e. the lattice is produced in a single step from the vector-quantized data); 2) VIT, 2-step (Viterbi decoder and word preselector, i.e. recognition is performed in 2 steps); 3) FORW, 1-step (Forward decoding algorithm, no word preselection), and 4) FORW, 2-step (Forward decoding and word preselection). The parser used a language model covering about 700 words and the equivalent branching factor was about 35.

Results are summarized in Table 4.1. The rate of successful sentence recognition/understanding is around 78%, about the same for the four types of recognition algorithms, which places the system well in the state-of-the-art. However, there are considerable differences when efficiency is taken into consideration. Efficiency at the parser level is represented by the average number of DIs generated during an analysis. The data show that the FORW algorithm permits the parser to save about 1/3 of computational activity. There is no definite evidence that using the 2-step algorithm instead of the 1-step produces similar benefits, though of course the 2-step algorithm increases efficiency at the recognition level. For these reasons, only the FORW algorithm was used in the final experimentation program.

The failures were partitioned into six classes, according to their causes: *resources*, referring to excessive time required for parsing, *missing* correct *words* in the lattice, excessive *gaps* and excessive *overlaps* between words, *better-scored* consistent wrong *sentences* and finally errors due to the *non-optimality* of the strategy. The distribution varies significantly in relation to the recognition algorithm. Variations are in accordance with expectations. For instance, the 2-step algorithm appears to be proner to lose words than the 1-step one. The FORW algorithm is clearly better able to spot words than VIT, as it is seen from the reduction in failures due to gaps and overlaps.

Trying to overcome the causes of failure implied conflicting requirements. The suboptimal parsing strategy used in the experiments occasionally prevents a correct partial parse from growing, and lets other parses come to a complete but wrong interpretation. These events are kept rare while suboptimality offers a considerable gain in efficiency. The flexibility of the understanding stage has permitted some adjustments to be performed on the operator application, so as to better exploit adjacency constraints during the backward parsing steps.

Failures due to gaps and overlaps could be overcome simply by relaxing the thresholds. For instance, raising the gap threshold by 1/3 reduces the failures in the 1-step VIT algorithm from 7 to 2. However, this would also increase the combinability of word hypotheses, worsening efficiency and raising the probability that incorrect but better-scored interpretations are found first. Besides, relaxed thresholds would be redundant in general, since the occurrences of big gaps and overlaps are somewhat rare and mainly related to special inter-word phonetic phenomena. A promising approach would then consist in taking into account such phenomena at the parsing level.

The understanding system used for the above experiments was the version existing at the beginning of 1988, in which the KS had been defined 'by hand' since the compiler of dependency rules and conceptual graphs was not yet available. During 1988 and 1989 a new version of the system was produced. For this system the knowledge bases are automatically produced by the compiler; in addition, while the first version was written in LISP on a SYMBOLICS, the final version is written in C and is running on a SUN-4 workstation. Of course the LISP parser has not simply been translated into C but has been redefined, still maintaining the overall organization described here. The main differences concern the internal structures used to represent phrase hypotheses, the way to represent morphological, syntactic and semantic constraints (mainly coded using bit-positions) and the procedures for constraint propagation and checking.

Nevertheless, in the final version some adjustments on the parsing process have been done. They basically consist in a limitation of the *subgoaling* operator applicability, depending on the characteristics of the slot boundary relative to the goal to which the operator should be applied. Recently a new experimentation has been performed, using a set of 210 lattices, each one pertaining a different sentence, produced using only the 1-step FORW recognition algorithm. Results indicate a correctness rate of 88.5% and an average number of generated DIs of about 280. The most relevant result given by the change of the programming language and machine is the drastic reduction of the average parsing time: from about 40 seconds to 2-3 seconds.

4.8.2 Performance of the Short Word Treatment

The experimental results relative to the short word treatment described in Sect. 4.7 were obtained at the end of the first experiment program and refer to the 1-step VIT and FORW recognition methods (see also Table 4.1). A limited subsequent experimentation, performed with the latest version of the understanding stage, indicated a limited improvement over the data shown here; however, the discussion we present for the data obtained in the first experiment program are still valid on the whole.

Table 4.2 shows the number of jolly words that have been skipped by the parser and the number of jollies actually missing in the corresponding original lattices. This latter figure clearly shows the relevance of the problem: nearly 2/3 of the auxiliary verbs and more than 1/4 of the articles are not recognized on the average. Table 4.2 also shows that

		aux. verbs	*pron.*	*prep.*	*art.*	*refl. markers*
Parsed sentences	*skipped*	26	27	57	90	11
	present	13	1	18	0	0
Original lattices	*missing*	24	5	9	25	0
	present	15	23	66	65	11

Table 4.2: Jolly word detection.

missing jolly words per sentence	1	2	3
n. of sentences	40	18	4
successfully parsed	35	15	3
average n. of generated DIs	318	440	563

Table 4.3: Successful parsing

the number of skipped jollies is higher than the number of jollies missing in the lattice, indicating that many words, albeit present, have been discarded by the operator *jverify* because of their bad acoustic scores or their scarce contribution to constraint propagation.

The most relevant advantage of the short word treatment is the dramatic increase of the number of sentences that can be analyzed successfully. Table 4.3 displays the number of lattices, corresponding to the sentences containing at least one word of jolly type, in which at least one of such words is missing. It is seen that about 75% of them have been successfully understood. This figure does not change substantially as the number of missing jollies per sentence increases, and hence indicates robustness. The computational load, given by the number of generated DIs, is somewhat affected by the number of missing jollies. However, this is mainly due to the fact that sentences with many jollies are also longer and syntactically complex. The actual efficiency can be better estimated from Fig. 4.36, where the average number of generated DIs is plotted as a function of the threshold on the width of the jolly temporal 'hole'. (In practice, the width of the whole, that should be defined by sj, is made 'fuzzy' by the presence of the tolerance on gaps ϵ_d. Thus, the values on the abscissa in Fig. 4.36 are actually given by $sj + \epsilon_d$ to take gap tolerance into account.) The figure displays also the amount of parsing failures related to jolly problems (failures due to other reasons have been ignored for simplicity). The curve indicates that raising the threshold (sj) does not change much the number of generated DIs (the relative oscillations of the values are small). This means that the relaxation of

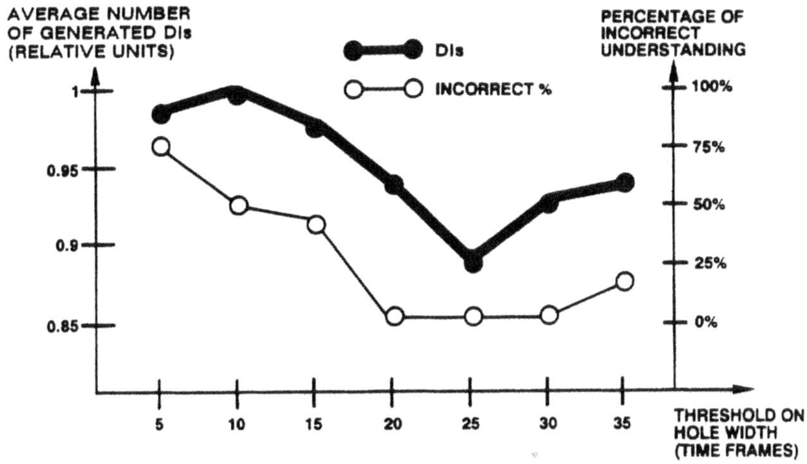

Figure 4.36: Performance vs. width threshold

constraints during the application of *jverify* is not a source of inefficiency. Moreover, there is a large range of values for which the parsing failure rate remains low.

The curve also shows that relaxing constraints may even speed up the parsing. This can be easily explained. When the threshold is low, no jolly is skipped, and failure occurs when jollies are missing from the lattice. When the threshold is raised, skipping begins to work: good-scored false jollies are no more a source of disturbance, and correct but bad-scored jollies are skipped, thus avoiding delaying the parsing; as a consequence the overall number of DIs decreases. Further enlarging the threshold reverses this tendency, since the too-much-relaxed constraints allow the aggregation of words that would have been discarded with stricter constraints; failures occur when one of such aggregations makes up a complete parse scoring better than the correct one.

In conclusion, experiments show that the presence of jolly slots solvable as described above, besides permitting successful analysis of a much greater quota of word lattices, also speeds up parsing by preventing it from being misled by false jollies. This well compensates for the growth of inferential activity due to the relaxed temporal constraints in the DIs containing 'holes'.

This is a novel improvement over systems that, to our knowledge, only admit one single skippable word and use a more rigid linguistic knowledge representation [38] or recognize any configuration of missing words but do not distinguish cases in which the information content of an absent word can be ignored [18]. An attractive feature of the present parsing technique is that the KS activities are modularized into a set of operators. Consequently, it remains open to 'local' improvements on single operators as well as to overall heuristic adjustments on the score-guided control strategy. As an example, the response of the predicate *jolly-type* of the operator *jverify* may be rendered more 'intelligent' by exploiting further information, such as estimates of the expected word length, that has not been taken into consideration in the present implementation.

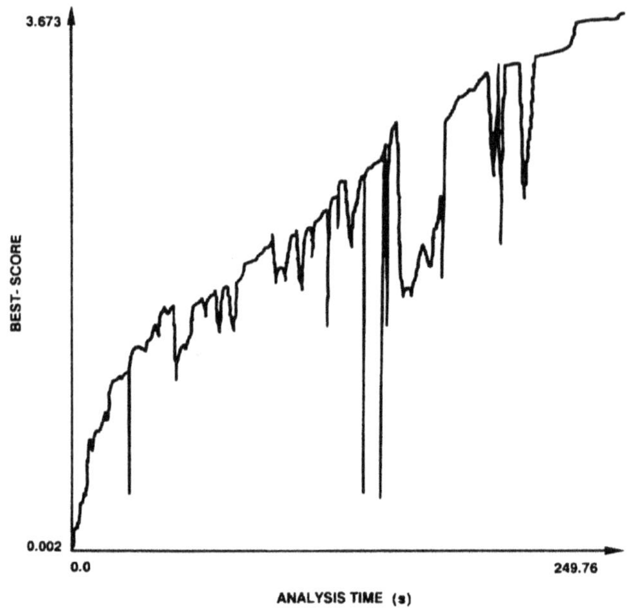

Figure 4.37: Best-score vs. time

4.8.3 Optimality and Efficiency

As was said in Sect. 4.5, optimality is a desirable feature of a parser for speech, as long as inefficiencies intrinsic in any optimal search strategy do not counterbalance the advantages offered by optimality itself. In the experiments described above, a *suboptimal* configuration has been used: it provided a considerable parsing speed-up at the expense of a very limited decrease in correct understanding rate. The behavior of the system in the suboptimal set-up can be seen in Fig. 4.37. The score of the best-score item selected at each cycle is plotted vs. the time at which the cycle started. If the system were thoroughly optimal, then the curve should be monotonically increasing. Therefore, a non-optimality makes the curve bend down; this happens when, starting from a DI having a certain score, a better (i.e. lower) scored DI is generated and selected at the subsequent cycle. Non-optimalities appear in Fig. 4.37 as spikes, showing that the amount of non-optimality (and the related risk of erroneous understanding) stays low. Moreover, they tend to appear late in the analysis, when the solution has already been found.

Given the shape of the above curve, it is apparent that in order to increase efficiency the solution should be found before the analysis enter the 'slow' region in which the worsening rate of the best score is low (and the curve is therefore nearly horizontal). In the previous sections it has been pointed out that in a thoroughly optimal analysis the computational load should depend on the average score of the solution, other conditions

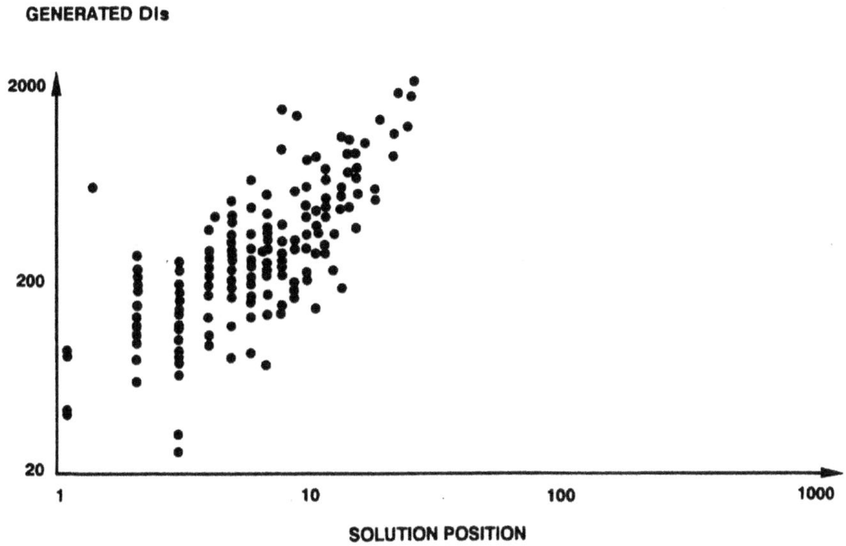

Figure 4.38: Generated hypotheses vs. solution position

being equal for comparison, in a simple bottom-up analysis strategy the load depends on the score of the worst-scored word hypothesis making up the solution. In the suboptimal strategy the optimal situation is no longer reached, but the shift from it is limited. This is seen from Figs. 4.38 and 4.39. In Fig. 4.38 each dot represents the number of generated DIs versus the so-called *solution position*. The solution position is the number of word hypotheses in the lattice whose score is better than the average quality factor of the solution, and approximately represents the amount of word material that an optimal analyzer should treat before coming up with the solution. Figure 4.39 is similar, but the abscissa represents the so called *worst position*, i.e. the number of word hypotheses having a score better than the score of the worst word hypothesis that makes up the solution. A comparison of the two figures shows that the dependency on the worst score is weak, whereas the one on the solution score is stronger (this is also reflected by the correlation coefficients, which are about 0.3 and 0.7 in the former and latter case respectively). The scattering of dots in Fig. 4.38 is high because the analyzed lattices differed in other parameters beside the solution score, including the lattice density (the number of hypotheses per unit of score) in the high-score regions of the lattices, the number of words making up the solution, and the number of short words that are skipped by the parser. A more focused analysis was not possible since the lattices were too few to be partitioned into classes each having all these features nearly equal. Nevertheless, the solution position is so preponderant in determining the computational load that its effect is still evident in Fig. 4.38.

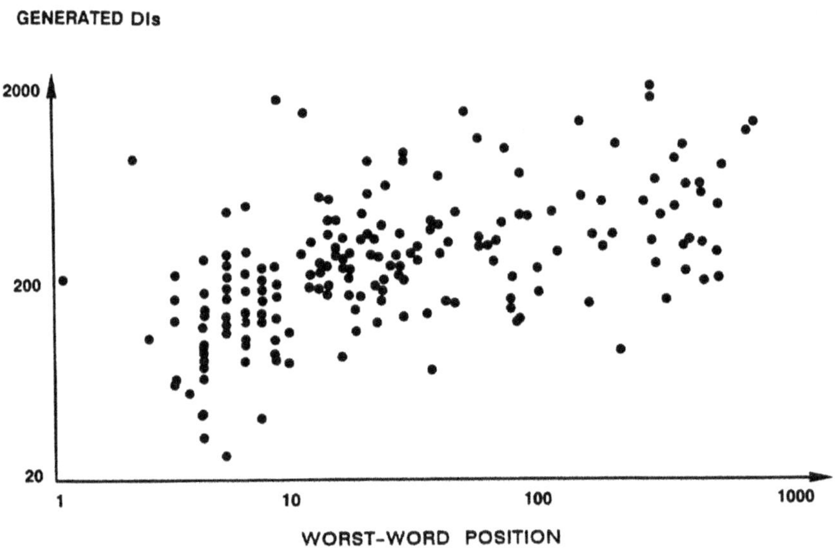

Figure 4.39: Generated hypotheses vs. worst word position

4.8.4 Some Specific Problems

Not all of the problems detected during the experiment programs have been solved, of course. Some of them overlay complex issues in speech understanding and require long research so that deeper insight may be gained into them. In the following we discuss the extant problems through a series of example failures observed in the experiments.

Excessive gaps and overlaps

A typical case of error arises when two correct words are too far apart from one another or overlap too much . Using higher thresholds would be of no use, as remarked in Sect. 4.8.1. As a matter of fact, such problems generally correspond to precise phonetic conditions. For instance, gaps relative to word pairs in which the second word begins with a plosive are substantially wider than in other cases. Similarly, overlaps are wider when a word ends with a vowel that is also the beginning vowel of the subsequent word. However, it has been observed that threshold values as small as 5 frames for ϵ_s and 10 for ϵ_d are sufficient to take into account all these phenomena; the majority of the failures due to excessive gaps or overlaps are not actually due to them. For instance, of the five failures due to excessive gap that have been observed in the VIT 1-step and FORW 1-step experiments, none corresponds to the case of a second word beginning with a plosive. Similarly, the five failures due to excessive overlap do not correspond to cases of adjacent words ending and beginning with the same vowel. On the other hand, the causes of failures due to excessive gap/overlap are not completely unpredictable. At least two regularities have been observed:

1. The Viterbi algorithm used at the recognition level appears to be less robust with regard to the Forward one as far as the correct detection of vocalized consonants. Consider the couple of words "piu' lungo". In the lattices generated with the Viterbi decoder, the end point of the hypothesis "piu'" is always estimated to lie in the point where the /u/ of "lungo" actually ends, because the /l/ is not detected. As a consequence, the word hypotheses "piu'" and "lungo" are largely overlapping. Three failures in the VIT 1-step experiments have been caused by this case only. A similar case happens with the words "quale regione", in which the /r/ is not detected causing the final /e/ of "quale" to extend over the first /e/ of "regione". On the other hand, the Forward algorithm always segments such words correctly. This suggests the need for a better investigation of the reasons for their respective weakness and robustness towards this phenomenon.

2. The present language model does not take into account inter-word coarticulation phenomena. Consider for instance the sentence "Quanto e' lungo il Po". The word "quanto" is not generally uttered as such, but as if it were the word "quante", because the final /o/ is assimilated to the /e/ of the word "e'". As a consequence, the word hypothesis "quanto" may be missing from the lattice or its end point may be incorrectly located, thus causing a parsing failure. Taking into account inter-word coarticulations has implications at both recognition and understanding levels, and would require a tighter cooperation between the two corresponding subsystems.

The other cases of gap/overlap could be best treated by adopting 'softer' ways of deciding if two word hypotheses can or cannot be considered adjacent than a rigid threshold method can permit. A study has been carried out in which the gaps/overlaps are represented by a statistical model obtained through measures on the lattices. Whenever two supposedly adjacent word hypotheses have to be joined, instead of checking the adjacency with the threshold 'filter', the parser assigns a penalty to the overall word agglomerate on the basis of the observed gap/overlap and of such statistical model. Results are still preliminary and are partly reported in [9]. They indicate that even a simple method like this one can be effective.

Non-optimality

In the present version of the system, some non-optimal variations to the standard parsing algorithm are used. Let us examine two cases.

1. One non-optimal variation involves a limitation of the application of the *subgoaling* operator. The *subgoaling* operator expands an empty field of a DI (i.e. a goal) according to the compositional part of some KSs. In the non-optimal case, the operator is applied only if the field is adjacent to some word hypothesis already present in the phrase hypothesis. This heuristic is used because it greatly improves the efficiency by reducing search. Of course it also may be the cause for failure; this happens when a correct DI is 'frozen' and incorrect DIs lead to a solution before the correct one is resumed. This has been experimentally proven a rare event (two failures in the first experiment phase).

2. Another non-optimal variation is the following. Whenever a complete DI is generated, it is compared to the other complete PHs referring to the same KS and, if it is sufficiently similar to one of them, it is 'deleted'. By 'sufficiently similar' we mean, in the present condition, that the two PHs have (i) the same semantics, (ii) the same grammatical relation, and (iii) the same time interval. This heuristic cuts some PHs that would have nearly the same behavior in the future parsing steps, but occasionally it prevents the generation of the correct solution. Three failures on 210 examined lattices have been caused by this heuristic. Removing it causes the number of generated PHs per lattice to rise by about 10% on the average, so that it may be eliminated without excessive loss.

Jolly words

The parameters for the jolly word treatment (threshold sj, classification of jollies into short, long and unknown, see Sect. 4.7) have been optimized in the last experiment phase. Still 5 lattices out of about 200, however, are not parsed correctly because the threshold sj for default solving is too low. It has been experimentally proven that raising such thresholds from the present value would increase the number of failures by allowing the generation of wrong but better-scored sentences with large 'holes' in them.

Bibliography

1. P. Baggia, M. Poesio: "Using conceptual graphs in the development of knowledge bases". Thesis extract, unpublished

2. D. Bigorgne, A. Cozannet, M. Guyomard, G. Mercier, L. Miclet, M. Querre, J. Siroux: "A versatile speaker dependent continuous speech understanding system". *Proc. of the ICASSP '88*, pp. 303-306, New York, NY, Apr. 1988

3. P.G. Bosco, E. Giachin, G. Giandonato, G. Martinengo, C. Rullent: "A parallel architecture for signal understanding through inference on uncertain data". *Proc. of PARLE - Parallel Architectures and Languages Europe*, Eindhoven, The Netherlands, June 1987, Springer-Verlag, Lecture Notes in Computer Science, vol. 258, pp. 86-102

4. R. Brachman, J. Schmolze: "An overview of the KL-ONE knowledge representation system". *Cognitive Science*, vol. 9, pp. 171-216, 1985

5. A. Brietzmann, U. Ehrlich: "The role of semantic processing in an automatic speech understanding system". *Proc. of COLING '86*, pp. 596-598, Bonn, Fed. Rep. Germany, Aug. 1986

6. R. Comino, R. Gemello, G. Guida, C. Rullent, L. Sisto, M. Somalvico: "Understanding natural language through parallel processing of syntactic and semantic knowledge: an application to data base query". *Proc. of the 8th IJCAI*, pp. 663-667, Karlsruhe, Fed. Rep. Germany, Aug. 1983

7. J. Courtain: "Algorithmes pour le traitement interactif des languages naturelles". These, Universite' Scientifique et Medicale de Grenoble, Grenoble, France, 1977

8. M. Danieli, F. Ferrara, R. Gemello, C. Rullent: "Integrating semantics and flexible syntax by exploiting isomorphism between grammatical and semantic relations". *Proc. of the 3rd Conf. of the Europ. Chapter of the ACL*, pp. 278-283, Copenhagen, Denmark, Apr. 1987

9. M. De Mattia, E. Giachin: "Experimental results on large vocabulary continuous speech understanding". *Proc. of the ICASSP '89*, pp.691-694, Glasgow, UK, Apr. 1989

10. L.D. Erman, F. Hayes-Roth, V.L. Lesser, D. Raj Reddy: "The Hearsay-II speech–understanding system: integrating knowledge to resolve uncertainty". *Computing Surveys*, vol. 12, pp. 213-253, June 1980

11. C.J. Fillmore: "The case for case". Bach, Harris (eds.): *Universals in Linguistic Theory*. Holt, Rinehart, and Winston, New York, 1968

12. L. Fissore, E. Giachin, P. Laface, G. Micca, R. Pieraccini, C. Rullent: "Experimental results on large-vocabulary speech recognition and understanding". *Proc. of the ICASSP '88*, pp. 414-417, New York, NY, Apr. 1988

13. R. Gemello, E. Giachin, C. Rullent: "A knowledge-based framework for effective probabilistic control strategies in signal understanding". *Proc. of GWAI '87*, pp. 104-113, Sept. 1987

14. E. Giachin, C. Rullent: "A control strategy for a knowledge-based approach to signal understanding". *Proc. of the 4th Esprit Technical Week*, pp. 836-849, Bruxelles, Belgium, Sept. 1987

15. E. Giachin, C. Rullent: "Robust parsing of severely corrupted spoken utterances". *Proc. of COLING '88*, pp. 196-201, Budapest, Hungary, Aug. 1988

16. E. Giachin, C. Rullent: "A parallel parser for spoken natural language". *Proc. of the 11th IJCAI*, pp. 1537-1542, Detroit, Mich., Aug. 1989

17. A. Giordana, P. Laface, A. Kaltenmeier, H. Mangold, R. Pieraccini, F. Raineri: "Algorithms for speech data reduction and recognition". *Proc. of the 2nd Esprit Technical Week*, pp. 845-853, Brussels, Belgium, Sept. 1985

18. G. Goerz, C. Beckstein: "How to parse gaps in spoken utterances". *Proc. of the 1st Conf. Europ. Chapt. ACL*, 1983

19. P.J. Hayes, A.G. Hauptmann, J.G. Carbonell, M. Tomita: "Parsing spoken language: a semantic caseframe approach". *Proc. of COLING '86*, pp. 587-592, Bonn, Fed. Rep. Germany, Aug. 1986

20. D.G. Hays: "Dependency Theory: a formalism and some observations". Memorandum RM4087 P.R., The Rand Corporation, 1964

21. F. Jelinek, R.L. Mercer, L.R. Bahl, J.K. Baker: "Perplexity - a measure of difficulty of speech recognition tasks". 94th Meeting of the Acoustical Society of America, Miami Beach, FL, 1977

22. P. Laface, G. Micca, R. Pieraccini: "Experimental results on a large lexicon access task". *Proc. of the ICASSP '87*, pp. 809-812, Dallas, Tex., Apr. 1987

23. A.L. Lepschy, G. Lepschy: *La lingua italiana*. Bompiani, Milano, 1986

24. L. Lesmo, P.Torasso: "Weighted interaction of syntax and semantics in natural language analysis". *Proc. of the 9th IJCAI*, pp. 772-778, Los Angeles, CA, Aug. 1985

25. F. Kubala et al.: "Continuous speech recognition results of the Byblos system on the Darpa 1000-Word Resource Management Database". *Proc. of the ICASSP '88*, pp. 291-294, New York, NY, Apr. 1988

26. W. Mann (chairperson): "Text generation". *American Journal of Comp. Linguistics*, vol. 8, no. 2, 1982

27. L.G. Miller, S.E. Levinson: "Syntactic analysis for large vocabulary speech recognition using a context-free covering grammar". *Proc. of the ICASSP '88*, pp 271-273, New York, NY, Apr. 1988

28. H. Ney, D. Mergel, A. Noll, P. Paeseler: "A data-driven Organization of the dynamic programming beam search for continuous speech recognition". *Proc. of the ICASSP '87*, pp. 833-836, Dallas, TX, Apr. 1987

29. G.T. Niedermair: "Merging acoustics and linguistics in speech understanding". *Proc. of the NATO ASI Conference*, pp. 479-484, Bad Windsheim, Fed. Rep. Germany, July 1987

30. M. Poesio, C. Rullent: "Modified caseframe parsing for speech understanding systems". *Proc. of the 10th IJCAI*, pp. 622-625, Milano, Italy, Aug. 1987

31. J.R. Rohlicek, Y.L. Chow, S .Roucos: "Statistical language modeling using a small corpus from an application domain". *Proc. ICASSP '88*, pp. 267-270, New York, NY, Apr. 1988

32. R. Schank: *Conceptual Information Processing*. North-Holland, New York, 1975

33. R.F. Simmons: *Computations from the English*. Prentice-Hall, 1984

34. A.R. Smith, L.D. Erman: "Noah - a bottom-up word hypothesizer for large-vocabulary speech understanding systems". *IEEE Trans. on Pattern Analysis and Machine Intelligence*, vol. 3, pp. 41-51, Jan. 1981

35. N.K. Sondheimer, B.Nebel: "A logical-form and knowledge-base design for natural language generation". *Proc. of the AAAI '86*, pp. 612-618, Philadelphia, PA, Aug. 1986

36. M.M. Sondhi, S.E. Levinson: "Computing relative redundancy to measure grammatical constraint in speech recognition tasks". *Proc. of the ICASSP '78*, pp. 409-412, Tulsa, OK, Apr. 1978

37. J.F. Sowa: *Conceptual structures*. Addison Wesley, Reading, MA, 1984

38. M. Tomita: "An efficient augmented–context–free parsing algorithm". *Computational Linguistics*, vol. 13, pp. 31-46, Jan.-June 1987

39. M. Tomita, J.G. Carbonell: "The universal parser architecture for knowledge–based machine translation". *Proc. of the 10th IJCAI*, pp. 718-721, Milano, Italy, Aug. 1987

40. M. Weintraub, H. Murveit, M. Cohen, P. Price, J. Bernstein, G. Baldwin, D. Bell: "Linguistic constraints in Hidden Markov Model based speech recognition". *Proc. of the ICASSP '89*, pp. 699-702, Glasgow, UK, Apr. 1989

41. T. Winograd: *Language as a Cognitive Process*, Vol. 1. Addison-Wesley, Reading, MA, 1984

42. W.A. Woods: "Optimal search strategies for speech understanding control". *Artificial Intelligence*, vol. 18, pp. 295-326, May 1982

Chapter 5

Implementation of a Parallel Logic + Functional Language †

Gian Paolo Balboni, Piergiorgio Bosco, Carlo Cecchi, Riccardo Melen, Corrado Moiso, Giorgio Sofi (CSELT)

5.1 Overview

This chapter presents the main features of the integrated logic plus functional languages IDEAL and K-LEAF along with their implementation on sequential and parallel architectures.

Our approach to integration is based on two levels [4]. The upper level is IDEAL (an Ideal DEductive and Applicative Language), a higher-order logic plus functional language [5] designed in order to offer in a unified coherent environment the most appealing features of Prolog and of modern functional languages: full invertibility, non-determinism, higher-orderness, lazy evaluation and type structures. The huge gap between the powerful computational model of IDEAL and the quite simple structure of the low-level underlying abstract machine is filled through a two-step compilation process:

1. an IDEAL program is transformed into a flat set of K-LEAF clauses. K-LEAF [6] is a well-founded first-order logic-functional language whose execution mechanism, based on SLD-resolution, provides a complete and efficient conditional equation solver, equivalent to conditional narrowing [8].

2. A K-LEAF program is compiled into K-WAM, a parallel extension of the WAM devised to efficiently implement the outermost resolution strategy (so as to achieve conditional narrowing and fully invertible lazy evaluation of functional calls).

The description of our parallel testbed PIPES (Packet Interconnected Processing ElementS) concludes the chapter.

5.2 Applications

The need for better ways to force computers to do exactly what we have in mind is still felt in the programming community. Even though, mainly for portability reasons, low-level imperative implementation languages (like C or parallel-C) are considered by many people the ultimate answer to the problem, we don't feel we are living in the "best possible world" and hope that an evolution will always be possible here as in any other scientific discipline. Our experience mainly comes from the world of complex real-time telecommunication systems, where typical code for call handling is more and more integrated with code implementing special services, such as speech access to information databases, intelligent network supervision and reconfiguration. The latter applications, as well as others of our interest, like robotic vision, require complex AI algorithms, which once prototyped, with substantial effort, in a conventional sequential environment, must be accelerated by one or two orders of magnitude to achieve the real-time performance. Therefore suitable linguistic tools should be available to ease both the early specification phase (by providing powerful structuring and control constructs, and reducing the possibility of introducing errors) and the move to parallel architectures in order to speed up the execution without complete reprogramming.

Our position is that *declarative* logic and functional languages are good candidates for such a role, being a good compromise between formal simplicity, expressive power, suitability for parallel execution and the possibility of optimization in order to approximate the efficiency of "normal" imperative languages. All of these targets have not been achieved yet, but the incoming results seem to suggest that all the technical problems can be solved soon. This position is not in contrast with the trend in "practical programming environments" towards the use of "fourth generation" languages, or *application generators* and *expert system builders*. Our impression is that such tools, which are emerging in a somewhat uncontrolled way (sometimes too complex and with unclear semantics), could be rephrased in a more compact and sound way in logic-functional languages (provided that a more appealing syntax than the bare mathematical one is offered to the user).

5.3 Languages

5.3.1 The Language K-LEAF

The K-LEAF syntax is based on Horn Clause Logic with Equality, extending pure Prolog in order to express *non-terminating conditional term rewriting systems with constructors*. Given a set V of variables, C of constructors, F of functions and P of predicates, a K-LEAF *program* consists of a set of clauses whose syntax is defined by the following grammar:

Clause	::=	Head :- Body. / Head.		
Body	::=	Atom / Atom, Body		
Term	::=	x / k(Term,...,Term)	$x \in V$ *and* $k \in C \cup F$	
Data-term	::=	x / c(Data-term,...,Data-term)	$x \in V$ *and* $c \in C$	
Atom	::=	p(Term,...,Term) /	$p \in P$	relational atom
		Term ≡ Term		strict-equality test
Head	::=	f(Data-term,...,Data-term)=Term /	$f \in F$	functional head
		p(Data-term, ...,Data-term)	$p \in P$	relational head

no multiple occurrences of the same variable in the head arguments,

in functional heads *var(rhs)* is included in *var(lhs)*

In order to guarantee confluency in function definition, the following additional constraint must be satisfied by a K-LEAF program: for each pair of functional heads of the form $f(d1,\ldots,dn) = t$ and $f(d1',\ldots,dn') = t'$, $(d1,\ldots,dn)$ and $(d1',\ldots,dn')$ are not unifiable. The clause $f(d) = t$:- b. means that $f(d)$ can be reduced to t if b is true (i.e. has a refutation). A K-LEAF *goal* is a conjunction of atoms (either relations, as in Prolog, or equations between terms) to be proved with respect to a program. An equation of the form $t \equiv R$, where t is ground (i.e. there are no occurrences of variables) and R is a variable, represents the evaluation of t to its normal form to which R will be bound.

The following set of clauses is a correct K-LEAF program:

```
plus(0,x)= x.
plus(s(x),y) = s(plus(x,y)).
nat(x) = cons(x,nat(s(x))).
odd(cons(x,cons(y,z))) = cons(x,odd(z)).
sqrlist=cons(0,sqrlist1(0,odd(nat(s(0))))).
sqrlist1(x,cons(y,z)) = cons(plus(x,y),sqrlist1(plus(x,y),z)).
p(x,y,z) :- p1(x,y,z,sqrlist).
p1(x,y,z,w) :- sqr(z,w) ≡ plus(sqr(x,w),sqr(y,w)).
sqr(0,cons(x,y)) = x.
sqr(s(x),cons(y,z)) = sqr(x,z).
```

The example defines some non-terminating functions, such as nat (the list of naturals), sqrlist (the list of squares), odd(s(0)) (the list of odd numbers): these *infinite* functions can be handled because of the non-strict (i.e. lazy) semantics of K-LEAF [6]; the relation $p(x,y,z)$ denotes all the triples $< x, y, z >$, such that: $x^2 + y^2 = z^2$. Possible goals (along with their answers) are:

?- p(x,s(y),z).	x = y := s(s(s(0))) z := s(s(s(s(s(0))))) ; ...
?- plus(s(s(0)),s(0)) ≡ r	r := s(s(s(0)))
?- plus(s(s(0)),s(x)) ≡ s(s(s(0)))	x := 0

5.3.2 The Language IDEAL

From the syntactic point of view IDEAL extends K-LEAF with the possibility to handle *λ−abstractions*, for the definition of higher-order functions/predicates. Moreover, the programs are constrained to be well typed with respect to a *polymorphic type system* [31]. The previous abstract K-LEAF syntax is enriched with the following rules:

Term	::=	Term **where** Program /	local definitions
		Term @ Term /	functional application
		λ (Abstr)	λ-abstraction
Abstr	::=	Data-term.Cond-term / Data-term.Cond-term ; Abstr	definition by cases
Cond-term	::=	Term / Term :- Body / :- Body	
Atom	::=	Term @ Term.	curried atom

As syntactic sugar, a clause of the form: $f = \lambda(x_1.\lambda(x_2.\ldots\lambda x_n.expr)\ldots))$. $n \geq 1$ can be written as: $f @\ x_1\ @x_2 @ \ldots @\ x_n = expr$.

Along with predefined constructors (for lists, numbers, etc.), new constructors can be introduced via polymorphic type declarations like the following one, which defines the type *tree* along with the associated constructors *leaf* and *node*:

tree(X) ::= leaf(X) ; node(X,tree(X),tree(X)).

The following example of functional program taken from [37] is a good introduction to the syntax of the functional subset of IDEAL (the type information inferred by the system is listed after the ":" symbol).

foldr @ Op@ Z = g
where g@ [] = Z.
    g @ [A | X] = Op @ (A, g @ X). $:(\alpha\#\beta \rightarrow \beta) \rightarrow \beta \rightarrow (list(\alpha) \rightarrow \beta)$.

An alternative definition of *foldr* (where the local definition g is replaced by a λ-abstraction) is:

foldr @ Op @ Z = λ[] . Z ; [A | X].´Op @ (A, foldr @ Op @ Z @ X)).

product = foldr @ (*) @ 1. :list(int) → int.
sum = foldr @ (+) @ 0. :list(int) → int.
?- sum @ $[1,2,3] \equiv R$. $R := 6$

Definitions of predicate combinators and lexically scoped predicate definitions are possible. This greatly improves conciseness and *modularization* of logic programs:

comb @ P @ (X,Y) :- P @ (Z,X),P @ (Z,Y).
    $:(\alpha\#\beta \rightarrow$ **truth-value**$) \rightarrow \beta\#\beta \rightarrow$ **truth-value.**

non-disjoint = comb@member.
    where member@ (X,[X | L]).
        member@ X,[Y | L]) :- member@(X,L)
            :list(α) # list(α) → **truth-value.**

```
person ::= (a;b;c;d;...).             % declares type person

  brothers =  comb@parent
    where     parent@(a,b).
              parent@(c,d).            :person # person → truth-value.
```

where *non-disjoint* succeeds when two lists are not disjoint, while *brothers* succeeds when its two arguments have a common parent.

5.3.3 Parallel IDEAL and K-LEAF

Due to their declarative nature, logic and functional languages are widely recognized as good candidates for a "natural" move from sequential programming of algorithms to the situation where several processing elements can be made to cooperate in parallel for the solution of a problem. We deliberately exclude here, although they are very important, other aspects of parallel/concurrent programming typical of embedded systems, where the program must cope with "simultaneous" and "continuous" changes of an external world. While the latter scenarios require real concurrency, i.e. at least some explicit notions at the user language of *merging* or *synchronization*, in the first case it is just a matter of programming style to introduce or not an explicit notion of parallelism.

With logic and/or functional languages we can, at least in principle, exploit the sources of parallelism already *implicitly* present in their operational semantics (based on resolution or on reduction), with no need of explicit constructs to spawn/synchronize parallel computations. A first consequence of this fact is that the same semantic characterization fits both the sequential and the parallel versions of such languages: concepts like *logic consequence*, *computed substitution answer*, *resolution step*, *normal form*, *rewriting step* are independent of the (sequential/parallel) way they are proved or computed.

The computational model of logic languages suggests two aspects that can, in principle, be handled concurrently, the search rule (OR-parallelism) and the computation rule (AND-parallelism) used to implement SLD-resolution. OR-parallelism introduces the possibility of performing in parallel all possible resolutions of the selected atom against the set of clauses defining it. AND-parallelism is related to the use of parallelism in the computation rule: two or more atoms in a goal can be resolved in parallel. On the other hand, the computational mechanism of functional languages, i.e. reduction, can be applied in parallel to different redexes in a term. As will be explained in detail in the following sections, we adopted (a version of) SLD-resolution on *flattened* programs as a uniform computational mechanism for our integrated logic+functional languages: this implies that in our case, OR-parallelism is essentially similar to that in logic languages, while AND-parallel execution of atoms derived from *flattening* of functional nestings into functional atoms is equivalent to parallel functional reduction.

The main problem in designing parallel algorithms is to achieve the "right" balance of parallel vs. sequential execution according to the actual features of the physical machine, such as number and power of processors, throughput and latency of the network. Several things are needed for a good compromise. On one side we have to guarantee enough parallelism for exploiting the processing power. On the other side we must ensure that each process has a suitable lifetime to compensate the overhead due to process creation and communication. The first constraint should primarily be solved by designing inherently parallel algorithms (for example, *quicksort* is "more" parallel than standard *bubblesort*).

The declarative and symbolic nature of logic and functional languages encourage the development of *divide-and-conquer* and *search-based* programs with great potential for parallelism. Other program analysis and transformation techniques, such as abstract interpretation [14], and partial evaluation, can further increase the degree of exploited parallelism without resorting to speculative parallelism (i.e. parallel evaluation of sub-problems which are not guaranteed to eventually contribute to the solution of the overall problem, e.g. parallel evaluation of both branches of if-then-else) which is usually more difficult to manage. In many cases the resulting fine granularity of potential parallelism does not fit the second constraint. A way to fix larger sequential parts of computation is needed. This can hardly be achieved by automatic tools. Our choice was to leave to the programmer such a responsibility, by providing him/her with simple annotations which characterize sequential and parallel parts: thus, our approach to parallelism could be called *controlled implicit parallelism.* The two forms of parallelism in IDEAL/K-LEAF are controlled in the following way:

- OR-parallelism: we must declare the procedures (i.e. function or predicate definitions) whose activations must be resolved in an OR-parallel way (in the current implementation of the language OR-parallel annotation is achieved by prefixing parallel procedures with $).
- AND-parallelism (for parallel functional computations, through flat SLD-resolution): annotations on functional calls (with the form *f*(...)//) denote that the resolution of the corresponding functional atoms must be performed by AND-parallelism.

As a simple example let us consider the IDEAL program:

```
sum_list([],F) = 0.
sum_list([E | L],F) = F @ E+sum_list(L,F).

$goal(L,R) :- sum_list(L,f) ≡ R.
$goal(L,R) :- sum_list(L,g) ≡ R.

f@X = long_computation(X).
g@X = short_computation(X).
```

A reasonable parallel modification of the program could be:

```
sum_list([],F) = 0.
sum_list([E | L],F) = (F@E)//+sum_list(L,F)//.

sum_list_seq([],F) = 0.
sum_list_seq([E | L],F) = F@E+sum_list_seq(L,F).

$ goal(L,R) :- sum_list(L,f) ≡ R.
$ goal(L,R) :- sum_list_seq(L,g) ≡ R.

f@X = long_computation(X).
g@X = short_computation(X).
```

Here *$goal* has to be executed by OR-parallelism; due to the "complexity" of f it is worthwhile to generate an AND-process for each application of f to a list element, while, for g, we estimate that sequential execution is better.

The IDEAL parallel procedures are translated into equivalent K-LEAF parallel procedures (see next sections). At the K-LEAF level additional constraints are put on the interaction among the OR-parallel part and the sequential one (executed by backtracking) in order to enable a better programmer intuition about the evolution of the parallel computation and to guarantee efficiency in sequential execution. Along with the static distinction between OR-parallel and sequential procedures (a procedure is a set of clauses defining the same predicate/function), this discipline imposes that a sequential procedure cannot directly invoke an OR-parallel one, but this interaction should take place only through a **parsetof** construct, an OR-parallel version of the built-in Prolog predicate **setof**. Moreover a parallel procedure can access the sequential component through an **all-solutions** construct (denoted by {...}); its related semantics is: first, all the solutions of the sequential goal are computed in a backtracking way, and, then, as many OR-parallel processes as the computed solutions are spawned to resolve the continuation.

```
app([],L2) = L2.
app([E | L1,L2]) = [E | app(L1,L2)].

$rev([]) = [].
$rev([E | L]) = app($rev(L),[E]).

g(L1,S) :- parsetof($rev(L2) = L1, L2, S).
```

- $rev is a parallel function, while *app*, and *g* are sequential functions and predicates;
- g(L0,[R]), returns in R the list whose reverse is L0 itself: if L0 = [1,2,3], then R := [3,2,1];
- a sequential call can occur in a parallel clause (e.g. *app* in the second clause for $rev), but it is implicitly embedded in an *all_solutions* construct;
- in a sequential clause, a parallel call is made by a *parsetof* construct (e.g. *$rev* in the clause for *g*).

IDEAL and K-LEAF actually inherit the constructs to control parallelism from a Parallel Prolog [24] developed in CSELT, which already provides the above primitives, together with other constructs for a better control of OR-parallel process spawning (**guards**), sound communication among independent OR-parallel processes (**assertion of lemmas**) and implementation of best-first search (user-defined **priorities** of OR-processes).

5.4 Models of Computation

The computational model of IDEAL consists of two parts:

1. a transformation to "compile" higher-order IDEAL programs into first-order K-LEAF ones;
2. a resolution-based execution mechanism (complete with respect to the K-LEAF declarative semantics) to efficiently execute K-LEAF programs. The (parallel) virtual machine for the execution of K-LEAF is described in Sect. 5.5.

5.4.1 Compiling IDEAL into K-LEAF

An IDEAL program is compiled into K-LEAF through *lambda-lifting* techniques for generating *fully lazy supercombinators* as recent compilers for functional languages [35]. In the logic programming framework, this technique can be seen as a partial evaluation of IDEAL programs with respect to given interpreter of IDEAL written in K-LEAF. As an example, consider the definition *twice*:

twice @ F @ X = F @ (F @ X).
plus1 @ X = s(X).

The result of its partial evaluation, in presence of the IDEAL interpreter (consisting of a first-order axiomatization of β-reduction written in K-LEAF), is the K-LEAF program:

twice @ F = twice1(F).
twice1(F) @ X = F @ (F @ X).
plus1 @ X = s(X).

where @ is an infix K-LEAF function symbol and *twice1* is a new constructor generated during the process of partial evaluation.

Lambda-lifting has been extended to cope with the aspects related to logic+functional integration, namely existential variables, conditions and predicates. For example, the IDEAL program:

$p@Z = 1 : -q@X, r@(\lambda y.X, Z).$
$r@(F, X) : -F@0 \equiv X.$

is transformed into the K-LEAF program:

$p@Z = 1 : -q@X, r@(p1(X), Z).$
$p1(X)@Y = X.$
$r@(F, X) : -F@0 \equiv X.$

A detailed description of the mapping of IDEAL into K-LEAF is reported in [7].

5.4.2 Execution of K-LEAF: Flattening and Outermost SLD-Resolution

The computational methods that have been proposed for the execution of languages based on Horn clause logic with equality are, in general, linear refinements of resolution and completion (i.e. SLD-resolution and narrowing, respectively). Among them we find conditional narrowing [20] [23] and SLDE-resolution (i.e. SLD-resolution with syntactic unification replaced by a semantic unification algorithm [25] [36] [29]).

Though the operational semantics of K-LEAF is *conditional narrowing*, which in [6] has been proved equivalent to a denotational semantics, its execution mechanism underlying the implementation, on the other hand, is *outermost SLD-resolution on homogeneous* (also called *flat*) *form* [6]. As described in [8] (basic-)conditional narrowing can be efficiently recast into SLD-resolution on a transformed program where functional nestings are eliminated by recursively replacing each functional call $f(t1,\ldots,tn)$ with a fresh variable v, named the *produced variable*, and adding the functional atom $f(t1,\ldots,tn) = v$ in the antecedent of the clause: the $=$ symbol is considered as an ordinary predicate and the axiom $x = x$ is added to the transformed program. For instance, the K-LEAF program in Sect. 5.3.1 is transformed into:

```
plus(0,x) = x.
plus(s(x),y) = s(v) :- plus(x,y) = v.
nat(x) = cons(x,v) :- nat(s(x)) = v.
odd(cons(x,cons(y,z))) = cons(x,v) :- odd(z) = v.
sqrlist = cons(0,v1) :- sqrlist1(0,v2)=v1, odd(v3)=v2, nat(s(0))=v3.
sqrlist1(x,cons(y,z)) = cons(v1,v2) :- plus(x,y) = v1, sqrlist1(v3,z)=v2, plus(x,y) =v3.
p(x,y,z) :- p1(x,y,z,w), sqrlist = w.
p1(x,y,z,w) :- sqr(z,w) = v1,v1 : v2,sqr(x,w) = v3,sqr(y,w)= v4,plus(v3,v4) = v2.
sqr(0,cons(x,y)) = x.
sqr(s(x),cons(y,z)) = v :- sqr(x,z) = v.
```

SLD-resolution on transformed programs seems to be more adequate than (conditional) narrowing:

- SLD-resolution was shown to be semantically equivalent to narrowing [8], with a considerable gain in efficiency (elimination of redundant solutions and, more generally, reduction of the search space);
- the full (relational + functional) language can be supported by a single inference mechanism;
- conditional equations can be easily handled, without need of extensions;
- call-by-need is obtained for free: a functional term is evaluated at most once.

Moreover, in spite of some independent effort in "direct" implementation of (conditional) narrowing, WAM (an abstract machine for Prolog [38]) can directly support it without extensions. An innermost strategy can be easily realized through the usual leftmost selection rule of Prolog, as long as the literals are put in the right order by the transformation. But, in general, the unlimited possibility of resolving functional atoms with $x = x$ has, as a serious drawback, a large amount of useless computation. From a theoretical point of view, the elimination of the reflexive clause causes the loss of completeness, unless functions are constrained to be *everywhere-defined* [23]. The introduction of an *outermost* selection strategy according to which a functional atom is resolved only when its produced variable would be bound to a non-variable term, eliminates all redundant resolutions against $x = x$. As the choice of functional atoms to resolve is *dynamically* performed (during unification), we must extend the WAM with a suspension/ reactivation mechanism for functional atoms (more complex than *freeze* [15]).

The outermost strategy we considered for the sequential implementation, reported in the following section, is specified by the following rules:

- all the *relational atoms* and *strict-equality tests* must be resolved: even if their selection order is immaterial, we adopted as default strategy for these atoms the left-to-right Prolog one;

- a functional atom is resolved only if the resolution of an atom A against a clause $H : -B$ *requires* its produced variable v (i.e. v is unified with a non-variable). To achieve better efficiency, through earlier detection of failures, the atoms required by a relational head or the *lhs* of a functional head are resolved before the atoms in B; those required by the *rhs* of a functional head are evaluated immediately after B;

- if a produced variable does not occur in the current goal to be resolved (and, thus, it can no longer be required by a resolution), its producer is eliminated (*elimination rule*).

Let us consider the following program (which is already in homogeneous form):

```
1) p(1,2) :- q(0).          3) f(1)=1.
2) q(0).                    4) f(2)=1.
```

and the goal ?- $p(f(x), x)$. transformed into ?- $p(v, x), f(x) = v$.

The only relational atom in the goal is $p(v, x)$ which is resolved with (1) to obtain ?-$f(2) = 1, q(0)$ with the most general unifier $\sigma = \{v := 1; x := 2\}$; σ requires the value of v , because it binds v to 1. The resolution of $\sigma(f(x) = v)$, i.e. the producer of v, is performed before the resolution of the relational atom $q(0)$ occurring in the body of (1). Its resolution with (4) succeeds and, then, the overall computation succeeds after the resolution of $q(0)$ against (2).

In some cases, e.g. during the resolution of strict-equality tests, the resolution of a functional atom is required even if the produced variable is still unbound. In these situations we want that, after the resolution of such a functional atom, its produced variable is bound either to a non-produced variable or to a term whose outermost functor is a constructor (i.e. it is in *head-normal-form*, hnf in the following). For sake of efficiency and maximization of possible parallelism, the general outermost strategy should allow "inner" sub-resolutions on *strict arguments*, i.e. those terms whose evaluation will eventually be required in the course of the computation. The above strategy is accordingly modified as follows: before the resolution of a functional/relational atom A, we must resolve (to hnf) all the functional atoms producing variables which are strict in A.

5.4.3 Parallel Outermost Strategy

In the implementation of the outermost strategy, parallelism can be exploited according to the parallel annotations; during the flattening, the AND-parallel annotation is inherited by functional atoms from the corresponding functional call:

- OR-parallelism: the resolution of an (OR-parallel annotated) atom $\$p(d1, \ldots, dn)$ or $\$f(d1, \ldots, dn) = d$ (where di are data terms) generates as many processes as the unifiable clauses defining $\$p$ or $\$f$. Each process executes the corresponding clause, and, if it succeeds, continues the evaluation of the goal; otherwise the process fails (and it is killed);

- AND-parallelism: the synchronous AND-parallel construct *and_par* (see Sect. 5.5.6) spawns n processes to resolve n parallel atoms, and waits for their termination. *And_par* is used to evaluate in parallel arguments of a function/predicate call: before the resolution of an atom A, all the AND-parallel functional atoms producing variables which are strict in A are evaluated in parallel by this construct; the same construct can be used to resolve in parallel the AND-parallel atoms dynamically required during unifications.

The flat K-LEAF form of the IDEAL program in Sect. 5.3.3 concludes this section:

```
sum_list([],F)          =  [].
sum_list([E | L],F)     =  R :- and_par ( F@E=R1, sum_list(L,F)=R2 ),
                               R1+R2 = R.

sum_list_seq([],F)      =  [].
sum_list_seq([E | L],F) =  R :- F@E=R1,sum_list_seq (L,F)=R2, R1+R2 = R.

$goal(L,R)              :-  sum_list(L,f)=R1, R1 ≡ R.
$goal(L,R)              :-  sum_list_seq(L,g) =R1, R1 ≡ R.
```

5.5 Language Implementation and Execution

In the previous section an overview of the outermost strategy was provided and it was pointed out that its implementation cannot be achieved only by directly compiling K-LEAF into Prolog. This is due to the fixed left-to-right atom selection order of Prolog, which does not allow a convenient expression of the more complex control of computation needed for outermost strategy, where the choice of functional atoms to be resolved must be performed within the unification algorithm. Therefore, we extended (the parallel version of) the WAM, so as to achieve a (parallel) implementation of K-LEAF. The choice of the WAM as a starting point for the abstract machine has been dictated not only by the resolution-based nature of the execution models of our integrated languages, but also by technological considerations: although we could re-invent a new abstract machine to support the logic-functional integration in the optimal abstract way, by choosing the WAM-approach we intended to capitalize on the extensive experience already available for sequential and parallel Prolog, and to be ready to incorporate all the new incoming optimizations.

The WAM is an abstract machine equipped with:

- a Code Area, containing object code (i.e. WAM instructions) produced by the compiler.
- a Symbol Table, allowing dynamic linkage/loading of procedures.

- a Stack, recording two kinds of information:
    1. *Choice points* (storing the status of the machine at the moment of the invocation of a multiple-clause procedure; that status will be restored upon backtracking, before starting the computation of the next alternative solution);
    2. *Stack frames* (storing the "environment" for the execution of the current clause, i.e. some control information and the bindings for logical variables (actually, permanent variables)).
- a Heap, where composite terms (i.e. structures and lists) are constructed (according to the so-called *structure-copying* technique).
- a Trail, storing information about the bindings to be *undone* upon backtracking.
- a PDL (Push Down List), used as a stack to hold operands and results of arithmetic instructions and also during recursive invocations of the unification algorithm.
- some Status Registers (e.g., P, program counter; E, environment pointer, referencing current stack frame; B, backtracking pointer, referencing most recent choice point; and lastly, the *X/A Register Array*, used both as temporary registers for storing temporary variables and as argument registers for parameter passing).

Prolog programs are compiled into WAM instructions, which can be classified into five groups:

- *procedural* instructions allocate/deallocate frames on the stack and call/return from a procedure.
- *get* instructions, corresponding to formal parameters of the head of the source clause, receive the parameters passed by the caller through argument registers and store them conveniently (temporary variables in temporary registers, permanent variables in locations of the stack frame). They define the *read/write-mode* for unification instructions, according to the instantiation state of the actual parameters.
- *put* instructions, corresponding to the arguments of body atoms, load actual parameters of the caller into argument registers. They set the *write-mode*.
- *unify* instructions, corresponding to the arguments of composite terms in a clause, are used either to build or access structures and lists, according to the *read* or *write* mode of execution, respectively, which is defined by *get* and *put* instructions.
- *indexing* instructions, allowing to reduce the non-determinism of a Prolog call selecting only the clauses (of a many-clause procedure) which actually unify with a given call.

The main reasons for the good performance of the WAM are: unification in most cases is not a general procedure, but is broken into simple test and assignment statements; an appropriate classification of variables (permanent, temporary and void) allows the cutting down of the number of memory accesses; some space-saving techniques (last call optimization and trimming [38]) contribute to reduce the amount of memory required at a given time, which is often a serious problem in logic programming; the technique of indexing on the first argument is an essential tool to reduce and in some cases eliminate the burden of handling backtracking.

The extensions of the WAM to execute K-LEAF are confined to the dereferencing primitive and to the instruction set, while the architecture is left unchanged (Sects. 5.5.2 and 5.5.3).

5.5.1 The Parallel Virtual Machine for K-LEAF

The parallel virtual machine we have designed takes into account both an OR-parallel (for the relational and equational part of K-LEAF) and an AND-parallel (for the achievement of functional parallel computations) component. The well-known problems discovered in implementing virtual machines supporting full AND/OR parallelism could be overcome by adopting the following restriction: an AND-parallel goal is constrained to be *one-solution* (i.e. it either fails or has only one answer substitution). Note that this is a natural restriction in the context of K-LEAF, where ground functional terms can be flattened into *one-solution* goals. More general conjunctions can be made to run in parallel by means of specialized primitives based on *setof* operators (see Sect. 5.5.6).

As our target is the achievement of a coarse-grain parallel implementation, where annotations control the creation of parallel computations, the design of the parallel virtual machine for IDEAL/K-LEAF can be split into two (almost) orthogonal parts:

- an efficient sequential implementation for K-LEAF execution (Sects. 5.5.2 and 5.5.3);
- a set of primitives to spawn and control/synchronize parallel computations (Sect. 5.5.5).

At the level of architectural abstraction the virtual machine for IDEAL/K-LEAF is a set of fully connected PCMs (i.e. a processor + a private memory + a local memory + a communication unit), whose local memories form a distributed global address space. Each PCM carries on, in a multiprogrammed way, non-trivial pieces of computations in a way substantially similar to a sequential K-LEAF execution, according to an imperative (K-WAM) code and acting on a number of internal data structures, which are the distributed pieces of the same structures of the sequential machine plus new ones for supporting the OR-binding conflicts.

The following aspects characterize parallel K-LEAF computations with respect to sequential ones:

- a computation can split into several either AND or OR parallel parts, and, thus, new processes must be created (and eventually eliminated);
- as the data structures of the whole parallel computation are distributed onto different local memories, at some point of the computation a process could require to access some location of the structures allocated in another local memory;

- OR-parallelism needs to cope with the problem of multiple-bindings (the same logical variable can be bound to different values in different OR-computations);

- the implementation of parallel functional reduction through AND-parallelism requires a synchronization mechanism similar to *read-only* variables in concurrent logic languages: a consumer P may require the value of a produced variable (which in this case must be implemented as a read-only variable) whose producer is one of the processes AND-related to P. If this value is not yet computed, P suspends and waits for it. Therefore, when a process binds a read-only variable, it must awake all the processes waiting for the value.

Intelligent local memories can be introduced to efficiently perform (in a mutually exclusive way) some complex memory accesses. The most relevant ones are variable dereferencing, and the requiring of a read-only variable. In the former case the operation could migrate on different local memories, by following the reference pointers, while in the latter, the operation could cause the suspension of the requiring process if the (dereferenced value of the) variable is an unbound variable. Multiprogramming is defined to be supported by processors, to cope with both physical (i.e. remote memory access) and logical (i.e. synchronization on *read-only* variables) latency.

5.5.2 Basic Compilation Scheme for Outermost Strategy

The efficiency of K-LEAF execution is related to the efficiency in recognizing produced variables and finding their producers: this is realized through a new WAM-type of term, called *prodvar*, which denotes the occurrence of a produced variable and *links* it to its producer. Moreover, by means of these links between variables and producers, the elimination rule is no longer considered, because a functional atom is implicitly eliminated as soon as in the current goal there are no longer occurrences of its produced variable. A *prodvar*(T,V,C), whose actual implementation in our K-WAM is shown in Fig. 5.1, joins the (internal representation of the) functional atom T and the produced variable V, while C is a control flag needed to avoid the duplication of the resolution of T=V (it is bound if T=V has been already resolved); *prodvars* can also be seen as a logical version of the functional closures.

The syntactic unification algorithm is replaced by a version *extended* to handle *prodvars*: when a *prodvar pv* must be unified with a term *t* then:

1. the algorithm performs the unification between *t* and the produced variable associated with *pv*;

2. if *t* is not a variable, *pv* is inserted into a global list, called *force-list*. In case of unification success, such *required* functional atoms are resolved to their *hnf* (through a *meta-call predicate*).

The basic compilation scheme of K-LEAF as well as the K-WAM, has been formally derived from the partial evaluation of the K-LEAF interpreter written in Prolog with respect to actual K-LEAF clauses. The compilation scheme for K-LEAF is an extension of the one for Prolog: most changes concern the compilation of the single clause, where

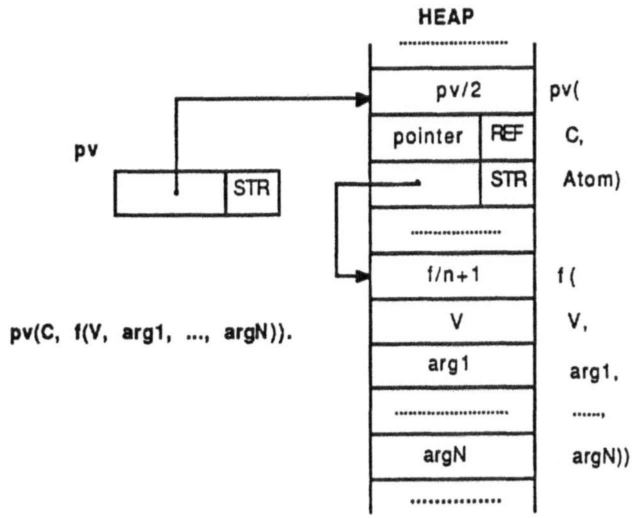

Figure 5.1: Actual representation of a *prodvar*

the instructions to deal with extended unification for outermost strategy and evaluation to hnf must be introduced. The other aspects (i.e. backtracking, indexing, environment management and procedure activation/return) are inherited from Prolog. An n-ary function g is compiled into an $n+1$-ary procedure g_f of K-WAM (i.e. $g(d1, \ldots, dn) = t$ is transformed into $g\_f(d1, \ldots, dn, t)$). Original WAM unification instructions have been enhanced to collect into the *force-list* the functional atoms required by the unifier. The force list is pointed by two state registers (namely, **hd** and **tl**), and is built on the heap. The *prodvars* collected into a force-list are resolved through the non-deterministic system predicate *force*.

Compared with the Prolog compilation scheme of the clause, the K-LEAF compiler adds some instructions that initialize (*initreg* instructions) and return (*close* instructions) the *force-lists*, and the calls to the predicate *force*. The activations of the required functional atoms are performed when the unification of the head is completed. Therefore required atoms need to be temporarily collected during unification. In clauses defining functions, a second *force-list* may be necessary, since *the prodvars* met in the unification of the functional-head rhs must be forced after the execution of the body. Clauses with relational heads need only one force-list "forced" before the body, while the general compilation scheme of those with functional heads is:

```
init_reg1                           initialize hd and tl
<lhs unification>                   get & unify instructions operate on tl
init_reg2                           hd → hd1, reset hd and tl
<rhs unification>                   get & unify instructions operate on tl
close1                (a1)          hd → a1 : put rhs force-list in a1
get_var_y             yj a1         save a1 in permanent register
close2                (a1)          hd1 → a1 : get lhs force-list
call                  force/1       force lhs prodvars
<compilation of the body>
put_value_y           yj a1         get rhs force-list from permanent register
call                  force/1       force rhs prodvars.
```

Several optimizations are performed in some cases to eliminate the instructions initializing/ returning/forcing the *force-list.*

The compilation of the rhs of functional heads must guarantee that the resolution of a functional atom binds its produced variable at least to hnf. It depends on the type of the rhs:

- function call: it is directly flattened as the last atom in the body;
- constructor: no further instructions are needed as in this case the *rhs* is already in *hnf* ;
- variable (e.g. X): the special instructions *get_result* or *execute_hnf are generated* if *X* occurs as an argument in the *lhs*, or if X occurs within a structure, respectively.

The second aspect peculiar to the compilation of K-LEAF programs is the generation of *prodvars*: if a function call is met in the body (respectively in the rhs of a functional head, but not at the outermost position), the compiler generates the instruction *put_prodvar* (respectively *get_prodvar*). We show the compilation of the standard *naive-reverse* function. K-WAM code is preceded by an abstract Prolog-like description:

```
rev([])=[].
rev([E | L])=app(rev(L),[E]).

rev(A1,R) :- initlist,A1=[],R=[],closelist(L),force(L).
rev(A1,R) :- initlist,A1=[E | L],closelist(FL), force(FL),app(pv(rev(L)),[E],R).
```

```
rev     switch_on_term     $ cl1 , $ cl2 , $ fail
$ ch1   try_me_else        $ ch2
$ cl1   init_reg1
        get_nil            a1
        put_nil            a3
        get_value_x        x2 , a3
        close1
        execute            force/1
$ch2    trust_me_else_fail
$ cl2   init_reg1
```

```
unify_var_y      y3
get_var_y        y2 , a2
close1
call             force/1 , 3
put_prodvar      rev_f/2 , a1
unify_value_y    y3
put_list         a2
unify_value_y    y1
unify_nil
put_value_y      y2 , a3
```

11	allocate	3	24	deallocate	
12	get_list	a1	25	execute	app_f/3
13	unify_var_y	y1			

5.5.3 The Actual Compilation Scheme

The outermost strategy implemented by the basic compilation scheme is optimal with respect to failures. On the other hand, even with sharing of functional calls, it is very inefficient for normal function evaluation: if a *prodvar* is required by n alternative resolutions of a call, then its evaluation is performed n times in n independent OR-branches. We developed a compilation scheme to overcome this drawback, without losing the advantages of outermost strategy. It can be applied to a wide class of functions defined by cases. We adopted a version of the pattern compilation proposed by Augustsson for Lazy-ML [2]. It is obtained by generating new clauses which evaluate (to their hnf) the arguments *always-required* by a procedure, once for all, before the attempt of unifying the computed values against the corresponding patterns in the clause heads (this method is recursively applied to the newly produced clauses). A simple characterization of always-required arguments is:

Given a procedure (with more than one clause) defining a predicate/function, its i-th argument is *always-required* if all the i-th formal arguments in the head patterns are non-variable terms.

For instance, *rev* always-requires its first argument:

```
rev(A1,R) :- eval_hnf(A1),aug_1(A1,R).     aug_1([],[]).
                                           aug_1([E | L],R) :- app(pv(rev(L)),[E],R).
```

aug_1 is similar to the original definitions of *rev*. But when it is invoked, the first argument is always at least in hnf; therefore its unification against [] or [.|.] never requires a functional atom.

The introduced scheme avoids that always-required arguments are evaluated as many times as the number of unifiable clauses but, on the other hand, it introduces another overhead, because it doubles the number of resolutions even if no functional atoms must be evaluated. We further optimized this compilation scheme to avoid this drawback, at least when the patterns of all the clauses in a procedure require the first argument. This optimization is based on an extension of the *switch-on-term* instruction, called *switch-on-prodvar*: an additional case is added to deal with an unevaluated *prodvar*:

```
proc/n :  switch_on_prodvar   $const $list $struct $pv
  $var:   try_me_else         $ else
          ...
  $pv:    save n                                 (allocate +) save a1 toan on the stack
          force_switch                           eval to hnf a1
          restore                                restore a1 to an (+ deallocate)
          execute proc/n
```

The new branch is entered only when we actually have to force a *prodvar*, otherwise the original clauses are directly activated.

A further improvement can be obtained by avoiding the building of *prodvars* for *always-required* function calls. If a function call $f(t1,\ldots,tn)$ occurs as an *always-required* argument in a context $C[f(t1,\ldots,tn)]$, the corresponding *prodvar* is not generated, but the context is transformed into the conjunction $(f(t1,\ldots,tn,V),C[V])$. If a variable X occurs as an *always-required* argument in a context $C[X]$, the context is transformed into the conjunction $(eval\_hnf(X),C[X])$, where $eval\_hnf$ is compiled into the K-WAM instruction $call\_hnf$. This transformation is performed to guarantee that, when a call is evaluated, all its *always-required* arguments are already in hnf (it is avoided if X is the first argument since this case is dealt by the *switch-on-prodvar* scheme).

The *context-sensitive* flattening avoids building *prodvars* immediately required as soon as the outer call is evaluated. In certain situations, it must introduce auxiliary functions, as it is shown in the following example. Let $q2$ be a predicate which does not *always-require* its first argument.

```
q1(X) :- q2(rev(g)).
```

is abstractly compiled into:

```
q1(X) :- q2(pv(str1_0)).

str1_0(R) :- g(R1),rev(R1,R).
```

Note that the flattening $q1(X) : -g(R), q2(pv(rev(R)))$ is not correct, with respect to the K-LEAF non-strict semantics, because the resolution of $g(R)$ would always be performed even if *rev* is not evaluated.

To summarize, the developed compiler for K-LEAF is based on the three presented schemes:

- context-sensitive flattening when an argument is *always-required*;
- compilation à la Augustsson for "complex" *sequential* patterns [27], with the *switch-on-prodvar* optimization);
- the most general scheme, based on run-time accumulation and resolution of required functional atoms, to deal with non-sequential patterns, one-clause procedures and rhs's of functional heads.

An example which requires the use of all three schemes is:

```
f(0,X) = X.
f(1,[]) = [].
f(1,[E|F]) = [E|f(g,F)].
```

```
f(Pv,L,R) :- prodvar(Pv),!,eval_hnf(Pv),f(Pv,L,R).
f(0,A2,R) :- initlist,get_result(A2,R),closelist(L),force(L).
f(1,A2,R) :- eval_hnf(A2),aug_0(A2,R).
```

```
f      switch_on_prodvar   $c_bl , $fail , $fail , $cl3
$ch1   try_me_else         $ch2
$cl1   init_reg1
       get_cons            &0 , a1
       get_var_x           x1 , a3
       get_result_x        x2 , a1
       close1_skip_execute
       execute             force/1
$ch2   trust_me_else_fail
$cl2   init_reg1
       allocate            2
       get_cons            &1 , a1
       get_hnf             y1 , a2
       get_var_y           y2 , a3
       close1_skip_call
       call                force/1 , 2
       put_value_y         y1 , a1
       put_value_y         y2 , a2
       deallocate
       execute             aug_0/2
$cl3   save                3
       force_switch
       restore
       execute             f_f/3
$c_bl  switch_on_cons      4 ,$fail
                           &0 , $cl1 | &1 , $cl2
```

```
aug_0([])=[].
aug_0([E|F])=[E|str1_1(F)].
```

```
str1_1(F)=f(g,F).
```

```
aug_0([],[]).
aug_0([E|F],R) :-  initlist,R=[E|pv(str1_1(F))],
                   closelist(L),force(L).
```

```
str1_1(F,R) :- g(R1),f(R1,F,R).
```

```
aug_0  switch_on_term      $cl1 , $cl2 , $fail
$ch1   try_me_else         $ch2
$cl1   get_nil             a1
       put_nil             a3
       get_value_x         x2 , a3
       proceed
$ch2   trust_me_else_fail
$cl2   init_reg1
       get_list            a1
       unify_var_x         x3
       unify_var_x         x4
       get_list            a2
       unify_value_x       x3
       unify_var_x         x5
       get_prodvar         str1_1/2 , x5
       unify_value_x       x4
       close1_skip_execute
       execute             force/1
```

```
str1_1 allocate            3
       get_var_y           y1 , a1
       get_var_y           y2 , a2
       put_var_y           y3 , a1
       call                g_f/1 , 3
       put_unsafe_value    y3 , a1
       put_value_y         y1 , a2
       put_value_y         y2 , a3
       deallocate
       execute             f_f/3
```

The described compilation is independent of the possible way arguments are characterized as *always-required*: user annotations (already exploited by the compiler) as well as complex strictness-analysis tools developed for functional languages [14] can be adopted. A detailed description of the sequential version of K-WAM can be found in [12].

5.5.4 C-Emulation of Sequential K-WAM and Benchmarks

An original C-emulator of WAM was upgraded in order to support K-WAM. For ease of implementation we chose to represent *prodvars* as standard Prolog terms, instead of more efficiently adopting a dedicated tag. The dereferencing primitive has been accordingly enhanced to deliver the ultimate value of an already evaluated *prodvar*-term. The operation of *forcing* a *prodvar* (i.e. evaluating a suspension) is implemented partly in K-WAM code and partly in C-language by means of a specialization of the original Prolog *meta-call* primitive. These extensions of the C-emulator are conservative, in the sense that any Prolog program compiled into WAM code can be executed by the K-WAM emulator (with an average overhead of 10%, which can be dropped to 5% by using a *prodvar* reserved tag).

Simple benchmarks have been run on the IDEAL/K-LEAF system

1. in order to get a flavor of the overhead imposed by lazy evaluation in the context of K-WAM code, and
2. in order to compare the performance of outermost resolution with lazy evaluation of functions. The C-emulated K-WAM was run on a VAX 8700. Run-times are expressed in milliseconds.

Examples:

revN:	Naive reverse of a list of N elements, in functional style
revIN:	Naive reverse (functional) of a list of N elements in inverted mode
fibN:	Nth fibonacci number
walkN:	Non-naive functional rev of a list of N elements repeated N times

The execution times in Fig. 5.2 refer to outermost compilation, with simple strictness analysis. In the compilation of *fib*, the first argument could not be inferred as *always-required*, due to the absence of a constructor in the head of the recursive case. By adding a strictness annotation to the *fib's* argument, which is equivalent to inferring the *always-required* status, *fib* is compiled as in Prolog, hence the ratio is close to 1.

A more significant benchmark showing good performance with respect to conventional languages has been represented by an *invertible* event-driven logic simulator described in [9].

We experimented with the expansion in C-code of the extended WAM instructions as an approximation to a native code generation in order to estimate the achievable performance on several machines.

Query	Time K-LEAF	Time Prolog	Ratio
rev40	82.5	28.5	2.9
rev20	2.25	0.75	3.0
revI40	1200	397	3.01
revI20	165	60	2.75
fib15	285	135	2,1
fib10	22.5	12.7	1.77
walk300	3037	2850	1.06
walk200	1282	1275	1.00

Figure 5.2: K-LEAF vs. Prolog execution time

Figure 5.3 reports the execution times (in seconds) of some C-expanded K-LEAF programs, and compares them with corresponding programs in Quintus Prolog 2.2, Lazy ML [2], and C-language, under Sun3/280. The programs are *queensN* (the C implementation is imperative and makes use of arrays), *fibN*, *walkN* and *revN* (with the list constructor strict in both arguments).

Even though we tested this approach on a small set of programs and we do not yet have a well engineered system for C-code generation, the above figures suggest that the approach of C-compilation is viable, with the great advantage of portability and scalability to new RISC machines. The drawback of the C-expansion approach, besides a significant increase of the total compilation time, is the occupation of the generated code: in fact, the average size of the C-object file is double that of the code produced by the Quintus Prolog compiler.

5.5.5 Execution of OR-parallel K-LEAF

As already explained in Sect. 5.3.3, OR-parallel K-LEAF extends the syntax of an OR-Parallel Prolog towards the capability to express functions and equations. To obtain an OR-parallel system for K-LEAF the required extensions to WAM are the (almost disjoint) union of those required by OR-parallel Prolog and K-LEAF outermost strategy: the same extensions made to sequential WAM to support sequential K-LEAF can be inserted, almost without changes, in an OR-parallel version of WAM. The extensions to the WAM for OR-parallel Prolog have to do with:

- OR-nondeterminism: primitives are introduced to create and spawn processes when a parallel procedure is invoked: in our solution *first-par-clause*, *par-clause* and *spawn* instructions replace *try*, *retry* and *trust* instructions used in sequential procedure to handle backtracking.
- multiple-bindings: new data structures (namely, *binding arrays* and *binding lists* in our solution) are added to cope with the problem of multiple-bindings: they

Query	K-LEAF	Quintus	Lazy ML	C
queens9	5.5	13.18	7.5	1.0
fib29	22.4	46.7	11.8	4.8
walk300	0.6	1.0	0.55	0.4
5*rev300	1.75	1.65	//	//

Figure 5.3: C-expanded K-LEAF performance

introduce a different variable representation which influences both the dereferencing and the binding algorithms.

- OR-parallelism and backtracking interaction: new primitives (e.g. *setof* constructs, ...) to allow this interworking are introduced.

The OR-parallel version of WAM we considered is the one developed for the language described in [24]: it is a variant of the SRI *multisequential* model [39] which was developed to keep as much as possible the WAM approach in parallel architectures with common address space. The main differences are:

1. *eager spawning of processes* with respect to *retroactive parallelization*: when a parallel atom must be resolved, the unification attempts against different clauses are performed sequentially, before spawning as many processes as successful unifications;

2. an original mix of *binding arrays* and *binding lists*, to implement multiple environments: binding arrays are used to bind stack variables, and binding lists to bind heap variables, shared among different OR-processes;

3. a discipline in the interaction between backtracking and OR-parallelism: a *parsetof* construct is used to start a parallel computation in a sequential context, while an *all-solutions* construct allows invocation of a sequential computation in a parallel context. All the details of this abstract machine, along with the description of its implementation on the architecture described in Sect. 5.6, can be found in [30].

As regards process creation, our preference for eager spawning was dictated by the need to possibly support a search strategy more general than depth-first (as a matter of fact, a best-first score-guided approach can profitably be used in many AI applications). Moreover, also from a philosophical viewpoint we share the opinion of [26], who have studied a large realistic OR-parallel application of natural language processing and postulate that process creation should be determined solely by annotations in the program, not by any run-time considerations. Compared with the multisequential model, where the number of parallel activities is naturally limited to the number of physical processors, in our model the combinatorial explosion typical of unwise brute-force OR-parallel search is controlled in a different way, but with comparable effectiveness, by carefully exploiting the control constructs and annotations provided by the language (e.g. distinction between parallel and sequential procedures, *guards, if-then-else parallel construct*, etc.). Besides, it is to be stressed that eager spawning of a process does not necessarily imply an immediate execution of it; namely, a new process is not started soon, but for some time is forced to stand *quiescent* in a scheduling queue, so occupying a minimal amount of memory and preventing eager creation of further processes. Stated in different terms, if the language primitive to modify process priority is not used, the resulting search strategy is *multi-depth-first*, that is, the same strategy as the multisequential model (the only overhead to pay for our more general scheme is that the actual representation of a quiescent process is a bit less concise than an ordinary *choice point*).

For the problem of implementing a structure of *Multiple Environments*, through which alternative OR-parallel subcomputations can assign different values to variables which are still unbound when the computation forks at a split node, none of the solutions proposed in the literature (for a broad and comparative discussion see [18] [40]) seemed to fully satisfy the requirements of our model. The most interesting methods can roughly be grouped under two classes: *Binding Array* (BA) and *Hash Windows* (and variations of these).

Very briefly, the former technique consists in transforming a reference to an unbound variable into the index of an array. Multiple bindings can be created in a straightforward way by assigning each process its own BA copy. Hash Windows are an optimization of a more elementary concept, the *Binding List* (BL), which in turn is a generalization of the Trail of the sequential model. An assignment to a variable of a *shared frame* (i.e. allocated before the most recent split point) cannot be done directly and, instead, a pair *<var_address, value>* is added to BL; the dereference algorithm is modified accordingly: if the variable cell contains the value *unbound* and belongs to a shared frame, BL is scanned sequentially until a binding for that variable is found; only if the search fails, that variable is actually still unbound. These two methods are opposite in behaviour: BA allows a fast constant-time access to a variable cell, but at the expense of a remarkable overhead for process creation or switching (in our model, each process requires an individual BA copy, which implies that at the moment of spawning a new process the parent's BA is duplicated); BL introduces no overhead at the time of process creation (the list is just turned into a tree reflecting the same topology of the search tree, where ancestor arcs are shared among descendant processes), but at the expense of a search of unpredictable length (Hash Window optimization can only slightly alleviate this shortcoming). Incidentally, it should be noticed that unless appropriate measures are taken, each BA always grows, as its lifetime has the same nature as the lifetime of Heap (deallocation taking place only upon backtracking and not upon procedure success); of course, naively copying an ever-growing data structure is not recommendable practice.

Starting from these considerations, we have devised an original technique combining BA and BL, which tries to keep the advantages of both methods while avoiding drawbacks as much as possible. Our strategy is based on a compile-time classification of potentially unbound parallel variables, which are distinguished into two classes: *local* (roughly, *stack* variables, i.e. not occurring in a structure) and *global* (*heap* variables). Only local variables are inserted into BA, while multiple bindings for global variables are built through BL; the resulting benefit is that in this way the Binding Array ceases to be an ever-growing structure and can shrink upon procedure success.

To illustrate the interworking of OR-parallelism, backtracking and outermost strategy and to mirror the underlying WAM-based implementation, we describe a kernel of an interpreter for OR-parallel K-LEAF, written in OR-Parallel Prolog [24]. By using a parallel version of Prolog as specification language, we abstract from the implementation details only related to OR-parallel resolution.

The first aspect concerns the representation of suspended functional atoms: in OR-parallel K-LEAF there are two types of *prodvars*: *s_prodvar/3* and *p_prodvar/3* denote suspensions of sequential and parallel functional calls, respectively. For example, the internal form of the second clause for *$rev* defined in Sect. 5.3.3 is:

$rev([E | L]) = s_prodvar(app(p_prodvar($rev(L),V1,C1),[E]),V2,C2).

The interpreter consists of four main parts. Here, we report the most relevant ones.

1. Resolution of sequential goals: *eval* uses *eval_atom* to describe the resolution of a functional/relational atom. It mirrors the sequential implementation detailed in Sect. 5.5.2.

2. Resolution of parallel goals: calls of strict-equality test and sequential atoms (e.g. A) are evaluated through the *all_solutions* construct ({...}).

```
$eval(true).

$eval( (G1,G2) ) :- $eval(G1),$eval(G2).

$eval(parsetof(G,T,S)) :- parsetof(G,T,S).

$eval(G) :- { eval(G) }.

$eval( X≡Y ) :- { strict_equality_builtin(X,Y) }.

$eval(atom(A)) :- sequential(A) → { eval_atom(A) }; $eval_atom(A).
```

Resolution of parallel functional/relational atoms: clause selection and extended unification are performed inside an *all_solutions* construct, to point out that the unification attempts against different clauses are performed sequentially, before spawning as many processes as unification successes (see below the conjunctions $kclause(H = R, B), unifyarg\ (Argh, Arg, [], Larg), unifyres(D, R, [], Lres)$ for functional atoms and $kclause(H = R, B)$, $unifyarg(Argh, Arg, [], Larg)$ for relational ones).

```
$eval_atom(T=D) :-  functor(T,F,N),T=..[_ | Arg],functor(H,F,N),H=..[_ | Argh],
                    { kclause(H=R,B),unifyarg(Argh,Arg,[],Larg),unifyres(D,R,[],Lres) },
                    $force(Larg),$eval(B),$force(Lres),$eval_hnf(D).

$eval_atom(P) :-    functor(P,F,N),T=..[_ | Arg], functor(H,F,N),H=..[_ | Argh],
                    { kclause(H,B),unifyarg(Argh,Arg,[],Larg) },
                    $force(Larg),$eval(B).
```

The following actions are performed during the resolution of the parallel functional atom $T = D$ with the clause $H = R : -B$ (compare them with the compilation scheme reported in Sect. 5.5.2):

- unifyarg(Argh,Arg,[],Larg): The enhanced unification between the arguments of the functional call T and the pattern in H is attempted and the list *Larg* holding all the required producers occurring in T is built;
- unifyres(D,R,[],Lres): The enhanced unification between D, the requested value for T, and R, the rhs of the functional head, is performed, and the required producers are inserted in the list *Lres*;
- '}': if the enhanced unification succeeds, a new process is created to continue the computation;
- $force(Larg): The producers in list *Larg* are resolved;
- $eval(B): The atoms in the body B are resolved;
- $force(Lres): The producers in list *Lres* are resolved;
- $eval_hnf(D): The value of D is examined: if it is a not- yet-evaluated *prodvar* then it is resolved.

3. Activation of required functional atoms: *$force* and *$eval_hnf* are the parallel counterparts of the sequential *force* and *eval_hnf*. The peculiar aspect of these predicates is that they must be able to resolve sequential (resp. parallel) functional atoms even in a parallel (resp. sequential) context:

   - a sequential functional atom in a parallel context is resolved through the *all_solutions* construct;
   - a parallel functional atom in a sequential context is resolved through the *parsetof* construct;

```
$force([]).
$force([s_prodvar(T,V,C) | L]) :- (var(C) → C=done,{eval_atom(T=V)};true), $force(L).
```

```
$force([p_prodvar(T,V,C) | L]) :- (var(C) → C=done, $eval_atom(T=V) ; true),
$force(L).

force([]).
force([s_prodvar(T,V,C) | L]) :- (var(C) → C=done,eval_atom(T=V) ; true),force(L).
force([p_prodvar(T,V,C) | L]) :- (var(C) → C=done,force_aux(T=V) ; true),force(L).

force_aux(G) :- parsetof($eval_atom(G),G,L),force_member(L,G).
force_member([X | _],X).
force_member([_ | L],X) :- force_member(L,X).
```

force_member is a version of member predicate that non-deterministically unifies G against an element (i.e. an instance of G) in the list L computed by the *parsetof* construct.

4. System primitives (e.g. *unifyarg, unifyres,...*): they are those of the sequential implementation enhanced to work with two types of *prodvar* terms. Both must be dealt with as *prodvars* in the sequential case, as it can be noted in the definition of kind (which can be thought as the specification of the system *dereferencing* primitive):

```
kind(lvar(L),K,Y) :- !,(var(L) → K=var,Y=lvar(L) ; kind(L,K,Y)).

kind(X,pv,X) :- (X=s_prodvar (_,_,C) ; X = p_prodvar(_,_,C) ), var(C), !.

kind(X,K,Y) :- (X=s_prodvar(T,V,C) ; X =p_prodvar(T,V,C) ),!,kind(V,K,Y).

kind(X,constr,X).
```

The WAM-based implementation of the above model is a first actual approximation to the implementation of the virtual machine (described in Sect. 5.5.1).

Some constraints are inherent in this implemented model: the execution of the constructs which allows evaluating a parallel (resp. sequential) atom G in a sequential (resp. parallel) context (i.e. namely, *parsetof* and *all_solutions*) terminates only when the whole search-space for G has been explored. This could be prevented by extending the *all_solutions* construct (so that it eagerly spawns the computed solutions) and by introducing a new version of the *parsetof* construct whose partially computed list is immediately made available for a consumer. Moreover, as regards the use of the *parsetof* construct to resolve parallel functional atoms required in a sequential computation, we have to remark that this could imply performing expensive copies of data structures (representing the functional atom along with its arguments). We are currently investigating a variant of the *parsetof* construct to be used in this case, so as to eliminate (or, at least, to reduce) the amount of data to be copied.

The following is the K-WAM code resulting from the compilation of the *naive-reverse* OR-parallel definition reported in Sect. 5.3.3:

```
$rev([])=[].

$rev([E | L])=app($rev(L),[E]).


$ rev(Pv,R) :- prodvar(Pv),!,$ eval_hnf(Pv),$ rev(Pv,R).

$ rev([],[]).

$ rev([E | L],R) :- $rev(L,R1),{app(R1,[E],R)}.
```

```
$rev   switch_on_prodvar   $cl1, $cl2, $fail, $cl3
$ch1   first_par_clause    $cl1
       par_clause          $cl2
       spawn
$cl1   get_nil             a1
       put_nil             a3
       get_value_x         x2 , a3
       save_process
       proceed
$cl2   par_allocate        4
       get_list            a1
       unify_var_y         y1
       unify_var_y         y4
       get_var_y           y2 , a2
       save_process
       put_value_y         y4 , a1
       put_var_ba          y3 , a2
       call                $rev$f/2
       all_solution
       spawn
       put_value_y         y3 , a1
       put_list            a2
       unify_value_y       y1
       unify_nil
       put_value_y         y2 , a3
       call                app$ f/3
       save_process_as
       par_deallocate
       proceed
$cl3   par_save            2
       par_force_switch
       par_restore
       execute             $rev$ f/2
```

A detailed description of the OR-parallel abstract machine for K-LEAF and the related compilation schemes is in [10] [11].

5.5.6 Mapping AND-parallelism into OR-parallelism

The kind of AND-parallelism necessary to fully exploit the parallelism available in functional reduction (namely the possibility to run in parallel arguments and function bodies) requires the introduction of synchronization primitives (similar to *read-only* variables) and a somewhat different memory management (see Sect. 5.5.1); on the other hand, as discussed in [7], we realized that the set primitives already present in the OR-parallel implementation allow simpler, but quite useful, forms of AND-parallelism, namely those related to the parallel evaluation of arguments of functions and data structures.

As a matter of fact a similar approach has been also recently taken by Argonne researchers in the context of the GigaLips project [16], in contrast with more complicated AND/OR schemes requiring a highly sophisticated and complex virtual machine like in [41]. It amounts to (syntactically) converting a problem of AND-parallelism into a program which only makes use of OR-parallel set primitives.

```
and_par(F,G) :-
              parsetof( (parsetof(F,F,Fset) $;
```

```
parsetof(G,G,Gset) ), (Fset,Gset),[(Fs,Gs),(Fs,Gs)]),
force_member(Fs,F),force_member(Gs,G).
```

Here F and G are evaluated in parallel (by the outermost *parsetof*) and each of them can lead to several OR-parallel solutions (computed by the innermost *parsetof* constructs). After this the computation becomes sequential, and calls of *force_member* (specified in the previous subsection) perform the join of solutions in a backtrack fashion (of course, an OR-parallel version of *force_member* will compute the join in an OR-parallel way). The major drawbacks of this method are the one previously mentioned for the interworking of outermost strategy and OR-parallelism, namely non-termination for infinite solution spaces and substantial copying. Anyway, we share with [16] the feeling that the method should suffice for many practical cases.

5.5.7 The Actual Parallel Implementation

A parallel C-emulator system for IDEAL/K-LEAF has been implemented on PIPES, our physical Transputer-based parallel machine described in the following section, by extending the already existing parallel Prolog system [30]: along with OR-parallel computations this emulator performs parallel functional computations through the *and_par* construct. The present system also includes a compiler from parallel K-LEAF to parallel K-WAM code and a lambda-lifting procedure from the parallel version of IDEAL to parallel K-LEAF. With respect to a sequential emulator the parallel implementation is obtained by:

- Vectorizing the sequential emulator, that is, a set of WAM states is used to represent several parallel WAM processes. Each processor holds a queue of (pointers to) such ready processes (one of them is actually running).
- Introducing a load-balancing policy.
- Implementing a fixed-length block memory management to grow the WAM data structures.
- Enriching the WAM instruction set and run-time primitives with those for handling OR-parallelism and outermost strategy.
- Introducing procedures and message passing for remote access triggered by software-based address control and for detecting distributed termination of OR-processes.
- Adding some graphic display of figures about degree of parallelism, processor load, etc.

One of the main issues in a distributed multiprocessor is the implementation of an efficient parallel scheduling policy, able to achieve a fair load balancing of processes among processors. Two principles have been followed:

a) In a physical architecture like ours, lacking a true common memory, a distributed policy is in any case to be preferred to a centralized one.

b) In order not to compromise locality of reference, processes can be transferred only once. After a process has been inserted into the scheduling queue of a processing node, further migrations are forbidden (a process in the course of execution may have generated a lot of data structures in the local memory of a Transputer, which we don't want to move).

Our load-balancing mechanism operates on a demand basis, so being able to distribute processes only when really needed. It works as follows:

a) When new processes are spawned in a processor, they are inserted into the scheduling queue of that node, ordered on the grounds of priority information.

b) All the time an *idle_mask* message circulates through a virtual ring network connecting all processors (realized via the Delta network). A bit set in a given position of the mask indicates that the corresponding processor is *idle* (i.e. the number of processes in that node is below a programmable threshold).

c) When a processor T_i receives the *idle_mask* message, if there are idle processors and if T_i has *sendable* processes (a process is sendable only if it hasn't been started yet. This measure is intended to reduce the cost of transmission and minimize non-local accesses). T_i sends one process to each idle processor, as long as there are sendable processes; at the end, *idle_mask* is propagated to T_{i+1} (note that if T_i is idle, before the propagation the corresponding bit in the mask is switched on). The above scheduling policy, which is able to favour locality of reference by preferably allocating a new process in the same node of the parent process, with minor adjustments and appropriate tunings, has shown a very satisfactory load-balancing performance in a wide range of situations, and has been essential to achieve the speed-up figures reported in Sect. 5.7.2.

5.6 Hardware Architecture

The parallel architecture which runs the IDEAL language is PIPES (Packet Interconnected Processing ElementS), an experimental multiprocessor machine designed and developed with the explicit target of being a suitable testbed for real-time AI applications.

The real-time constraint requires the availability of adequately high processing power, so parallel architectures targetted to this task must be easily upgradable, while a common feature of many AI applications is the need for an efficient support of irregular, strongly data-dependent communication among several processes cooperating for solving the problem.

The need for a good solution for this communication problem was the driving force in the early phase of PIPES design: several interconnection structures were analysed, from the point of view of efficiency, complexity and suitability for large numbers of interconnected nodes, and the choice of employing a packet-switched multistage interconnection network was made as the best trade-off amongst the various factors.

The chosen solution is suitable for architectures up to several hundreds of processing nodes, and in this sense PIPES is a very good testbed for parallel applications and experiments which, in a short or medium timescale, need an easy-to-use and efficient support to a generalized communication. Parallel architectures based on multistage networks are probably not the best solution when thinking about machines comprising thousands or millions of processing nodes: for such a perspective different approaches should be taken into account [1].

It is worth mentioning, before going into greater details, that PIPES is a transputer-based architecture as several others proposed in the same period [33], [34], [17]. The fundamental difference with respect to those relies on the fact that, while using a transputer for several interesting reasons explained in the following, PIPES does not just make available to the user the INMOS point-to-point communication capabilities, which allows for the construction of any static graph of fixed degree among the computational nodes of the system (although programmable and rearrangeable as in the Supernode), but it gives a direct, complete and efficient (since totally hardware-based) interconnection capability among all its computational nodes. PIPES is so oriented to extend the range of applicability of transputer-based systems to irregular, dynamic, communication-intensive algorithms.

5.6.1 Architectural Overview

The PIPES architecture is based (see Fig. 5.4) on a number of processing nodes which comprise a Processing Element (PE), built around the 32-bit microprocessor transputer T414/T800, a (local) page of memory (MM) of about 2 Mbytes, and a Processor-Memory to Network Interface (PMNI) circuit.

The processing nodes in PIPES are connected by means of two distinct interconnection networks, one for local (i.e. near neighbours) communication (LCN: Local Communication Network), and the other for non-local communication (NLCN: Non-Local Communication Network). These two networks are different in implementation techniques, conveyed traffic and purpose.

The LCN is a point-to-point network implemented by directly wiring in a fixed topology the high-speed serial links provided by the transputer. In the PIPES machine the chosen topology is a bidirectional ring, which is easy to manage and anyway adequate to support the traffic, as this network is mainly used for downloading of programs and data at bootstrap-time and flowing of a small amount of control information from/to the host at run-time.

The NLCN is realized by means of a buffered multistage packet-switched interconnection network with Delta topology, whose characteristics and related performance will be detailed in the next section. It support unidirectional communication, so each processing node must be connected to both one input and one output of the network. The NLCN provides a fast transfer mode at the physical machine level to support efficiently requests and responses to/from remote MMs, i.e. remote read or write operations to non-local

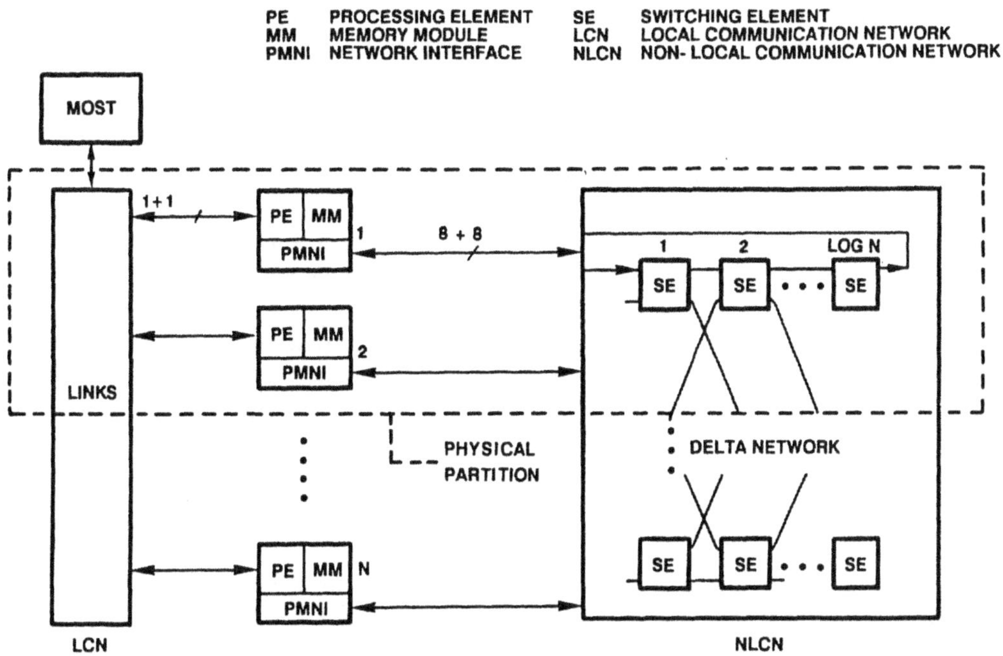

Figure 5.4: PIPES general architecture

pages of memory which are logically shared but physically distributed among the processing nodes. That allows PIPES to be classified as an evolution of the shared memory architectures because it retains the global adressability of all its storage locations (i.e. "remote pointers" are a basic data type supported at the virtual machine level) according to a concept pionereed by the CM* development and now used in a number of contemporary machine (e.g. Alice, Rediflow, BBN Butterfly and IBM RP3).

The function which is responsible for giving the visibility of the global memory space of the machine to each single processing node is implemented in the PMNI block. This block is able to recognize remote memory access, formatting a packet of remote request, sending it through the NLCN, handling the answer obtained and giving in turn the answers to requests of the same kind coming through the NLCN from other computational nodes and it is implemented partially in software within the PE and partially in hardware using the commercially available INMOS link adaptor to interface the PE to the NLCN.

When implementing the aforementioned "global distributed storage" concept, the problem of dealing with the latency time for remote memory accesses inevitably arise, and a solution must be find to avoid wasting CPU power in long series of "wait" cycles. In PIPES this solution consists in multi-micro-tasking the processing element of the node among a (small) number of virtual processors, and in supposing a very fast context switch, when a virtual processor refers to a remote data structure, to a different virtual processor able to proceed with its processing using local data. This technique exploits one of the peculiar features of the transputer device, i.e. its sub-microsecond switching time that makes it a very good approximation of the ideal processor for this kind of architecture.

PIPES is hosted by an IBM PC AT equipped with an INMOS evaluation board. The host is connected to the LCN as though it were the $N + 1$ processing nodes, and its run-time role is mainly collecting information from the system and giving the user some monitoring facilities.

5.6.2 The Non-Local Communication Network

As already stated, the NLCN function is implemented by a binary Delta network. A network of this type, connecting N inputs to N outputs (N being a power of 2), consists of $N/2 \cdot log_2 N$ 2 x 2 switches arranged in $log_2 N$ stages interconnected by patterns of unidirectional links in such a way that any output can be reached from any input (see Fig. 5.5). Moreover, the following condition must be met: if any path is described by a bit string indicating the output chosen at each binary switch traversed, the descriptors of all the paths leading to the same network output are identical. Therefore this path descriptor is a string of $log_2 N$ bits which can be used to identify the output.

A large literature exists on Delta networks; see for instance [21] [22]. The definition above allows different topologies, all of which, however, can be shown to be equivalent in a MIMD environment characterized by uniform traffic. Therefore, the choice of a particular topology depends in this case only on implementation (i.e. "wireability") considerations.

The network operating mode is packet switching, where packets have variable length and begin with a field, called a "routing tag", which contains the path descriptor relative to its destination: this allows each binary switch to route the packets autonomously after having analysed the proper bit of the routing tag. To optimize the crossing delay, cut-through switching is employed, i.e. the first bytes of a packet can be sent to the next stage immediately after the routing decision has been taken, even before other bytes of the packet have been received at that stage.

The performance of a packet-switched Delta network can be reduced due to two different phenomena: routing conflicts and backpressure due to buffer filling. The provision of buffers inside each switching element allows, in the case of two packets competing for the same output port, to resolve the conflict by storing the loser and forwarding it at the following cycle.

Overflows are prevented by means of a backpressure mechanism consisting of a signal which blocks the transmission of new packets from the preceding stage when the buffer is full: therefore no loss of packets is possible within the NLCN. The evaluations described in the next section show that the operating conditions and the dimensioning of the network are such that the network throughput is not decreased by backpressure.

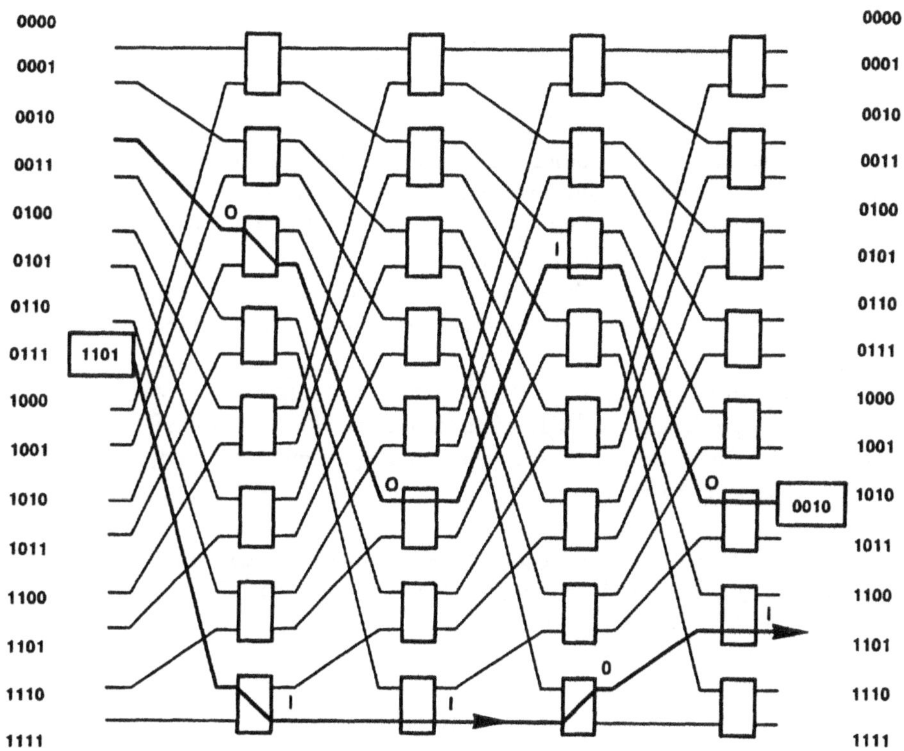

Figure 5.5: An example of a Delta network

5.6.3 Performance Evaluation

The overall performance of the multiprocessor depends on the interplay of widely different factors, related to different hierarchical levels, like the NLCN throughput and delay characteristics, the scheduling and process communication mechanisms, the load distribution strategy, the efficiency of the compiler and the potential for parallelism of the algorithm itself.

While for the higher-level, software-related issues the experimentation on a prototype is the only viable evaluation methodology, hardware and firmware-related issues can be studied by simulation or analytical modelling. In the following the NLCN performance and the efficiency of the solution adopted to overcome the memory latency problem are evaluated.

For what concerns the NLCN, an exact analytical model of a packet-switched Delta network is infeasible due to its huge state space. However, simulation runs taking into account all the characteristics of this implementation (finite length input buffers, back-pressure flow control, cut-through switching) have given several interesting results.

The first is that, for any fixed load below saturation, if the buffer size is increased starting from unity, a value is found for which the backpressure effect disappears and the delay is negatively affected only by routing conflicts inside the switches. Conversely, for any given buffer size, there is an operating region below saturation where the stages interact strongly through the backpressure mechanism, while, at lighter loads, the interaction is negligible. The buffers of the switching element in the NLCN are designed so that the network operates in this region for all the foreseeable loads in the intended application.

Because in the light loading region the behavior of a single switch with infinite buffers can be taken to represent, with little approximation, the behavior of the network as a whole (for instance the network average delay can be assumed to be $log_2 N$ times the delay of a switch), it is useful to develop an analytical model of the 2 x 2 switch.

A model for the switch with output buffers, under the assumption of synchronous operation, uniform traffic distribution and independent arrivals of unit-length packets, has been described in [28]. The throughput/delay characteristic is expressed by the following formula:

$$d = 1 + p/[4(1-p)] \qquad (5.1)$$

where p is the throughput, i.e the probability that a unit-length packet occupies a time slot on a link.

Under the same assumptions an excellent approximated model for the input buffering case can be developed by analysing the behaviour of one of the buffers, influenced only by the probability that the other one is empty or full (and not by its complete state description: see Fig. 5.6). The resulting formula for the throughput/delay characteristic is:

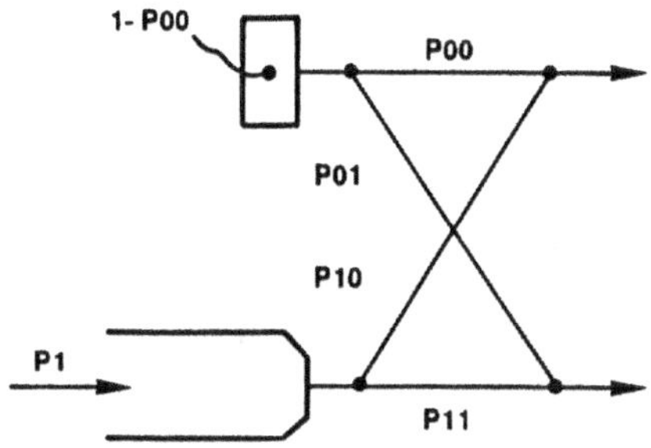

Figure 5.6: Model with independent buffers

$$d = 2(1-p)/[\sqrt{1-p} + 1 - 2p] \tag{5.2}$$

The above evaluations give a positive assessment of the NLCN performance in the expected operating conditions ($p < 0.1$). However, when N becomes large, a fast interconnection network cannot totally overcome the memory latency problem, but only make an architectural solution somewhat cheaper. In the present case such a solution consists in multiprogramming each processing element and performing a context switch at any remote memory request.

The efficiency of the solution can be evaluated with the simple queuing network of Fig. 5.7 In this model the single server represents the CPU, and its associated queue the ready list, while the delay unit (a parallel infinite server) models the remote memory access time. The variable n represents the number of processes waiting for a response from the system; if N is the multiprogramming level, there are $N - n$ processes in execution or in the ready list. t_E is the CPU processing time and t_T the remote memory access time.

The main hypotheses behind this much simplified model are that a remote memory server is employed (the CPU time otherwise necessary to handle remote memory access requests is not considered) and that uniform behaviour of the multiprocessor is assumed, because a single processor is taken to represent the whole system. Moreover, t_E includes also the context switch time, therefore the actual processor utilization is penalized by the factor $(t_E - t_{CS})/t_E$, which clarifies the need for a fast context switch mechanism. The solution of the model in an operational framework [19] can be conveniently expressed by means of $E(\tau, N)$, the Erlang-B formula with load factor $\tau = t_T/t_E$. The results for the utilization U, the average number of suspended processes nav and the response time T (time to complete a cycle) are:

$$U = 1 - E(\tau, N) \tag{5.3}$$

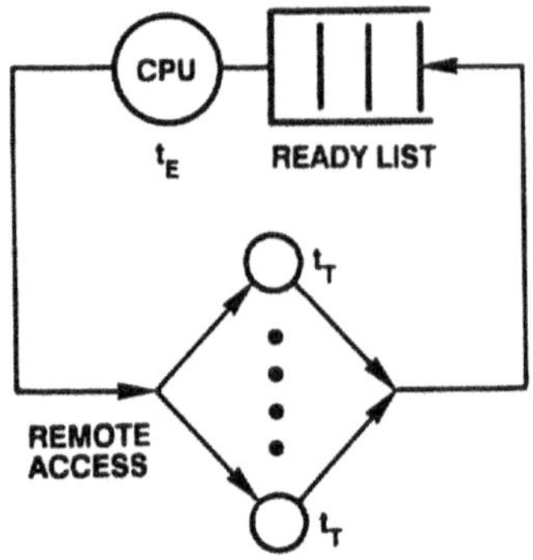

Figure 5.7: A simple model of a processing element

$$nav = \tau(1 - E(\tau, N)) \tag{5.4}$$
$$T = t_E(N + \tau E(\tau, N) \tag{5.5}$$

t_E is a parameter which can be estimated by executing the same algorithm on a uniprocessor, while t_T includes two network crossing times, which in their turn depend on *nav*. Thus, it is necessary to use the results obtained for the NLCN (equation 5.1 or 5.2 above) and solve the model iteratively.

To obtain results which are largely technology-independent it is convenient to express the times involved in terms of switching element cycles. Let $t_E = C$ cycles; if $D(nav)$ is the network crossing delay, t_T can be expressed as $t_T = 2D(nav) + C/K$, where the term C/K represents the remote memory handler operating speed. If M is the number of processors, the performance results can be given as a set of curves with parameters C, K and M. Figure 5.8 gives U versus N for $M = 256$ (a rather large multiprocessor), $C/K = 10$ and $C = 50$, 20 and 10 (50 is a reasonable value, while 20 and 10 represent a very slow network). It is readily seen that a multiprogramming level of 3 is enough to obtain a very high utilization in all reasonable situations.

It is also clear from (5) above that increasing N beyond this limit, while it does not give significant improvements in terms of utilization, is harmful for the response time of the multiprocessor, in addition to the increment of related hardware costs.

5.6.4 The Switching Element

A key point for being able to build PIPES machines with a significantly high number of processing nodes (e.g. up to some hundreds) is the availability of an integrated 2 x 2 packet router for the multistage network, the Switching Element (SE). Such a device has been defined, study and designed up to layout level at our laboratories, and silicon was provided by SGS-Thomson in the framework of a National Grant for microelectronics.

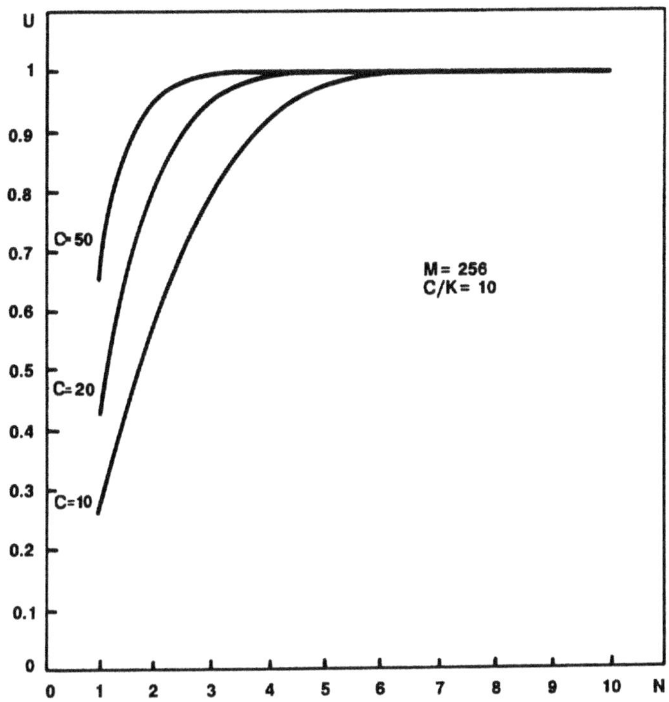

Figure 5.8: CPU utilization versus multiprogramming level

The basic function implemented by the SE is the switching of data coming from one or both of its two input ports towards one of the two output ports, chosen properly on the basis of one bit of a tag information included in the data itself. In such a way it allows the construction of the self-routing networks described above. The normal routing is made on the basis of the indication given by the most significant bit of the tag. If it is 1 the packet is sent through the lower output port of the SE, if it is 0 through the upper one. A mechanism is provided in the output ports to rotate left the tag indicator in order to maintain in the most significant position the bit which will be used in the next stage.

The SE has two internal buffers, one for each input port, to temporarily store up to 64 bytes (that could be either an entire packet, several packets or only a portion of a packet) if some routing conflicts arise inside the component or the next stage of the network is unable to accept further bytes. It routes the packets according a "cut-through" policy: that means that, when the internal input buffer is completely empty, the bytes constituting the packet cross the component as soon as they enter the input port, without waiting for the storing of the entire packet in the buffer before sending it to the next stage of the network. This is a way to minimize the crossing delay of the network and can lead to a scattering of the bytes of one packet through different stages of the network as shown in Fig. 5.9.

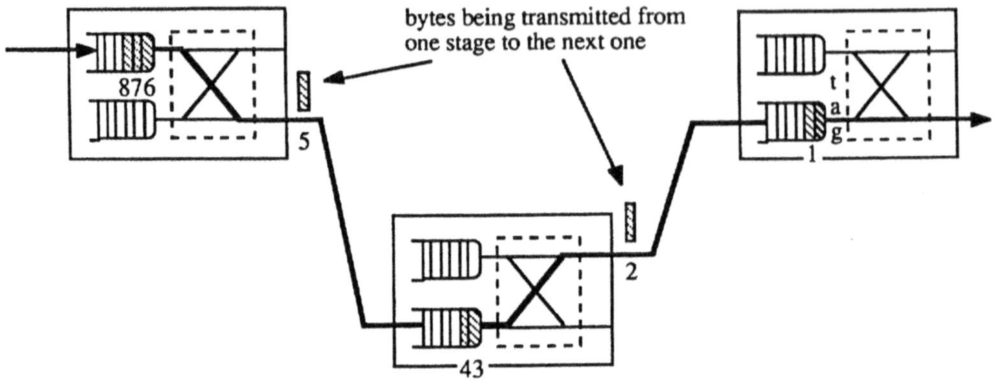

Figure 5.9: An example of cut-through switching

There are two ways to indicate the length of is a packet; either of them can be selected by the user by means of some external configuration pins.

- A counter can be embedded in the packet, containing the number of the following data byte.
- A ninth bit can be transmitted, in parallel to every byte, which always has the value 0 and rises to logical 1 only when the last data byte of the packet is sent.

Additional features, again selectable by pin configuration, are one-to-all broadcasting mode and an error detection (CRC-based) mechanism.

The SE has an internal synchronous structure and it is designed to be driven by an external clock source of any frequence from 100 kHz up to 20 MHz.

The input/output ports use a quite classical four-phases protocol operating with two control signals Request and Acknowledge (see Fig. 5.10). These control signals can be internally double- sampled at the internal clock frequency (to minimize the probability of metastable states), and allow for the construction of interconnection networks in which any device has its own, independently running clock, without imposing the constraint of distributing a single system clock.

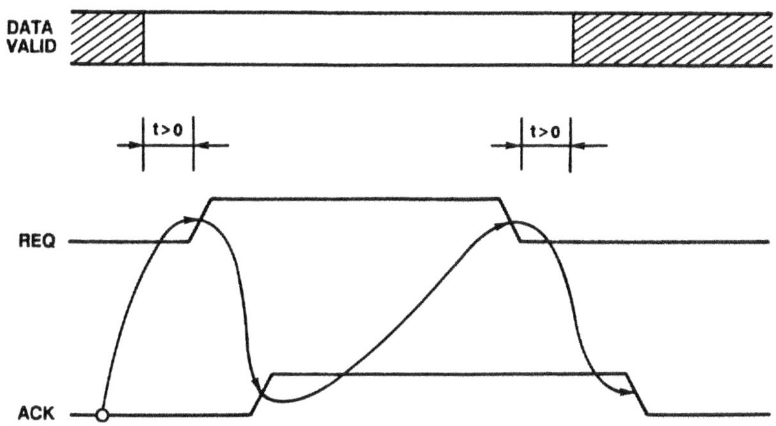

Figure 5.10: Input/output protocol of the SE

The peak input-output rate of any port of the component is one byte every 100 ns if an internal double-sampling mechanism is disabled, or one byte every 200 ns if it is enabled. These figures refer to an SE embedded into an ideally fast environment, able to respond simultaneously to any stimulus, and gives a global throughput of 320 Mbit/s for the four I/O ports of the device. The minimun crossing delay for the component is 300 ns when measured for the tag of the packet in a "buffer empty" condition.

The number of transistors for the SE is about 35000. The technology used is the 3-micron C-MOS, a current SGS process. The physical dimension of the die of the component is 6.7 x 5.7 millimeters, and it is packaged in an 84-pin grid array support; the power dissipation is very small (about 200 mW at 20MHz) thanks to the extensive use of internal "domino" structures.

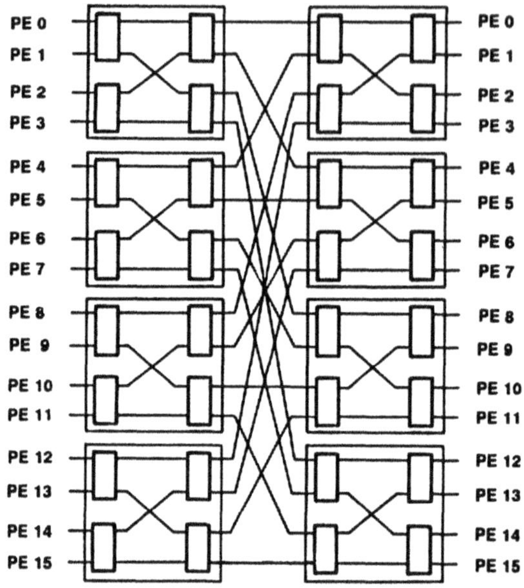

Figure 5.11: Physical partitioning in PIPES 16 machine

5.6.5 The Physical Prototypes

Two different kind of prototypes of the machine have been designed and built: PIPES 16 and PIPES 64; these two prototypes differ not only in the number of processing nodes, as suggested by their names, but in the implementation of the NLCN subsystem too.

PIPES 16 is built around an NLCN whose SEs have an implementation based on off-the-shelf components as registers, FIFOs, PLDs, flip-flops, etc. This came from the need to have available a parallel machine for software development before the end of the design of the custom SE. Such an MSI implementation of the NLCN works at a basic frequency of 16 MHz, giving a minimun network crossing time of about 3 μs. It provided just the basic switching function (e.g. broadcast and CRC generation are not implemented), and drove the physical partitioning of the whole system (see Fig. 5.11): the machine is made up from eight CPU cards, each containing two processing nodes, and eight NLCN cards, each containing four 2 x 2 SE functions arranged to form an equivalent 4 x 4 SE to minimize the number of off-boards connection. PIPES 16 has been operating since December 1987.

PIPES 64 is instead based on the integrated version of SE. The availability of this custom circuit allows this parallel system to be built in a highly modular fashion, being built around a single type of board. Thanks to the freedom in chosing NLCN topology on the basis of wireability and partitionability considerations, we arrange a slicing of the whole system as suggested in Fig. 5.4, and any board of PIPES 64 thus contains four processing nodes and a corresponding slice of the NLCN , according an approach that again minimizes the off-board connections. The custom-based implementation of

NLCN works at a basic frequency of 20 MHz, giving a crossing time of 350 ns each stage, corresponding to 2.1 μs for the whole network.

5.7 Conclusions

5.7.1 Experience with Programming Style

One might wonder how much the mix of OR-parallelism, backtracking and lazy evaluation is effective for parallelism exploitation. We tested it with the invertible simulator turned into a fault-finder described in [9], whose elegant IDEAL implementation relies heavily on such a combination of features. The fault-finder run by our OR-parallel K-LEAF system exhibits a great amount of parallelism, in spite of the lazy constructors. As in the sequential case where lazy evaluation performs an intelligent control of backtracking, in the OR-parallel execution the same intelligent policy is applied to parallel split points, starting only parallel computations which are relevant to the final solution.

Also standard algorithms, requiring problem decomposition, for example sorting, simulation, etc., can be naturally expressed for AND-parallel execution. A general unavoidable difficulty remains in modifying recursive programs in order to keep granularity of non-trivial grain. While in *fib* definition this is easily obtained with the change of the recursion base, in other cases, e.g. in programs recurring over lists, the needed changes could produce a less "abstract" program:

fib(N) = sequential_fib(N) :- N < X.
fib(N) = fib(N-1)// + fib(N-2)// :- N ≥ X.
    where X = suitable_value_for_sequential_execution.

5.7.2 Speed-up

We report in Fig. 5.12 some figures obtained on PIPES about the speed-up of the following benchmarks:

- 8-Queens: all-solutions of OR-parallel version of 8-queens problem;
- Fib25: AND-parallel computation of *fib(25)*;
- Image: the image understanding program described in [32].

The above figures confirm the good performance of the model on such a distributed architecture. At present an overhead of about 30-60% is imposed by the execution of the parallel model on a single processor with respect to pure sequential execution. This is mainly due to the software emulation of memory management (demand-driven block allocation produces an overhead of 30% with respect to pure sequential execution), software handling of non-local accesses, and lack of micro-context switching on remote read operations.

Acknowledgement
This work has been partially supported by ESPRIT Project No. 415.

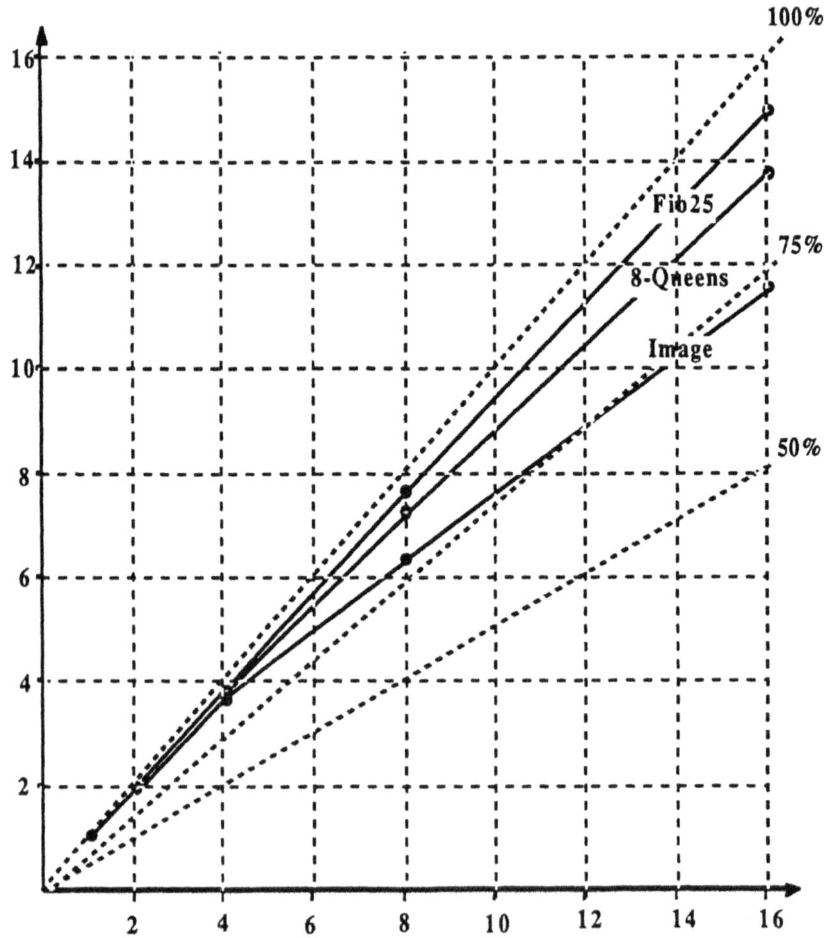

Figure 5.12: Speed-ups on PIPES

Bibliography

1. W.C. Athas, C.L. Seitz: "Multicomputers: message-passing concurrent computers." *IEEE Computer*, pp.9-24, August 1988

2. L. Augustsson: "Compiling lazy functional languages." PhD Thesis, Chalmers University of Technology, Goteborg, 1987

3. G.P. Balboni, G. Giandonato, R. Melen: "A parallel architecture for AI-based real-time applications." *Proceedings of 1987 AFCEA European Symposium*, Rome, 1987

4. M. Bellia, P.G. Bosco, E. Giovannetti, G. Levi, C. Moiso, C. Palamidessi: "A two-level approach to logic plus functional programming integration." *Proceedings of the PARLE Conference*, Lecture Notes in Computer Science 258, Springer-Verlag, pp.374-393, 1987

5. P.G. Bosco, E. Giovannetti: "IDEAL: An Ideal DEductive Applicative Language." *Proceedings of the 1986 Symposium on Logic Programming*, IEEE Comp. Society Press, pp.89-94, 1986

6. P.G. Bosco, E. Giovannetti, G. Levi, C. Moiso, C. Palamidessi: "A complete semantic characterization of K-LEAF, a logic language with partial functions." *Proceedings of the 1987 Symposium on Logic Programming*, IEEE Comp. Society Press, pp. 318-327, 1987

7. P.G. Bosco, C. Cecchi, C. Moiso: "Feasible computational models for logic plus functional programming integration." ESPRIT Project 415, Subproject D, Deliverable D3, 1987

8. P.G. Bosco, E. Giovannetti, C. Moiso: "Narrowing vs. SLD-resolution." *Journal of Theoretical Computer Science*, vol. 59, no. 1-2, North-Holland, pp.3-23, 1988

9. P.G. Bosco, C. Cecchi, C. Moiso: "Exploiting the full power of logic plus functional programming." *Proceedings of the 5th Conference and Symposium on Logic Programming*, MIT Press, pp.3-17, 1988

10. P.G. Bosco, C. Cecchi, C. Moiso, G. Sofi: "The abstract parallel machine for IDEAL/K-LEAF." ESPRIT Project 415, Subproject D, Deliverable D5, 1988

11. P.G. Bosco, C. Cecchi, C. Moiso: "Compilation tools for IDEAL/K-LEAF." ESPRIT Project 415, Subproject D, Deliverable D6, 1988

12. P.G. Bosco, C. Cecchi, C. Moiso: "An extension of WAM for K-LEAF: a WAM based compilation of conditional narrowing." *Proceedings of 6th Conference on Logic Programming*, MIT Press, 1988

13. T.H. Brus, M.C. Van Eekelen, M.O. Van Leer, M.J. Plasmeijer: "CLEAN: a language for functional graph rewriting." *Proceedings of the 3rd Conference on Functional Programming Languages and Architecture*, Lecture Notes in Computer Science 274, Springer-Verlag, pp. 364-374, 1987

14. G.L. Burn: "Abstract interpretation and the parallel evaluation of functional languages." PhD Thesis, University of London, 1987

15. M. Carlsson: "Freeze, indexing and other implementation issues in the WAM." *Proceedings of the 4th Conference on Logic Programming*, MIT Press, pp. 40-58, 1987

16. M. Carlsson, K. Danhof, R. Overbeek: "A simplified approach to the implementation of AND-parallelism in an OR-parallel environment." *Proceedings of the 5th Conference and Symposium on Logic Programming*, MIT Press, pp. 1565-1577, 1988

17. R.S. Cok: "A medium grain parallel computer for image processing." *Proceedings of 8th Technical Meeting of OCCAM User Group*, pp. 113-124, March 1988

18. J. Crammond: "A comparative study of unification algorithms for OR-parallel execution of logic languages." *IEEE Transactions on Computers*, pp. 911-917, October 1985

19. P.J. Denning, J.P. Buzen: "The operational analysis of queuing network models." *ACM Computing Surveys*, pp. 225-262, September 1978

20. N. Dershowitz, D.A. Plaisted: "Logic programming cum applicative programming." *Proceedings of the 1985 Symposium on Logic Programming*, IEEE Comp. Society Press, pp. 54-66, 1985

21. D. M. Dias, J.R. Jump: "Analysis and simulation of buffered delta networks." IEEE Trans. on Computers, pp. 273-282, April 1981

22. D. M. Dias, J.R. Jump: "Packet switched in N logN multistage networks." *Proceedings of Globecom*, pp.114-120, 1984

23. L. Fribourg: "SLOG: A logic programming language interpreter based on clausal superposition and rewriting." *Proceedings of the 1985 Symposium on Logic Programming*, IEEE Comp. Society Press, pp. 172-184, 1985

24. G. Giandonato, G. Sofi: "Parallelizing logic programming based inference engines." *Proceedings of the 3rd International Conference on Supercomputing*, International Supercomputing Institute Inc., pp. 282-287, 1988

25. J.A. Goguen, J. Meseguer: "Equality, types and generic modules for logic programming." In: D. DeGroot and G. Lindstrom (eds.): *Logic Programming: Functions, Relations and Equations*, Prentice-Hall, pp. 295-364, 1986

26. L. Hirschman, C. Hopkins, R. Smith: "OR-parallel speed-up in natural language processing: a case study." *Proceedings of the 5th Conference and Symposium on Logic Programming*, MIT Press, pp. 263-279, 1988

27. J.W. Klop: "Term Rewriting Systems." *Notes for the Summer Workshop on Reduction Machines*, Ustica, 1985

28. C.P. Kruskal, M. Snir: "The performance of multistage interconnection network for multiprocessor." *IEEE Trans. on Computers*, pp. 1091-1098, Dec. 1983

29. A. Martelli, C. Moiso, G.F. Rossi: "Lazy unification algorithms for canonical rewrite systems." *Proceedings of the Colloquium on Resolution of Equations in Algebraic Structures*, Prentice-Hall, 1989

30. C. Merlo, C. Moiso, M. Porta, G. Sofi: "Parallel Prolog for signal- understanding parallel machines." ESPRIT Pilot Project 26, Deliverable 14b, 1988

31. R. Milner: "A theory of type polymorphism in programming." *Journal of Computer and System Sciences,* 17, 3, pp. 348-375, Dec. 1978

32. C. Moiso, M. Porta, M. V. Oneto, G. Sofi: "Application of a logic parallel language: recognition of two-dimensional objects." *Proceedings of the 4th Italian Conference on Logic Programming,* Bologna, June 7-9, 1989

33. Parsys: "Supernode SN1000 series brochures." THORN EMI Central Research Labs, Dawley Road, Hayes, Middlesex UB3 1HH, UK, 1988

34. Parsytec: "Megaframe Supercluster brochure." PARSYTEC mbH, Julicher Str. 338, D-5100 Aachen, Fed. Rep. Germany, 1987

35. S.L. Peyton Jones: "The Implementation of Functional Programming Languages." Prentice-Hall, 1987

36. P.A. Subrahmanyam, J. H. You: "FUNLOG: A computational model integrating logic programming and functional programming." In: D. DeGroot and G. Lindstrom (eds.): *Logic Programming: Functions, Relations and Equations*, Prentice-Hall, pp. 157-198, 1986

37. D.A. Turner: "MIRANDA: a non-strict functional language with polymorphic types." *Proceedings of the 2nd Conference on Functional Programming Languages and Architectures*, Lecture Notes in Computer Science 201, Springer-Verlag, pp. 1-16, 1985

38. D.H.D. Warren: "An Abstract Prolog Instruction Set." Technical Note 309, SRI International (Oct.1983)

39. D.H.D. Warren: "The SRI model for OR-parallel execution of prolog. Abstract design and implementation." *Proceedings 1987 Symposium on Logic Programming*, IEEE Comp. Society Press, pp. 92-103, 1987

40. D.H.D. Warren: "OR-parallel models of Prolog." *Proceedings of the International Joint Conference on Theory and Practice of Software Development (TAPSOFT '87)*, Lecture Notes in Computer Science 250, Springer-Verlag, pp. 243-259, 1987

41. H. Westphal, P. Robert, J. Chassin, J.C. Syre: "The PEPSys model: combining backtracking AND- and OR-parallelism." *Proceedings of the 1987 Symposium on Logic Programming*, IEEE Comp. Society Press, pp. 436-448, 1987

Chapter 6

Conclusions and Future Developments

Alberto Ciaramella, Giancarlo Pirani, Claudio Rullent (CSELT)

The speech understanding system that has been described in the preceding chapters represents a first prototype that is still open to evolution and improvement. Although the final result of the ESPRIT Project P26 represented a good achievement of our goals, we do believe that both the recognition and understanding stage should get better performance in order to make the overall system more accurate and robust.

In particular, most of these new developments are being carried out in a project of the ESPRIT II program, which actually started in early 1989, including in the new consortium two partners of the previous project, i.e., CSELT and Daimler Benz (formerly AEG).
The new project is called SUNDIAL (Speech UNderstanding and DIALogue) and addresses the problem of speech based co-operative dialogue as an interface for computer applications in the information service domain, with particular reference to the telecommunications environment.

The following treatment will explain how we intend to make the developed system evolve from the state of a one-way speech understanding system to that of a real dialogue system which is capable of offering automated information services over the telecommunication network. Of course, a sketch is also given of the foreseen hardware updating which is necessary for an efficient implementation of the algorithms.
The developments that are foreseen assume that the general architecture of the speech understanding system described in Fig. 1.1 of Chap. 1 will not be substantially altered; only some feedback communication from the understanding stage to the recognition one will be added.

6.1 Recognition Algorithms

The improvements of the recognition stage should be introduced to optimize the global performance of the overall system in terms of correct sentence understanding rate. Anyway, as this global optimization is quite complex we are following the pragmatic approach of improving independently the performance of the recognition stage in terms of word

P.26 system	updated system
1K words	1K-2K words
high quality speech	telephonic speech
speaker trained	speaker independent
training with isolated words	training with continuous speech
application indep. training	mixed application indep. and dep. training
manual endpointing	automatic endpointing
no feedback from understanding	feedback from understanding

Table 6.1: Features of the P.26 system and goals of the new developments

accuracy (WA), defined as [1]:

$$\mathrm{WA} = 100 * \left(1 - \frac{substitutions + insertions + deletions}{correct\ number\ of\ words}\right)\% \tag{6.1}$$

where ***substitutions***, ***insertions***, and ***deletions*** are the number of substituted, inserted, and deleted words respectively.

In fact, we are convinced that the recognition stage has to achieve the highest degree of accuracy in order to have the best performance of the linguistic processor in terms of both correct understanding and time consumption.

As far as the general characteristics of the recognition system are concerned, the major developments that we intend to carry out for enlarging the scope of the system are summarized in Table 6.1.

The items that stem from this development program are dealt with in the following.

Lexicon. The choice of the lexicon is clearly dependent on the application. We do not foresee for the near future a dramatic increase of its size, deeming that an active vocabulary of 1K-2K words can be enough for many telecom applications.

Anyway, linguistic phenomena will be taken into account when function-phrase entries have to be defined in the lexicon to improve the recognizer performance; also non-word entries will be added to take into account phenomena like hesitations, coughs, etc.

Function phrases must be chosen according to either their degree of statistical relevance within the given application or their degree of difficulty when trying to identify each constituent word at the acoustic level. The latter is the case, for example, with several monosyllabic connectives.

However, that does not imply the need to impose a strict linguistic constraint at the recognition level, since each constituent word may maintain its own entry in the lexicon; simply, at the expense of a little increase of the lexicon size, function phrases

[1]K.F. Lee, H.W. Hon: "Large-vocabulary speaker-independent continuous speech recognition using HMM." *Proc. of the ICASSP '88*, pp.123-126, April 1988

help improving recognition performance. Non-word phenomena must also be modeled in some way.

Input speech quality. When telephonic speech is considered, the main problems are represented by the noise and distortion due to the telephone line and by the different telephone handsets distributed over the network.

These factors determine a high degree of variability and are a major source of performance degradation. We intend to contend with this impairment by replacing discrete HMM with continuous density hidden Markov models (CDHMM), where the number of states and of Gaussian mixtures will be optimized by means of intensive experimentation, and including also differential cepstral coefficients into the parameters of the observation vector. As far as noise is concerned, the training database implicitly models this phenomenon when the test conditions are homogeneous with the training ones. Furthermore, an explicit background noise model can be trained separately, while impulsive noise bursts should only cause minor degradation of the acoustic decoding performance.

Speaker independence. Selection criteria to be used for the training data base will take into account the major sources of variability at the speaker level; for example, samples for each of the main regional intonations will be collected, and speakers' ages will globally span a range as large as possible. Obviously, a proper balance between male and female speakers is to be guaranteed.

In order to carry out the experiments on the speaker independent system we have collected specific databases. A training database (MDB) and a testing database (FDB) have been collected relevant to 110 and 20 different speakers respectively, who uttered 80 sentences (MDB) and 60 sentences (FDB) each. Utterances were collected through a normal telephone set connected to a PABX. Speech, sampled at 16 kHz with a 16-bit linear quantizer, was stored on WORM optical disk, for a total of about 3.2 Gbytes.

Uttering style for training. When we collected the new speech database for training the models of the acoustic units, we chose to use continuously spoken sentences in order to be as close as possible to the uttering style that is employed by the user that addresses the understanding system. In this way, also the main coarticulation effects will be captured to be automatically included into the acoustic decoder, thereby improving its accuracy; for this purpose both intra– and inter–word coarticulation will be taken into account and properly represented by means of specific tree structures.

Application dependence of training. We will investigate the improvement attainable by means of a training speech database that is at least partially application oriented. In particular, each speaker belonging to the training set was asked to utter 40 application sentences and 40 phonetically balanced sentences. To what extent the degree of application dependence of the training database affects the recognition performance (and also the overall understanding system performance) will be carefully studied in the near future.

Sentence boundaries. Detecting the end of a sentence is a simply stated, but hard-to-solve problem. Integration of low level acoustic knowledge (energy-based endpointing) with high level acoustic knowledge (spectral modeling of silence and noise) and linguistic knowledge will be exploited.

Feedback from understanding While for the P26 system the connection between recognition and understanding is unidirectional, the introduction is foreseen of a linguistic feedback requesting either the acoustic evaluation of alternative parsed sentences, or the acoustic scoring of segments of the utterance with alternative, partially parsed word sequences. Also this process of recognition activation driven by linguistic analysis will have to be carefully designed to be able to cope with real time requirements.

6.2 Real-time Hardware Implementation

The next evolution step should imply an improvement of the system in ease of use, accuracy, and technology. As mentioned previously, the system will evolve to a full dialogue system, with remote speaker-independent telephone input without any console interaction and with telephone synthesized output. Some of these improvements are larger in scope than the recognizer in itself, and new specific blocks have to be added, such as for example the dialogue manager, which in each case is more related to the understanding stage, and the telephone network interface, which will be interposed between the telephone network and the recognizer A/D converter, with the capability of connecting and disconnecting the machine to and from a specific user call.

We will not deal in detail with these general system upgradings outside the recognition stage; we point out that some of these upgradings concern directly the hardware implementation of the recognition stage itself, which will provide:

- a more asynchronous behaviour, for reacting to the external telephone environment (connect, disconnect) and to less regular user interactions, as experimented within the dialogue;
- a higher accuracy, for handling a speaker-independent, telephone input application;
- more robust sentence endpointing, since console interactions are not possible.

These constraints require the addition of computational power in the recognition stage, for example the extraction of differential cepstral, but require also the improvement of accuracy in other algorithms, for example maximum-likelihood acoustic decoding. As efficient implementations of forward acoustic decoding algorithms and CDHMM require floating point representation, we intend to use floating point DSPs. This is a timely technological transition, since DSPs were still in embryonic form to comply with the time schedule of P26, but now they are a consolidated reality. Using modern floating point DSPs allows us not only to implement a larger set of algorithms, but also to gain other benefits like:

- more powerful DSPs;
- faster algorithm prototyping;
- faster programming prototyping.

As far as the first item is concerned, we point out that modern floating point DSPs have increased not only the computational accuracy, but also the memory available on the chip and addressable outside, the input/output capabilities, the internal number of registers and even the clock rate. The most severe limitation we faced in P26 was in fact the limited addressing capabilities of DSPs for our application: we chose the TMS32020 as the best for this aspect, which could externally address up to 64K words (of 16 bits): this was quite large for 1985 technology, and meanwhile we had to implement specific hardware around this DSP in order to increase this address capability through paging: now instead the native addressing capability of the new floating-point TMS320C30 is 16Mwords (of 32 bits), well appropriate for our requirements.

The technological transition from integer to floating-point DSPs is not straightforward, since we have to re-implement our DSP firmware, but in this case we can benefit by both algorithms and programming fast prototyping. Floating-point DSPs allow algorithm fast prototyping, given that they assure the same accuracy obtained in the off-line simulation, hence the phase of integer simulation between floating-point simulation and DSP programming is now completely skipped. Fast prototyping in programming is even more important in modern floating-point DSPs, since their architecture, more similar to traditional microprocessors especially in available memory address space and in the number of internal registers, allows us to use high level languages, for example C.

We have to point out however that the use of C alone is not efficient for DSPs, since it produces a decrease of speed, which is the most important benefit of the DSP architectures: an order of magnitude of difference in computational time can be easily demonstrated by comparing a C implemented versus an assembler implemented DSP algorithm. Hence, in order to obtain reasonably efficient implementations in a reasonable amount of programmer time, for floating-point DSPs we will observe the following guidelines:

- implement in C language only the program control section, which requires less computations;
- use C callable assembler optimized libraries for implementing comon use functions, such as FFT, filters, etc.;
- use assembler code only for implementing C-callable assembler optimized custom functions not available in the library used or for driving specific hardware related functions, as input/output.

Moreover, in our new project we will use as much as possible of the standards and facilities provided by vendors. Furthermore, confirming our choice of the powerful VME bus, we will host the whole system in a SUN 4, equipped with commercially available floating-point TMS320C30 DSP boards, using the C language and the SPOX library for firmware development.

6.3 Understanding Algorithms

An evaluation of what has still to be done to improve the speech understanding algorithms must start from the analysis of their limits in the current scenario (speaker dependent recognition algorithms and good quality voice signal) and from the target scenario where our speech understanding applications are envisaged (voice services over the telecommunication network).

The limits of the current system are more or less common to all existing speech understanding systems and will be examined in the following paragraphs.

Linguistic coverage and system performance. A first aspect is that there are two contrasting requirements in a speech understanding application: from one side to have a large linguistic coverage and from the other to have good efficiency and good correct understanding rate. Unfortunately a very large linguistic coverage reduces constraints with the result of increasing the error rate; furthermore the larger search space reduces efficiency.

We have seen that it is extremely dangerous to develop a large linguistic knowledge base by incrementally testing it only on written sentences (i.e. lattices containing only the correct word hypotheses): when testing it on real lattices efficiency problems are likely to occur and their solution can require a certain number of changes in the knowledge base.

We do not envisage magic solutions to this problem: a great effort has been made to increase efficiency both by devising an efficient control strategy and by optimizing and rewriting the final version into the C language.

A reasonable possibility is to carefully consider the selected application and to decide to accept a certain level of linguistic coverage. In this way there is the risk of not understanding a user utterance because it is not within the system linguistic coverage.

The problem of long sentences. What we have seen is that the effort of parsing very long sentences is not really worthwhile; in fact it increases enormously the search space while the probability of understanding them is not really good, both for the risk of errors at the recognition level (the longer the utterance, the higher the probability of errors) and for the risk of consuming the system resources (time and memory) before obtaining the solution.

On the other hand, if the user has to reach a goal that cannot be fulfilled without using a long sentence, it becomes necessary to allow the user to reach his/her goal through the use of a set of interconnected sentences, each of them followed by a system answer. In other words the system has to manage a dialogue with the user. From a certain point of view it seems that we are trying to solve a problem by introducing a greater one, but that is not true because for certain domains it is quite unnatural to interact with single utterances. That happens when the user goal has to be satisfied through a number of simple steps where each of them can depend on the results of the previous ones; in such case it is natural for the user utterances to refer to the contexts of the previous steps. A brief description of the what we are planning to do in the future in the dialogue field will be given later on.

From the current scenario to the future one. In the current scenario it has been possible to reach an acceptable rate of correct understanding (87%) and a good efficiency (two or three seconds) within the system linguistic coverage. From the point of view of the understanding algorithms, the new scenario (speaker independence and telephonic quality of the voice signal) is likely to lead to the generation of lower quality lattices.

With the term "lower quality" we mean mainly the fact that the number of word hypotheses could be larger (maintaining the same very low probability of a correct word missing from the lattice) and that the reliability of each word hypothesis (its score and its time limits) could be lower.

For the understanding algorithms a lower quality lattice is likely to mean a large search space and a lower reliability of the generated phrase hypotheses. That leads to a lower efficiency and a higher error rate. The problem of reduced efficiency can easily mean an increase of the error rate too, as the the system resources (time and memory) are predefined and if a solution is not found within such limits the analysis fails.

There are two main ways to preserve efficiency and correct understanding rate in the new scenario: to have faster algorithms with large amount of memory and to have more "reliable" algorithms. The first approach can reduce the error rate if the understanding algorithms are not able to find a solution within the available system resources. The second one is useful when the algorithms find an incorrect solution (i.e. a sentence that has not been uttered).

Both kind of phenomena will have to be dealt with, but what it is not still so clear is the relative weight of the two phenomena. By now we have the feeling that the second one is likely to be the most frequent one. We are envisaging two different activities for the two different phenomena.

Increasing performance through parallelism. In the first case the idea is to move towards parallel processing: the parsing algorithms studied have been shown to be easily parallelized and an experimental parallel LISP version (running on a pool of LISP machines) has been developed, starting from an intermediate LISP version of the parsing algorithms.

Now the idea is to develop a parallel C parser starting from the current parser; this parallel parser should be first tested simply in a multitasking environment (like UNIX) and then tested in a real multiprocessing environment.

At the moment this activity has not been started, as we prefer to evaluate the real lattices generated in the final scenario to see the relative relevance of the first phenomenon compared to the second one. Only at that point it will be possible to select, if necessary, the most suitable parallel machine (taking into account costs, number of processors, MIPS for each of them and finally the software environment available for parallel processing).

Interaction between recognition and understanding. For the second phenomenon the best results should be obtained from the research concerning the recognition algorithms. Nevertheless there are sources of errors that are on the border between recognition and understanding: they are connected to the way in which the word hypotheses are inserted into the lattice by the recognition algorithms and to the way they are used by the understanding algorithms. Particularly critical are the aspects connected to the treatment of short words, the co-articulation phenomena and the combination of the scores of word hypotheses to obtain the scores of phrase hypotheses.

A possible solution is to have a tighter but still highly controlled interaction between recognition and understanding algorithms. In particular the first idea is to have the possibility of a verification activity performed by the recognition algorithms on the solutions proposed by the understanding algorithms.

The verification procedure. The procedure is described through the following steps:

1. The understanding algorithms do not terminate as soon as a solution is selected by the Scheduler (the current situation) but the analysis continues until the system resources are used up. In this way none, one or more than one candidate solutions are obtained; their scores are not used to select the solution.

2. Each candidate solution will be analyzed by the recognition algorithms. Each of them is usually characterized by the presence of gaps where short words with low semantic content are supposed to be present, although not detected by the recognition algorithms (or ignored by the understanding algorithms). For each gap a set of candidate words is obtained, taking into account the syntactic knowledge and the solution parse tree. In such a way a graph of words (not of word hypotheses) is generated; such a graph can be seen as a very specific "grammar".

3. The recognition algorithms have to analyze each graph to find the best path, i.e. the most probable sequence of words. For each analyzed graph the result is then a sequence of words (the most plausible one) and a score. Note that the scores of the different graphs refer to the same time interval, so the critical problem of comparing scores pertaining to different time intervals is solved.

4. The understanding algorithms select the path (sequence of words) with the best score and accept it as the solution.

The advantage of the verification procedure. The main advantage of this approach is that in the first phase of the parsing a set of candidate solutions are found using constraints (linguistic and temporal constraints) that have been somewhat relaxed (think of the thresholds values, the missing short words, etc.) while the second phase makes a strict verification of these solutions. What is likely to happen is that at the end of the second phase it will be possible to select, say, the second solution instead of the first one because, for instance, the speech portion corresponding to a gap in the first candidate solution where a preposition was supposed to be, did not match such a preposition at all.

Note that a typical problem is, e.g., that in a sentence like "leggimi due messaggi di Rossi" ("read me two messages of Rossi") the word hypothesis for "due" can have a score worse than the score of the sentence; in such a case the first candidate solution could be "leggimi ? messaggi di Rossi" ("Read me the messages of Rossi") that is linguistically correct too. That results by assuming the presence of an article "i" where "due" should be. The second candidate solution could be the correct one.

When the verification of the first candidate solution is performed, the Markov chain for the sentence "leggimi i messaggi di Rossi" is matched against the speech signal while in the case of the second candidate solution the Markov chain is "leggimi due messaggi di Rossi". Even if "due" has not a good matching in that portion of the speech it is likely to have a better global score by matching it than to match a "i" and to require a different alignment of the two words "leggimi" and "messaggi" (as they must now be much more adjacent or the matching of "i" would be quite bad).

6.4 The Role of a Dialogue Manager

Different application domains tend to stress different linguistic phenomena: certain problems are much more crucial for certain domains than for others.

Mapping to pragmatic representation. An example is the complexity of the task of translating the semantic representation of an utterance into the pragmatic representation that is needed to execute the action able to satisfy the user requirement expressed through the utterance. As an example, consider the sentences "Ci sono messaggi per me?" ("Are there any messages for me?", "Qualcuno mi ha inviato dei messaggi" ("Did someone send me any messages?") and "Ho ricevuto dei messaggi?" ("Did I receive any messages?").

All these utterances have different semantic representations (their conceptual graphs refer to different concepts) but nevertheless the user goal is the same in the three cases; therefore, they must have the same pragmatic representation so that the system action can be the same as well. This problem was not so serious in the case of the geographical domain as it is likely to be, for example, in the case of voice access to electronic mail.

To deal with this problem means to find proper ways of performing a mapping from conceptual graphs to pragmatic representations. We are thinking of using frames to represent pragmatic knowledge and using abstractions to match the frame slots against the utterance conceptual graph.

Referring to previous contexts. When more than one sentence is used to reach the user goal, sentences need to refer to previous contexts and themselves can create new contexts. As an example, the utterance "Ci sono messaggi di Rossi?" ("Are there messages from Rossi?") can be answered with the sentence "Ci sono cinque messaggi di Rossi" ("There are 5 messages from Rossi"). At this point the user can say "Leggimi i primi due" ("Read me the first two."). Using this utterance she/he refers to the first two messages of Rossi.

Sometimes the reference to previous contexts is made in an implicit way (as in this case); in others there could be an explicit reference to one of the previous contexts through the use of pronouns. The main problem is to solve the reference problem, i.e. to find the most probable context in which the utterance has to be interpreted.
All the available knowlege should be used for this purpose: the reference markers in the case of an explicit reference, the available entities that are associated with the context (like Rossi's messages in the case of the example), the sentence given to the user by the system, etc.

Repairing sentence understanding failures. When the understanding of a user utterance fails, the utterance could be understood incorrectly or it could be not understood at all.

In the first case it is quite difficult to find a way of repairing the failure as the system is not conscious of not having understood correctly, unless in the rare cases where there are clear contradictions from which an incorrect understanding can be inferred.

In the second case something can be done: the system can try to take over the control of the dialogue, reducing the user freedom.

Note, however, that there are not miraculous solutions to understanding failures: reducing the user freedom means reducing the efficiency of the user interaction, as it is difficult to find a good position between the two extremes of a system-directed dialogue and a user-directed dialogue.

GPSR Compliance

The European Union's (EU) General Product Safety Regulation (GPSR) is a set of rules that requires consumer products to be safe and our obligations to ensure this.

If you have any concerns about our products, you can contact us on ProductSafety@springernature.com

In case Publisher is established outside the EU, the EU authorized representative is:

Springer Nature Customer Service Center GmbH
Europaplatz 3
69115 Heidelberg, Germany

Batch number: 10704763

Printed by Printforce, the Netherlands